Elements of Quantum Mechanics

Elements of Quantum Mechanics

Michael D. Fayer
Stanford University

New York • Oxford
OXFORD UNIVERSITY PRESS
2001

Oxford University Press

Oxford New York
Athens Auckland Bangkok Bogotá Buenos Aires Calcutta
Cape Town Chennai Dar es Salaam Delhi Florence Hong Kong Istanbul
Karachi Kuala Lumpur Madrid Melbourne Mexico City Mumbai
Nairobi Paris São Paulo Shanghai Singapore Taipei Tokyo Toronto Warsaw

and associated companies in
Berlin Ibadan

Published by Oxford University Press, Inc.
198 Madison Avenue, New York, New York, 10016
http://www.oup-usa.org

Library of Congress Cataloging-in-Publication Data

Fayer, Michael D.
Elements of quantum mechanics / Michael D. Fayer.
p. cm.
Includes bibliographical references and index.
ISBN 0-19-514195-4
1. Quantum theory. I. Title.

QC174.12.F38 2000
530.12—dc21 00-037518

Printing (last digit): 9 8 7 6 5 4 3 2 1

Printed in the United States of America
on acid-free paper

CONTENTS

PREFACE ix

CHAPTER 1 **Absolute Size and the Superposition Principle 1**

CHAPTER 2 **Kets, Bras, Operators, and the Eigenvalue Problem 7**

A. Kets and Bras 7
B. Linear Operators 11
C. Eigenvalues and Eigenvectors 13

CHAPTER 3 **Momentum of a Free Particle and Wave Packets 18**

A. Momentum States of a Free Particle 19
B. Normalization of the Momentum Eigenfunctions 20
C. Wave Packets 23
D. Wave Packet Motion and Group Velocities 29

CHAPTER 4 **Commutators, Dirac's Quantum Condition, and the Uncertainty Principle 34**

A. Dirac's Quantum Condition 34
B. Commutators and Simultaneous Eigenfunctions 36
C. Expectation Values and Averages 39
D. The Uncertainty Principle 42

CHAPTER 5 **The Schrödinger Equation, Time-Dependent and Time-Independent 45**

A. The Schrödinger Equation 45
B. The Equation of Motion of the Expectation Value 48
C. The Free-Particle Energy Eigenvalue Problem 49
D. The Particle in a Box Energy Eigenvalue Problem 51
E. Particle in a Finite Box, Tunneling and Ionization 57

CHAPTER 6 **The Harmonic Oscillator in the Schrödinger and Dirac Representations 66**

A. The Quantum Harmonic Oscillator in the Schrödinger Representation 67
B. The Quantum Harmonic Oscillator in the Dirac Representation 79
C. Time-Dependent Harmonic Oscillator Wave Packet 90

CHAPTER 7 **The Hydrogen Atom 93**

A. Separation of the Schrödinger Equation 93
B. Solutions of the Three One-Dimensional Equations 97
C. The Hydrogen Atom Wavefunctions 105

CHAPTER 8 **Time-Dependent Two-State Problem 112**

A. Electronic Excitation Transfer 116
B. Projection Operators 119
C. Stationary States 120
D. The Nondegenerate Case and the Role of Thermal Fluctuations 123
E. An Infinite System—Excitons 125

CHAPTER 9 **Perturbation Theory 133**

A. Perturbation Theory for Nondegenerate States 133
B. Examples—Perturbed Harmonic Oscillator and the Stark Effect for the Rigid Plane Rotor 139
C. Perturbation Theory for Degenerate States 145

CHAPTER 10 **The Helium Atom: Perturbation Treatment and the Variation Principle 152**

A. Perturbation Theory Treatment of the Helium Atom Ground State 152
B. The Variational Theorem 158
C. Variation Treatment of the Helium Atom Ground State 160

CHAPTER 11 **Time-Dependent Perturbation Theory 163**

A. Development of Time-Dependent Perturbation 163
B. Vibrational Excitation by a Grazing Ion–Molecule Collision 165

CHAPTER 12 **Absorption and Emission of Radiation 172**

A. The Hamiltonian for Charged Particles in Electric and Magnetic Fields 173
B. Application of Time-Dependent Perturbation Theory 178
C. Spontaneous Emission 186
D. Selection Rules 188
E. Limitations of the Time-Dependent Perturbation Theory Treatment 189

CHAPTER 13 **The Matrix Representation 193**

A. Matrices and Operators 193
B. Change of Basis Set 199
C. Hermitian Operators and Matrices 204
D. The Harmonic Oscillator in the Matrix Representation 205
E. Solving the Eigenvalue Problem by Matrix Diagonalization 208

CHAPTER 14 **The Density Matrix and Coherent Coupling of Molecules to Light 213**

A. The Density Operator and the Density Matrix 213
B. The Time Dependence of the Density Matrix 214
C. The Time-Dependent Two-State Problem 217
D. Expectation Value of an Operator 219
E. Coherent Coupling of a Two-State System by an Optical Field 221
F. Free Precession 226
G. Pure and Mixed Density Matrices 228
H. The Free Induction Decay 229

CHAPTER 15 **Angular Momentum 232**

A. Angular Momentum Operators 232
B. The Eigenvalues of $\underline{J}^2$ and $\underline{J}_z$ 236
C. Angular Momentum Matrices 240
D. Orbital Angular Momentum and the Zeeman Effect 242
E. Addition of Angular Momentum 246

CHAPTER 16 **Electron Spin 255**

A. The Electron Spin Hypothesis 256
B. Spin–Orbit Coupling 258
C. Antisymmetrization and the Pauli Principle 268
D. Singlet and Triplet States 278

CHAPTER 17 **The Covalent Bond 280**

A. Separation of Electronic and Nuclear Motion: The Born–Oppenheimer Approximation 280
B. The Hydrogen Molecule Ion 282
C. The Hydrogen Molecule 288

Problems 295

Physical Constants and Conversion Factors for Energy Units 314

INDEX 315

PREFACE

This book is the outgrowth of teaching the first graduate-level quantum mechanics course in the Department of Chemistry at Stanford University for over twenty years. The course has a broad student body, which includes students from chemistry, chemical engineering, biophysics, biology, materials science, electrical, mechanical, and other engineering fields, and physics. The book was developed from a set of lecture notes that was provided to the students each year. After each year, the notes were corrected, updated, and added to. Several years ago, I began turning the lecture notes into this book.

The book is intended to be used as a first graduate-level course, taught in one semester or in one quarter of extended lectures. It assumes that the reader has some familiarity with the concepts of quantum theory—for example, the concept of the Uncertainty Principle, quantized energy levels, and so on—but it is self-contained. It starts with the most basic concepts and develops quantum theory from them. The major distinctions between this book and the many fine books on quantum chemistry are (a) its emphasis on a general approach that does not focus mainly on the Schrödinger representation of quantum theory and (b) a more extensive consideration of time-dependent problems than is usually provided in a first graduate course. In developing the course that grew into this book, I could not find an appropriate book to use. Books that seemed to be at the correct level of presentation for a first graduate course focused either totally or mainly on the Schrödinger representation and included only a minimal amount of material on time-dependent quantum mechanics. Books that contained the desired material—that is, a general presentation that embraced other representations of quantum theory and included time dependence to a significant degree—were too advanced, requiring too much prior knowledge of quantum mechanics, classical mechanics, and mathematics, for a first course. This book is intended to fill the gap.

The book begins with a qualitative discussion of two key concepts put forward by Dirac, absolute size and the superposition principle. These are used to draw a distinction between quantum theory and classical mechanics and to motivate the use of vector algebra (linear algebra) as the basic mathematics that underpins quantum mechanics. The necessary linear algebra, in the context of quantum theory, is supplied in Chapters 2 and 13 and in other portions of the book, so that readers who have not had a formal course in linear algebra can successfully follow the material. Throughout the book, sufficient mathematical detail and classical mechanics background are provided to enable readers to follow the quantum mechanical developments and analysis of physical phenomena. For example, in Chapter 12, Absorption and Emission of Radiation, the coupling of molecules (treated quantum mechanically) to a radiation field (treated classically) is described in considerable detail. The final results—that is, the expressions involving the transition dipole bracket (matrix element)—are relatively simple. The transition dipole bracket is frequently presented as a fait accompli, or it is developed in a manner that is so terse that a vast knowledge of classical electricity and magnetism is required to understand the derivation. Here, all of the necessary material is supplied so that the development can be followed and the nature of the approximations inherent in the final results can be appreciated. Chapter 12 ends with a brief description of the quantum theory of radiation, in which the radiation field is treated quantum mechanically rather than classically. This discussion is used to show that spontaneous emission emerges naturally from a quantum theory of radiation, although it must be grafted onto semiclassical theory in which the matter is treated quantum mechanically but the radiation field is treated classically.

The book has a distinctive flavor throughout. This flavor can be illustrated by briefly considering the content of several of the chapters. Chapter 6 treats the harmonic oscillator, first in the Schrödinger representation and then in the Dirac representation (raising and lowering operator formulation). The differences in the two approaches provide a graphic illustration of a fundamental aspect of quantum theory; that is, different representations involve distinct mathematical approaches to the same problem but yield identical calculated values for observable quantities. The Dirac raising and lowering operator method is usually found in books in chapters on the matrix formulation of quantum mechanics, and it is not presented in many books that treat the harmonic oscillator only with the Schrödinger equation. The raising and lowering operator methods provide an alternative to the Schrödinger approach that is more mathematically tractable. In addition, the Dirac representation of the harmonic oscillator is the basis of the quantum theory of solids and the quantum theory of radiation that is discussed at the end of Chapter 12. Chapter 15, Angular Momentum, develops the quantum mechanical theory of angular momentum with raising and lowering operator methods. It is difficult to read very far into modern literature without encountering raising and lowering operator techniques. The last section of Chapter 6 discusses the time evolution of a harmonic oscillator wave packet in the context of modern ultrafast spectroscopy. The harmonic oscillator wave packet is used to illustrate dynamics of molecular bonds brought about by an ultrashort pulse of light through the preparation of a vibrational wave packet on an electronic excited-state surface. Thus, the harmonic oscillator, a standard problem of early quantum theory, is connected to topics of current research interest.

Chapter 8, Time-Dependent Two-State Problem, uses the problem of the time evolution of two coupled quantum mechanical states to illustrate basic features of quantum theory and as a vehicle to discuss the nature of several important physical phenomena, namely, electron transfer, vibrational energy flow, and electronic excitation transfer. First, the time evolution caused by the coupling between two states of identical energy is examined in detail. Then the nondegenerate case is discussed, and it is used to qualitatively describe the influence of thermal fluctuations of solute–solvent interactions on the time evolution of transfer processes. The distinction between coherent oscillatory probability flow and incoherent probability hopping is made. At the end of the chapter, the ideas developed for the two-state problem are extended to an infinite number of states on a periodic lattice. The infinite problem is used to introduce the concept of the band structure of solids. The Bloch theorem of solid-state physics is developed and used to find the eigenstates and energy eigenvalues (band structure) for a one-dimensional exciton band. The transport of exciton wave packets is described through the calculation of exciton group velocities, and coherent and incoherent exciton transport are discussed. The exciton band problem is used to briefly explain electrical conductivity and resistive heating. Thus, the chapter on the time-dependent two-state problem is used to develop quantum theory methods, such as projection operators, in the context of physical problems that occur in chemistry, biology, materials science, and physics.

Chapter 14, The Density Matrix and Coherent Coupling of Molecules to Light, develops the density matrix as a formalism for solving time-dependent quantum mechanical problems. Time-dependent density matrix calculations have become important in many areas of modern research. In this chapter, the use of the time-dependent density matrix is illustrated through its application to the problem of coherent coupling of a radiation field to a two-level molecular system. The optical Bloch equations are derived, and they are used to calculate the time evolution of the population of a system that begins in the ground state. In contrast to the time-dependent perturbation theory treatment of molecule-radiation field coupling (Chapter 12), it is shown that strong coherent coupling can produce oscillation of population between the ground and excited state. The role of the off-diagonal density matrix elements—that is, the system's "coherences"—is described and illustrated with examples of free precession, frequency beating for systems composed of species with two different frequencies, and the free induction decay that arises from an inhomogeneous distribution transition frequencies. The close relationship between pulsed nuclear magnetic resonance and coherent optical experiments is brought out. The chapter on the time-dependent density matrix develops the formalism in the context of modern time-dependent coherent spectroscopy.

The examples given above illustrate the nature of material that is found throughout the book. Many topics are presented along conventional lines—for example, the hydrogen atom, which is developed in considerable detail. Chapter 17 is used to explain the nature of the covalent bond, one of the most important contributions of quantum theory to chemistry. The Born–Oppenheimer approximation is presented, and the electronic and nuclear (vibrational) "wave equations" are discussed. Very simple analytical treatments are used to approximately solve the hydrogen ion molecule and hydrogen molecule problems. This chapter is not a chapter on electronic structure calculations, although it is an introduction. Rather, it is used to demonstrate that the covalent bond is an inherently quantum mechanical effect. Numerical methods are very important in modern quantum theory. However, a conscious choice was made in developing this book to use examples that have analytical solutions because they can be worked out in detail within the confines of the pages of the book.

Following the last chapter of the text is a section containing problems for each chapter. These are problems that have been used in the quantum mechanics course that spawned the book. The problems are tied closely to the text, and they are used to amplify and expand the topics that are covered. In many instances, the problems are applications of methods developed in the text to important physical problems. Problems for later chapters use concepts and methods from earlier chapters.

This book will provide a solid grounding in the fundamentals of many aspects of quantum mechanics, and it explicates a variety of physical problems that are key components to understanding broad areas of physical science. It will bring readers to the point where they can focus their future efforts on more specialized topics in quantum theory.

CHAPTER 1

Absolute Size and the Superposition Principle

By the late 1920s, several formulations of quantum theory had been put forward. Schrödinger and Heisenberg developed two mathematically distinct representations of quantum mechanics. There were many discussions, and considerable disagreement, about which form of the theory was correct. By 1928 Dirac had advanced a general approach to quantum theory. He was able to demonstrate that both the Schrödinger and the Heisenberg methods were subsets of a more general theory. Dirac based his developments on two fundamental concepts: absolute size and the superposition principle. These two ideas are at the heart of the differences between classical physics and quantum theory. Before developing these ideas in detail and applying them to physical problems, they will be discussed qualitatively.

First consider the concept of size. In classical theory, big and small are relative concepts. Large objects are explained in terms of smaller constituents. Small objects are described in terms of still smaller ones. There is no end to this procedure. An object is big or small only in comparison to some other object. An elephant is big when compared to a mouse, but a mouse is big when compared to an ant. Size is relative.

Science deals with observables. You can only make an observation of an object by letting it interact with an outside influence. An object is big if the disturbance accompanying an observation can be neglected. An object is small if disturbance cannot be neglected. If you examine a wall by throwing bowling balls at it, you will cause a nonnegligible disturbance. But you can examine the wall with low-intensity light, and the disturbance will be negligible. Therefore, the size of an object is related to the disturbance associated with the observation of the object.

What is the size of the disturbance caused by observing an object? In classical theory it is assumed that you can reduce the size of the disturbance to be as small as desired. No matter what is under observation, it is possible to find an experimental method that will cause a negligible disturbance. This implies that *size is only relative.* The size of an object depends on the object and on your experimental technique. There is nothing inherent in the concept of size. Any object can be considered to be large by observing it with the right method, a method that causes a negligible disturbance.

Dirac put forward the concept of absolute size. He said, "There is a limit to the fineness of our powers of observation and the smallness of the accompanying disturbance—a limit which is inherent in the nature of things and can never be surpassed by improved technique or increased skill on the part of the observer." This is the fundamental assumption that gives absolute meaning to size. If an object is such that the unavoidable limiting disturbance accompanying a measurement is negligible, then the object is big. If the limiting disturbance is not negligible, the object is *small in an absolute sense*. Classical mechanics is not set up to describe objects that are small in an absolute sense.

One of the features of classical mechanics that appears to be straightforward, causality, is lost in quantum theory. Absolute size is intimately related to the nature of causality in quantum theory. In classical mechanics there is a causal relationship between events. Motions can be described by trajectories. If the initial position and momentum of an object are given and the forces acting on it are taken into account, it is possible to describe the trajectory of the object and predict future observables such as its position. If you throw a rock, its trajectory can be determined and its location at any later time can be predicted. The predictions can be tested by direct observation. The motion will follow the determined trajectory as long as the rock is not disturbed. Of course, if a bird flies into the rock in the middle of the rocks trajectory, it is not surprising that this nonnegligible disturbance will result in subsequent observations not agreeing with the predicted trajectory. In quantum theory, it is also possible to predict trajectories. For an object that is small in the absolute sense, the trajectories will hold unless an observation is made to test the prediction. Any observation of a small object will cause a nonnegligible disturbance because this is "inherent in the nature of things." Causality only applies to undisturbed systems. The act of observation of a small quantum mechanical system, by definition of a small object, causes a nonnegligible disturbance. Therefore the results of one observation will not allow a causal prediction of the result of a subsequent observation. This is not surprising from definition of a small quantum mechanical system. You can tell what a system is doing as long as you don't observe it. Indeterminacy comes in the calculation of observables. While the precise outcome of a series of observations cannot be predicted for a small object, quantum theory provides a method for calculating the probability that a particular outcome of an observation will occur.

There is nothing strange or mysterious about the lack of causality in quantum theory. In quantum theory, predictions can be made about undisturbed systems. For a small system, the act of observation causes a significant disturbance. Therefore, the lack of causality comes directly from the concept of absolute size.

The second concept Dirac put forward is the superposition principle. It is one of the most fundamental and drastic of the quantum mechanical laws. First, two examples will be discussed before formulating the principle in detail—that is, polarization of photons and interference of photons.

Consider a crystal used as a polarizer. Light polarized along one crystal axis propagates right through. Call this the parallel direction. Light polarized along the perpendicular axis doesn't go through. Classical electrodynamics, which represents light as a wave, can describe possible outcomes of an experiment. Parallel light goes through the polarizer, and perpendicular light is absorbed or reflected. For light polarized at an angle α to the parallel crystal axis, a fraction $\cos^2 \alpha$ passes through the polarizer (see Figure 1.1).

In explaining the photoelectric effect, Einstein put forward that light is not composed of waves. Light is composed of individual photons. When polarized light is used in the photoelectric effect to eject electrons, the electrons come out preferentially in certain directions. Thus polarization must be ascribed to individual photons. A beam of plane-polarized light is composed of plane-polarized photons. So the beam of light impinging

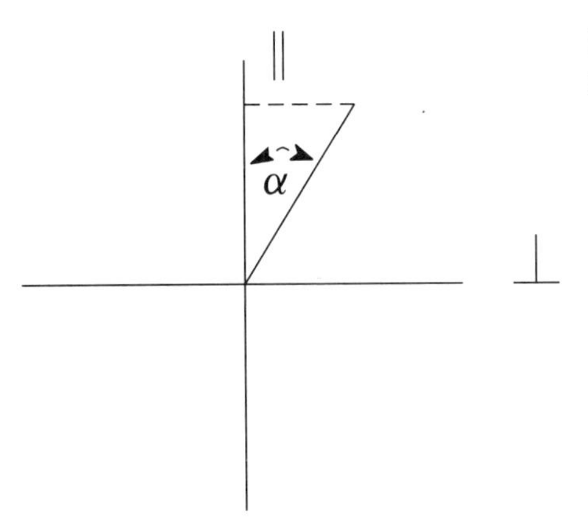

Figure 1.1 Projection of the electric field with polarization α onto the || light axis.

on the polarizer is composed of polarized photons. This does not create any conceptual problem if the photon polarization is parallel or perpendicular. Parallel photons go right through. Perpendicular photons are absorbed or reflected. However, for a photon polarized at some angle α, it is not clear what the outcome of an observation will be.

Quantum mechanics describes observables. Questions can only be asked about observables. Therefore, to discuss what happens, an experiment is needed. Consider an experiment in which photons are incident on the polarizer one at a time. Place a photodetector after the polarizer. The observable is the detection of a photon at the back side of the polarizer. Quantum mechanics predicts the result. Sometimes you observe a whole photon with the same energy as the incident photon. Sometimes no photon is observed. When a photon is found, it is always polarized parallel. The photon has somehow jumped from having polarization α to polarization parallel. If the experiment is conducted for a large number of photons, $\cos^2\alpha$ are observed at the back side of the polarizer.

To understand this result, it is necessary to consider a superposition of photon states. A photon of polarization α, P_α, is some type of superposition of the states of polarizations parallel, $P_\|$, and perpendicular, $P_\perp$. Any state of polarization can be resolved into, or expressed as, a superposition of two mutually perpendicular states of polarization,

$$P_\alpha = aP_\| + bP_\perp,$$

where a and b are coefficients that tell how much of each polarization is contained in the superposition that is P_α. (Standard notation will be introduced below.) When a photon meets the polarizer, an observation is being made as to whether it is polarized parallel or perpendicular. The act of observation forces the photon entirely into one or the other of the states that make up the superposition—that is, entirely parallel or entirely perpendicular. The act of observation causes the photon to make a sudden jump from being partially in each of the polarizations in the superposition to being in only one. For one photon, it is not possible to say which will occur, but the outcome of many observations yields the classical result: the fraction $\cos^2\alpha$ of the incident photons has parallel polarization.

A second qualitative illustration of the superposition principle involves the interference of light. An incoming light beam impinges on a 50% reflecting mirror (a beam splitter). Half of the beam is transmitted and half is reflected. Each beam reflects off of a subsequent mirror and is sent approximately back on itself. The two beams are made to overlap, crossing at a small angle. This produces an optical interference pattern in the beam overlap region. The apparatus and the interference pattern are illustrated in Figure 1.2.

The interference pattern is easily explained using classical electromagnetic (E & M) theory, which takes light to be a wave. In the classical picture, half of the wave reflects from the mirror and half of it passes through. The two waves come back and overlap to produce the interference pattern. Since the waves are crossed at an angle, the distance each of the waves travels to reach a particular point along the x axis (see Figure 1.2) varies with x. At some points, the distances traveled are such that the waves add in phase, resulting in a maximum in the pattern. At some points, the differences in the distances traveled are one-half of a wavelength. The waves are 180° out of phase, causing destructive interference and a null in the pattern. Each of the waves is described by a wavefunction, and it is straightforward to calculate the details of the interference pattern.

Difficulties with the classical description arose once it was recognized that the incoming beam of light is composed of photons, not waves. Initially, it was thought that nothing would be different if the beams of light were simply viewed as beams of particles instead of waves. Then at the beam splitter, half of the photons are reflected and half are transmitted. These two beams of distinct photons recombine to give the interference pattern. Photons in one beam were believed to produce the alternating regions of light and dark by either combining with or annihilating photons in the other beam. To make this consistent with the classical E & M description, it was thought that the classical E & M wavefunction described the number of photons in a region of space.

Defining the wavefunction to describe the number of photons in a region of space is a serious error. There are many problems with this description. One is that the pattern is independent of the intensity of the incoming light beam. If a piece of film is used to record the pattern, as the intensity of the light is reduced, it takes longer to obtain the pattern, but the pattern is identical. This is true even if the intensity is reduced such that only one

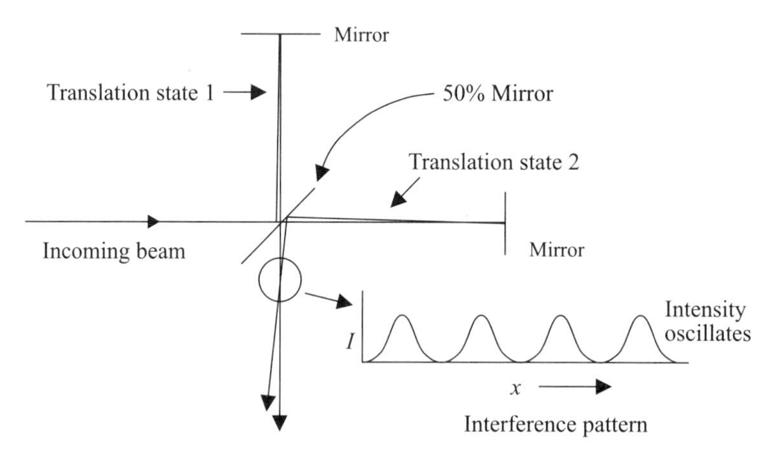

Figure 1.2 A schematic of an optical interference apparatus showing beam splitting and then spatial overlap of the resulting two beams. A blow-up is shown of the beam overlap region and the resulting interference pattern.

photon per hour impinges on the apparatus. In the context of the incorrect view that the wavefunction describes the number of photons in a region of space, any one photon goes into only one leg of the apparatus. With only one photon in the apparatus at a time, there is no photon in the other leg to interfere with. Thus the interference pattern should vanish.

The quantum mechanical analysis of this problem describes the situation in terms of a superposition of two photon translation states. A photon in one leg of the apparatus is in translation state 1, T_1; a photon in the other leg is in translation state 2, T_2. When a photon meets the beam splitter, there is equal probability of finding the photon in either leg of the apparatus. The state of the system is an equal superposition of T_1 and T_2. In some sense, the state of the system, T, is a superposition of the two possible translation states, that is,

$$T = T_1 + T_2.$$

Each photon is in both legs of the apparatus in that there is equal probability of finding the photon in each leg. The interference comes from each photon interfering with itself in the beam overlap region. Thus the interference pattern will be identical regardless of the number of photons that are in the apparatus at a given time. Each translation state of the photon is described with a classical E & M wavefunction. The fundamentally important point that comes out of the superposition state description of the light interference problem is that *the wavefunction gives the probability of finding a photon in a region of space*, not the number of photons in a region of space. The connection between light waves and photons is statistical in nature.

To formulate the superposition principle, it is first necessary to define the term "the state of a system." A system is composed of a collection of bodies that have certain properties such as mass, moment of inertia, or charge. The bodies interact according to specific laws of force. Certain motions are consistent with the properties of the bodies and the laws of force. Each such motion is a state of the system. Dirac gave the definition as follows:

❖ **Definition:** The state of a system is an undisturbed motion that is restricted by as many conditions as are theoretically possible without contradiction.

A state may be the state at one particular time or the state of the system throughout time. An example is a proton and an electron interacting through a Coulomb interaction to form a hydrogen atom. The lowest energy state of this system is the $1s$ state. There are other states—for example, the $2s$ and the $2p$. The nature of these distinct states is determined by the charges, masses, and the interaction. (The states of the hydrogen atom are discussed in detail in Chapter 7.)

Dirac put forward the superposition principle as a fundamental assumption of quantum theory.

❖ **Assume:** Whenever a system is in one state, it can always be considered to be partly in each of two or more states.

The original state is regarded as a superposition of two or more other states. Conversely, two or more states can be superposed to give a new state. This is nonclassical superposition. In mathematics, it is always possible to form superpositions, but they may not be physically useful. As the qualitative examples of the polarization of photons and the interference of photons showed, the superposition principle is essential for the understanding of quantum mechanical systems because it allows observables to be explained.

As will be discussed quantitatively below, the superposition principle is central to the calculation of quantum mechanical observables. Standard Dirac notation for states of a system will be introduced in the next chapter, but first consider a system having states A and B. Observation of system in state A yields the result a. Observation of the system in state B yields the result b. Observation on a superposition of A and B gives either a or b. An observation never yields any other result. The probability of obtaining result a or b in a single measurement depends on relative weights of A and B in the superposition. Quoting Dirac, "The intermediate character of the state formed by superposition thus expresses itself through the probability of a particular result for an observation being 'intermediate' between the corresponding probabilities for the original state, *not through the result itself being intermediate between the corresponding results for the original states.*" For example, if a hydrogen atom is placed in a superposition of its $2p$ and $3p$ states, and the energy is measured, on a single measurement, either the $2p$ energy will be observed or the $3p$ energy will be observed. An energy will never be observed that is between the $2p$ and $3p$ energies even though the state of the system is a superposition of the two atomic states. When making a series of observations on identically prepared atomic size systems, the result from one observation to the next in general will vary. By making enough observations, it is possible to map out a probability distribution for the results. Quantum mechanics provides the method for the calculation of such probability distributions.

CHAPTER

2 Kets, Bras, Operators, and the Eigenvalue Problem

A. KETS AND BRAS

In Chapter 1 the superposition principle was discussed qualitatively. To develop quantum theory, it is necessary to have a mathematical formulation that builds the superposition principle into the theory. Superposition implies that states are added together to get new states. The mathematical entities associated with quantum mechanical states must also be able to be added together to obtain a representation of the superposition state. Vectors have this property. In real three-dimensional space, any vector can be written as the superposition of the three basis vectors. Under appropriate combination rules, three vectors can be combined to get new basis vectors that are also capable of describing a general vector. Therefore, vectors have the necessary property to be used to represent quantum mechanical states. In quantum mechanical systems, there can be many more than three states. One vector is needed for each state. There will be a finite or infinite number of vectors depending on whether there are a finite or infinite number of states of the system.

Following Dirac, the quantum mechanical vectors are called ket vectors or just kets. For a general ket, the symbol is $|\ \rangle$. For a particular one, A, the symbol is $|A\rangle$. Kets can be multiplied by a complex number and added, that is,

$$|R\rangle = C_1 |A\rangle + C_2 |B\rangle, \tag{2.1}$$

where C_1 and C_2 are the complex numbers. Any number of ket vectors can be added,

$$|S\rangle = \sum_i C_i |L\rangle; \tag{2.2}$$

or if $|x\rangle$ is continuous over some range of x,

$$|Q\rangle = \int C(x) |x\rangle \, dx. \tag{2.3}$$

Ket vectors can be dependent or independent. A ket that is expressible linearly in terms of other kets is dependent on them:

$$|R\rangle = C_1 |A\rangle + C_2 |B\rangle. \tag{2.4}$$

$|R\rangle$ is dependent on $|A\rangle$ and $|B\rangle$. A set of kets is independent if no one of them is expressible linearly in terms of the others.

So far a set of vectors, the kets, have been defined which can be multiplied by complex numbers and added. To proceed, the following assumption is made.

❖ **Assume:** Each state of a dynamical system at a particular time corresponds to a ket vector, the correspondence being such that if a state results from a superposition of other states, its corresponding ket vector is expressible linearly in terms of the corresponding ket vectors of the other states.

This is the basic connection between the physical system and the mathematical formulation.

The order of superposition of kets does not matter:

$$|R\rangle = C_1\,|A\rangle + C_2\,|B\rangle = C_2\,|B\rangle + C_1\,|A\rangle\,. \tag{2.5}$$

If $|R\rangle = C_1\,|A\rangle + C_2\,|B\rangle$, then $|A\rangle$ can be expressed in terms of $|R\rangle$ and $|B\rangle$: $|A\rangle = b_1\,|R\rangle + b_2\,|B\rangle$. Because each state of a system corresponds to a ket, a state is dependent on other states if its ket is dependent on the corresponding kets of the other states.

The superposition of a ket with itself does not result in a new ket but is a ket corresponding to the same state as the original ket:

$$C_1\,|A\rangle + C_2\,|A\rangle = (C_1 + C_2)\,|A\rangle \Rightarrow |A\rangle \tag{2.6}$$

unless $(C_1 + C_2) = 0$; since a ket is a vector, it must have a finite length. In the same manner, a ket corresponding to a state multiplied by any complex number (other than 0) gives a ket corresponding to the same state. The important point is that a state corresponds to the direction of the ket vector. The length is irrelevant, and the sign is irrelevant. This represents a significant difference between classical and quantum mechanical superposition. Consider classical standing waves on a string with fixed ends (see Figure 2.1). In a classical system of this type, if two identical states are added together, the result is a standing wave with twice the amplitude. This is a different state of motion. In quantum theory, the addition of identical states does not result in a different state. In the classical system, it is possible to have the state of zero amplitude—that is, no motion of the string. In quantum mechanics, a ket cannot have zero amplitude, since the state of the system corresponds to the direction of the ket vector, and a vector cannot have zero length.

The ket vectors that correspond to the states of a physical system have been introduced. In mathematics, whenever there exists a set of vectors, another set of vectors, called dual vectors, can be formed. A number ϕ can be associated with a ket vector $|A\rangle$, with a linear function of $|A\rangle$, that is,

$$|A\rangle + |A'\rangle \Rightarrow \phi + \phi'$$
$$C\,|A\rangle \Rightarrow C\phi.$$

A number that is a linear function of a vector is obtained by taking the scalar product of the vector with some other vector. These other vectors are call bra vectors or just bras. For a general bra, the symbol is $\langle\,|$. For a particular one, B, the symbol is $\langle B|$. A bra vector and a ket vector combined as $\langle \text{bra}|\text{ket}\rangle$ is called a bracket. Any complete bracket, $\langle\,|\,\rangle$,

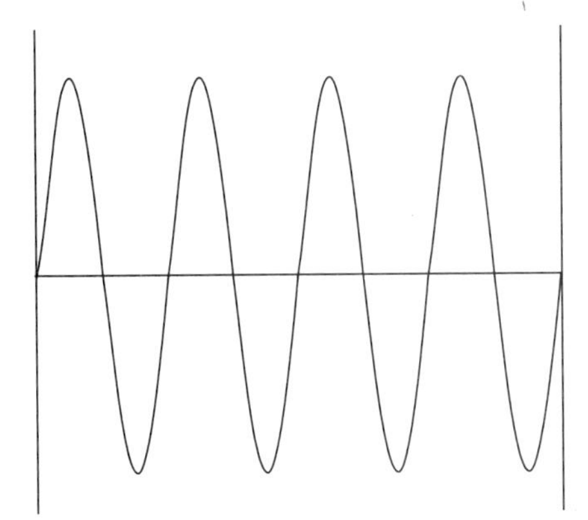

Figure 2.1 A standing wave on a classical string with fixed ends.

is a number. A bracket represents the scalar product of a bra vector and a ket vector. An incomplete bracket is a vector, and

$$\langle B| \left\{ |A\rangle + \left|A'\right\rangle \right\} = \langle B|A\rangle + \left\langle B|A'\right\rangle \tag{2.7}$$

$$\langle B| \{C\,|A\rangle\} = C\,\langle B|A\rangle\,. \tag{2.8}$$

A bra is completely defined when its scalar product with every ket is known. If $\langle P|A\rangle = 0$ for all $|A\rangle$, then $\langle P| = 0$. The addition of bras is defined by the scalar product with ket $|A\rangle$,

$$\left\{ \langle B| + \left\langle B'\right| \right\} |A\rangle = \langle B|A\rangle + \left\langle B'|A\right\rangle. \tag{2.9}$$

The multiplication of a bra by C, a complex constant, is defined by the scalar product with $|A\rangle$,

$$\{C\,\langle B|\}\,|A\rangle = C\,\langle B|A\rangle\,. \tag{2.10}$$

The bra vectors are different from the ket vectors. So far the only connection is the scalar product.

❖ **Assume:** There is a one-to-one correspondence between kets and bras such that the bra corresponding to $|A\rangle + \left|A'\right\rangle$ is the sum of the bras corresponding to $|A\rangle$ and $\left|A'\right\rangle$ and that the bra corresponding to $C\,|A\rangle$ is $\overline{C}$ or C^* times the bra corresponding to $|A\rangle$.

C^* or $\overline{C}$ is the complex conjugate of the complex number C. Thus the bra corresponding to $|A\rangle$ is the complex conjugate $\overline{|A\rangle}$. The bra corresponding to $|A\rangle$ is $\langle A|$. The state of a dynamical system can be specified by the direction of a bra as well as a ket. The theory is symmetrical between bras and kets.

The complex conjugate relationship between bras and kets implies

$$\langle B|A\rangle = \overline{\langle A|B\rangle}, \tag{2.11}$$

since $\langle A| = \overline{|A\rangle}$ and $\langle B| = \overline{|B\rangle}$. Setting $|B\rangle = |A\rangle$, it is clear that $\overline{\langle A|A\rangle}$ must be real, because a number equal to its complex conjugate is real. $\langle A|A\rangle$ is the scalar product of the ket vector with its corresponding bra. It is a real number. Assuming $\langle A|A\rangle > 0$, then the length of the ket vector is a positive real number unless $|A\rangle = 0$.

The scalar product of real space vectors gives a real number, and the scalar product is symmetrical with respect to interchange of the order of multiplication, that is,

$$\vec{A} \cdot \vec{B} = \vec{B} \cdot \vec{A}.$$

The scalar product of a bra and a ket gives a complex number,

$$\langle A|B\rangle = \overline{\langle B|A\rangle}$$

and interchange of the order gives the complex conjugate of the number.

A bra and a ket are orthogonal if their scalar product is zero, that is,

$$\langle B|A\rangle = 0.$$

Two kets, $|B\rangle$ and $|A\rangle$, are orthogonal if

$$\langle B|A\rangle = 0;$$

and two bras, $\langle B|$ and $\langle A|$, are orthogonal if

$$\langle B|A\rangle = 0$$

Two states of a dynamical system are orthogonal if the vectors representing them are orthogonal.

While it is the direction of a ket or bra vector, not the length, that defines the state of a system, it is frequently convenient to normalize the lengths of the vectors that correspond to the states of a system. The length of a bra, $\langle A|$, or a ket, $|A\rangle$, is $(\langle A|A\rangle)^{1/2}$. If

$$(\langle A|A\rangle)^{1/2} = 1$$

for all $|A\rangle$, the states are said to be normalized. Even when the direction and the length of a quantum mechanical vector are defined, it is still not completely determined, since it can be multiplied by $e^{i\gamma}$ with γ a real number. This changes neither the length nor the direction of a ket. For a normalized ket, $|A\rangle$, that is,

$$(\langle A|A\rangle)^{1/2} = 1,$$

if

$$|A'\rangle = e^{i\gamma}|A\rangle,$$

then

$$\langle A'| = e^{-i\gamma}\langle A|$$

and

$$(\langle A'|A'\rangle)^{1/2} = [e^{-i\gamma}e^{i\gamma}\langle A|A\rangle]^{1/2} = 1.$$

Therefore, multiplying by $e^{i\gamma}$ does not change the length of the ket. $e^{i\gamma}$ is called a phase factor. Phase factors in which γ is a function of time will be used in a number of time-dependent problems discussed in following chapters.

Ket vectors and bra vectors are used to mathematically represent states of a physical system. A number of assumptions have been made and rules put forward that relate states of dynamical systems to the abstract quantities, kets and bras. Development of quantum theory lets mathematical relations involving kets and bras go over to physical observables. The object is to be able to predict and understand factors affecting experimental observables. To go further, it is necessary to introduce the manner in which kets and bras are manipulated.

B. LINEAR OPERATORS

A ket $|F\rangle$ can be a function of a ket $|A\rangle$, that is,

$$|F\rangle = \underline{\alpha}\,|A\rangle\,, \tag{2.12}$$

where $\underline{\alpha}$ is an operator. (Operators are designated by an underline.) In quantum theory, the operators are linear. An operator is a linear operator if it obeys the two relations

$$\underline{\alpha}\left(|A\rangle + |A'\rangle\right) = \underline{\alpha}\,|A\rangle + \underline{\alpha}\,|A'\rangle, \tag{2.13}$$

and with C a complex number

$$\underline{\alpha}(C\,|A\rangle) = C\underline{\alpha}\,|A\rangle\,. \tag{2.14}$$

A linear operator is completely defined when the result of its application to every ket is known. Linear operators are additive:

$$[\underline{\alpha} + \underline{\beta}]\,|A\rangle = \underline{\alpha}\,|A\rangle + \underline{\beta}\,|A\rangle\,. \tag{2.15}$$

Multiplication of linear operators is associative:

$$[\underline{\alpha}\,\underline{\beta}]\,|A\rangle = \underline{\alpha}\,[\underline{\beta}\,|A\rangle] = \underline{\alpha}\,\underline{\beta}\,|A\rangle\,. \tag{2.16}$$

However, in general, multiplication of linear operators is not necessarily commutative:

$$\underline{\alpha}\,\underline{\beta}\,|A\rangle \neq \underline{\beta}\,\underline{\alpha}\,|A\rangle\,. \tag{2.17}$$

It will be seen in subsequent chapters that the noncommutivity of linear operators is extremely important in connection with the calculation of observables and the Uncertainty Principle. In special cases, linear operators may commute. If

$$\underline{\gamma}\,\underline{\delta}\,|A\rangle = \underline{\delta}\,\underline{\gamma}\,|A\rangle\,,$$

for all $|A\rangle$ the operators $\underline{\delta}$ and $\underline{\gamma}$ are said to commute. The statement that two operators commute will sometimes be expressed as $\underline{\gamma}\,\underline{\delta} = \underline{\delta}\,\underline{\gamma}$. This is a shorthand notation. It does not mean that $\underline{\gamma}\,\underline{\delta}$ is identical to to $\underline{\delta}\,\underline{\gamma}$, but rather that when $\underline{\gamma}\,\underline{\delta}$ is applied to an arbitrary ket, the result is the same as applying $\underline{\delta}\,\underline{\gamma}$ to the ket.

In working with kets and linear operators, the ket is placed to the right of the operator and the operator operates to the right. Linear operators can also operate on bras. The bra is placed to the left of the linear operator—that is, $\langle B|\,\underline{\alpha}$—and the operator operates to the left.

A linear operator applied to a ket vector yields a ket vector. A linear operator applied to a ket and left multiplied by a bra is a number. To see this consider

$$\langle B|\,\underline{\alpha}\,|A\rangle = \langle B|\,\{\underline{\alpha}\,|A\rangle\},$$

but $\{\underline{\alpha}\,|A\rangle\}$ yields a new ket $|Q\rangle$. So,

$$\langle B|\,\underline{\alpha}\,|A\rangle = \langle B|Q\rangle = C,$$

a complex number. A closed bracket is always a number.

In certain circumstances, a combination of a ket and a bra can act as a linear operator. $\langle A|B\rangle$ is a number, but $|A\rangle\langle B|$ is a linear operator. $|A\rangle\langle B|$ is not a closed bracket. To see that $|A\rangle\langle B|$ is a linear operator, apply it to an arbitrary ket, $|P\rangle$:

$$|A\rangle\langle B|P\rangle = |A\rangle\,\phi = \phi\,|A\rangle\,, \tag{2.18}$$

where ϕ is a number because $\langle B|P\rangle$ is a closed bracket. Operating on $|P\rangle$ gave $|A\rangle$, a new ket. The application of $|A\rangle\langle B|$ to a ket yields another ket. Therefore, $|A\rangle\langle B|$ is a linear operator. $|A\rangle\langle B|$ can also operate on a bra $\langle Q|$.

$$\langle Q|A\rangle\langle B| = \theta\,\langle B| \tag{2.19}$$

θ is a number since $\langle Q|A\rangle$ is a closed bracket. Operating on $\langle Q|$ yields a new bra, $\langle B|$. Linear operators of this type can play an important role in quantum theory. Linear operators of the form $|A\rangle\langle A|$ are called projection operators. They will be used in subsequent chapters.

To this point, an algebra involving bras, kets, and linear operators has been introduced.

1. The associative law of multiplying holds.
2. The distributive law holds.
3. The commutative law does not hold.
4. $\langle\ \rangle$, a closed bracket, is a number.
5. $\langle\ |$ and $|\ \rangle$ are vectors.

The bras and kets correspond to the state of a dynamical system at a given instant of time. The linear operators also have physical significance.

❖ **Assume:** The linear operators correspond to the dynamical variables of a physical system.

The dynamical variables are physical properties such as coordinates, components of velocity, momentum, angular momentum, dipole moment, and various functions of these. The linear operators correspond to the questions that can be asked about a system—that is, system observables. If the question is, "What is the energy?", there is an energy linear operator. If the question is, "What is the momentum?", there is a momentum operator. If the question is, "What is the location of a particle?", there is a position operator. For every observable, there is a linear operator. However, there is not a single mathematical form for a linear operator representing a particular observable. There are sets of linear operators for the dynamical variables that are internally consistent. Each such set forms a "representation" in quantum theory. Mathematically, the linear operators in different representations can be very different. However, the results of the calculation of an observable for a given system will be identical, independent of the choice of the representation. The prescription for selecting linear operators is given by Dirac's "Quantum Condition," which is presented in Chapter 4.

There is a fundamental difference between dynamical variables in classical mechanics and quantum mechanics. In quantum theory, linear operators, which represent the dynamical variables, are subject to an algebra in which the commutative law of multiplication does not hold. This is a consequence of developing a formalism that builds the superposition principle into the description of systems that are small in an absolute sense.

There are a number of useful relations involving kets, bras, and linear operators; these are called conjugate relations. It was seen above that $\langle B|A\rangle = \overline{\langle A|B\rangle}$. Furthermore,

$$\langle B|\overline{\underline{\alpha}}|P\rangle = \overline{\langle P|\underline{\alpha}|B\rangle}, \tag{2.20}$$

where $\overline{\underline{\alpha}}$ is the complex conjugate of the operator $\underline{\alpha}$. It is called the adjoint. It represents the complex conjugate of a dynamical variable. $\overline{\overline{\underline{\alpha}}} = \underline{\alpha}$, the adjoint of the adjoint equals the original operator. If $\overline{\underline{\alpha}} = \underline{\alpha}$, the operator is self-adjoint. It corresponds to a real dynamical variable. As mentioned above, this statement does not mean that $\overline{\underline{\alpha}}$ is algebraically equal to $\underline{\alpha}$, but rather that the application of either to an arbitrary ket gives the same result. The same statement applies to the following relations as well.

$$\overline{\underline{\beta}}\,\overline{\underline{\alpha}} = \overline{\underline{\alpha}\underline{\beta}} \tag{2.21}$$

$$\overline{\underline{\alpha}\underline{\beta}\underline{\gamma}} = \overline{\underline{\gamma}}\,\overline{\underline{\beta}}\,\overline{\underline{\alpha}} \tag{2.22}$$

$$\overline{|A\rangle\langle B|} = |B\rangle\langle A| \tag{2.23}$$

In general, the complex conjugate of any product of bras, kets, and linear operators is the complex conjugate of each factor with factors in reverse order.

C. EIGENVALUES AND EIGENVECTORS

Consider the equation

$$\underline{\alpha}|P\rangle = p|P\rangle \tag{2.24}$$

A linear operator $\underline{\alpha}$ is applied to a ket $|P\rangle$, and the result is to obtain the same ket back multiplied by a number, p. Equation (2.24) is called an eigenvalue equation. Usually the operator $\underline{\alpha}$ is known, and the problem is to find p and $|P\rangle$. This is called an eigenvalue problem. The kets, $|P\rangle$, that solve the equation are called eigenkets, and the associated numbers are called eigenvalues. The trivial solution, $|P\rangle = 0$, is ignored. In general, the same operator, $\underline{\alpha}$, applied to other kets will change both their length and direction.

The eigenvalue problem can also be written in terms of bras; that is,

$$\langle Q| \underline{\alpha} = q \langle Q| , \tag{2.25}$$

where $\langle Q|$ is an eigenbra of $\underline{\alpha}$ with eigenvalue q. Eigenbras and eigenkets are referred to as eigenvectors. Eigenvectors and eigenvalues only have meaning in reference to a particular linear operator. It is said that an eigenvector belongs to an eigenvalue of a linear operator.

The eigenvalue problem belongs to a branch of mathematics called linear algebra. Treated strictly as a mathematical problem, there is a great deal known about the properties of eigenvalue equations and the solution of eigenvalue problems. Some of the results of the mathematical analysis of eigenvalue problems have important physical meanings; these results are used in the analysis of quantum mechanical problems.

Observables in quantum theory are obtained from eigenvalue equations. Kets and bras represent the states of a dynamical system. Linear operators represent the dynamical variables. Observables are associated with real dynamical variables. Observables are eigenvalues of Hermitian linear operators. Hermitian linear operators are self-adjoint, that is,

$$\langle a| \underline{\gamma} |b\rangle = \langle a| \overline{\underline{\gamma}} |b\rangle ; \tag{2.26}$$

and if $|b\rangle = |a\rangle$, then

$$\langle a| \overline{\underline{\gamma}} |a\rangle = \langle a| \underline{\gamma} |a\rangle . \tag{2.27}$$

For every observable there will be a Hermitian linear operator and a corresponding eigenvalue equation. If the question is, "What is the momentum?", there is a momentum operator and a momentum eigenvalue problem. The eigenvalues that result are the possible observable values of the momentum of the system, and the eigenvectors represent the states of the system that correspond to the eigenvalues. If the question is, "What is the energy?", there is a different linear operator and a different eigenvalue problem. The eigenvalues of the energy eigenvalue problem are the observable energies of a system, and the eigenvectors are the associated representations of the states having the energy eigenvalues. As will be proven below, the eigenvalues of Hermitian operators are real numbers. Since laboratory observables are always real numbers, the linear operators that represent observables are Hermitian.

It is straightforward to prove a number of important relations concerning the eigenvalue problem and observables. The proofs are not difficult and will be presented because the results are used frequently.

If an eigenvector of $\underline{\alpha}$ is multiplied by any number C, then the product is still an eigenvector with the same eigenvalue:

$$\underline{\alpha} |A\rangle = a |A\rangle$$

$$\underline{\alpha}[C |A\rangle] = C[\underline{\alpha} |A\rangle] = C[a |A\rangle] = a[C |A\rangle]$$

Multiplying an eigenvector by a constant does not change the eigenvalue. This is the reason that only the direction of the ket matters, not its length. It may be convenient to normalize kets, but normalization does not change the values of the observables that are calculated. A ket can be multiplied by any complex number other than zero, and the ket will still represent the same state of the system because it will still generate the same observable eigenvalue.

Any superposition of several independent eigenkets of a linear operator having the same eigenvalues is also an eigenvector having the same eigenvalue:

$$\underline{\alpha}\,|P_1\rangle = p\,|P_1\rangle$$

$$\underline{\alpha}\,|P_2\rangle = p\,|P_2\rangle$$

$$\underline{\alpha}\,|P_3\rangle = p\,|P_3\rangle$$

$$\begin{aligned}\underline{\alpha}\,\{C_1\,|P_1\rangle + C_2\,|P_2\rangle + C_3\,|P_3\rangle\ \} &= C_1\underline{\alpha}\,|P_1\rangle + C_2\underline{\alpha}\,|P_2\rangle + C_3\underline{\alpha}\,|P_3\rangle \\ &= p\,\{C_1\,|P_1\rangle + C_2\,|P_2\rangle + C_3\,|P_3\rangle\ \}\,.\end{aligned}$$

Thus the eigenvalue of the superposition is the same as the eigenvalue of the eigenkets that compose the superposition. An example is the three different p orbitals of the hydrogen atom, usually written as p_x, p_y, and p_z, which all have the same energy. These are eigenkets of the energy eigenvalue problem for the hydrogen atom. As discussed in Chapter 7, the solution of the hydrogen energy eigenvalue problem in the Schrödinger representation yields three eigenvectors with the same energy, p_1, p_0, and p_{-1} where the subscripts are designations of an angular momentum quantum number. The commonly used p_x, p_y, and p_z are superpositions of p_1, p_0, and p_{-1}.

For observables, the eigenvalues associated with the eigenkets are the same as the eigenvalues associated with the eigenbras. Assume they are not the same. Since an observable is involved, the operator $\underline{\gamma}$ is Hermitian, that is, $\overline{\underline{\gamma}} = \underline{\gamma}$:

$$\underline{\gamma}\,|P\rangle = a\,|P\rangle$$

The equivalent equation, but for bras and assuming the eigenvalues are not the same would be the complex conjugate equation with a different eigenvalue, b. Since $\overline{\underline{\gamma}} = \underline{\gamma}$, it can be written

$$\langle P|\,\underline{\gamma} = b\,\langle P|\,.$$

Left-multiplying the top eigenvalue equation by $\langle P|$ and right-multiplying the bottom equation by $|P\rangle$ gives

$$\langle P|\,\underline{\gamma}\,|P\rangle = a\,\langle P|P\rangle$$

$$\langle P|\,\underline{\gamma}\,|P\rangle = b\,\langle P|P\rangle\,,$$

respectively. Subtracting,

$$0 = (a - b)\,\langle P|P\rangle\,.$$

But,

$$\langle P|P\rangle > 0$$

since this is the length of the ket vector. Therefore,

$$a = b.$$

For this reason, the theory is said to be symmetrical in bras and kets. An eigenbra, which is the complex conjugate of an eigenket, yields the same observable value.

The complex conjugate of any eigenket is an eigenbra belonging to the same eigenvalue, and conversely.

This can be proven in a similar manner, but it will not be done here.

The eigenvalues of Hermitian operators are real:

$$\underline{\gamma}\,|P\rangle = a\,|P\rangle\,.$$

Taking the complex conjugate of this equation yields

$$\langle P|\,\overline{\underline{\gamma}} = \langle P|\,\overline{a}.$$

Since the operator is Hermitian,

$$\langle P|\,\underline{\gamma} = \langle P|\overline{a}.$$

Left-multiplying the first equation by $\langle P|$ and right-multiplying the third equation by $|P\rangle$ yields

$$\langle P|\,\underline{\gamma}\,|P\rangle = a\,\langle P|P\rangle$$
$$\langle P|\,\underline{\gamma}\,|P\rangle = \overline{a}\,\langle P|P\rangle\,.$$

Subtracting gives

$$0 = (a - \overline{a})\,\langle P|P\rangle\,.$$

Since $\langle P|P\rangle > 0$ (the length of a vector), $(a - \overline{a})$ must be zero, and

$$\overline{a} = a.$$

Therefore, a is real because a number equal to its complex conjugate is real.

The next relationship is referred to as the Orthogonality Theorem.

Two eigenvectors of a real dynamical variable (Hermitian operator - observable) belonging to different eigenvalues are orthogonal:

$$\underline{\gamma}\,|\gamma'\rangle = \gamma'\,|\gamma'\rangle$$
$$\underline{\gamma}\,|\gamma''\rangle = \gamma''\,|\gamma''\rangle\,.$$

Forming the complex conjugate of the first equation yields

$$\langle\gamma'|\,\underline{\gamma} = \langle\gamma'|\,\gamma'$$

since $\underline{\gamma}$ is Hermitian, and the eigenvalue, γ', is real.

Right-multiplying by $|\gamma''\rangle$ gives

$$\langle\gamma'|\,\underline{\gamma}\,|\gamma''\rangle = \gamma'\,\langle\gamma'|\gamma''\rangle\,.$$

Left-multiplying the second equation by $\langle\gamma'|$ gives

$$\langle\gamma'|\,\underline{\gamma}\,|\gamma''\rangle = \gamma''\langle\gamma'|\gamma''\rangle.$$

Subtracting yields

$$\langle\gamma'|\gamma''\rangle(\gamma' - \gamma'') = 0.$$

But, $\gamma' \neq \gamma''$ because these are different eigenvalues. Therefore,

$$\langle\gamma'|\gamma''\rangle = 0;$$

$|\gamma'\rangle$ and $|\gamma''\rangle$ are orthogonal.

So far, kets, bras, linear operators, and eigenvalue problems have been introduced. The relationship of these to the states of physical systems has been discussed. A prescription for the calculation of observables associated with physical systems has been provided. While substantially more formalism will be presented in subsequent chapters, there is now enough formalism on hand to begin to address problems of physical significance.

CHAPTER 3

MOMENTUM OF A FREE PARTICLE AND WAVE PACKETS

In this chapter the momentum states of a free particle and the description of free particles in terms of wave packets will be discussed using the formalism and ideas presented in the previous chapters. Eigenvectors, linear operators, and eigenvalues have been introduced.

- **Hermitian linear operators** represent real dynamical variables, which are observable quantities.

- **Eigenvectors** represent the states of a system associated with a particular observable quantity.

- **Eigenvalues** are the values of the observables associated with a particular linear operator and its eigenvectors.

Using these, the momentum states of a free particle and the nature of the momentum eigenstates will be determined. Superpositions of momentum eigenstates will be formed to obtain wave packets, and some characteristics of wave packets and their motion will be described. The quantum description of a free particle will be contrasted with the classical mechanics picture. The free-particle problem is important because the concepts that are used to understand its quantum nature are central to the more complex problems that are addressed is subsequent chapters.

A free particle is a hypothetical entity. It is an object, such as an electron or a photon, that is traveling through space with no forces acting upon it. It is hypothetical in the sense that even in interstellar space, there are gravitational forces and electromagnetic forces acting on a particle. However, in many real situations, particles behave essentially as free particles. A photon moving in a vacuum or even through air is well-described as a free particle. An electron moving in a vacuum chamber toward a sample in a low-energy electron diffraction (LEED) experiment is also basically acting as a free particle. Therefore, describing a free particle properly is important for understanding realizable experimental situations. A free particle is the simplest particle to describe quantum mechanically. There are no forces acting on it; no potential energy enters into the problem. Classically, a free particle moves in a straight line. At a given instant of time, it has a well defined position and momentum. Once these are known, the position of the classical free particle can be determined for all time. This classical behavior is distinct from the behavior of a quantum free particle, and the differences illustrate the fundamental differences between the classical and quantum description of matter.

A. MOMENTUM STATES OF A FREE PARTICLE

To determine the momentum of a quantum mechanical free particle, it is necessary to solve the momentum eigenvalue problem. The solution will yield the momentum eigenvalues—that is the possible values of momentum that a free particle can have. Eigenvectors will give a description of the nature of the free particle. The treatment presented here is nonrelativistic. Therefore, there are no restrictions on the momentum that would arrive from considerations of the theory of special relativity. Because the free particle is moving in some direction in space, the coordinate system can be selected to make this a one-dimensional problem without loss of generality.

The momentum eigenvalue equation is

$$\underline{P}\,|P\rangle = \lambda\,|P\rangle\,, \tag{3.1}$$

where $\underline{P}$ is the momentum operator. The λ's are the momentum eigenvalues. The $|P\rangle$'s are the momentum eigenkets. They are the states of definite momentum, each having one of the λ's as its eigenvalue, the value of the momentum observable. To go farther, it is necessary to have a specific form for the momentum operator. As mentioned in Chapter 2, these can be obtained using Dirac's Quantum Condition, which will be introduced in Chapter 4. Different choices of operators define the different representations of quantum theory. Here the momentum operator from the Schrödinger representation will be used. It is the form that is probably the most familiar. The form of the Schrödinger momentum operator will be justified subsequently with Dirac's Quantum Condition.

In the Schrödinger representation the momentum operator has the form

$$\underline{P} = -i\hbar\,\frac{\partial}{\partial x}, \tag{3.2}$$

where the x axis has been chosen as the direction along which the particle is moving. $\hbar$ is Planck's constant divided by 2π, $h/2\pi$. Substituting equation (3.2) into equation (3.1) results in a very simple differential equation. Consider the function

$$|P\rangle \equiv c e^{i\lambda x/\hbar} \tag{3.3}$$

as a solution. c is a constant other than zero that sets the length of the vectors. Since the momentum is an observable—that is, it is a real dynamical variable—the eigenvalues associated with the momentum operator are all real:

$$\lambda = \text{real numbers.}$$

This was proven in Chapter 2. If the λ weren't all real, then $e^{i\lambda x/\hbar}$ would blow up at either positive or negative infinity. This would not be well-behaved.

In connection with the photon interference problem (Chapter 1), the idea was put forward that the function representing the state of a system gives the probability of finding a photon (particle) in a certain region of space, not the number of photons (particles) in the region. Therefore, it is not permissible to have a function representing the state of a system become infinite; that is, the probability of finding a particle can't become infinite anywhere. The explicit function representing a state of a system in a particular coordinate system and in a particular representation is usually called a wavefunction or eigenfunction. The above reasoning in terms of probabilities means that wavefunctions must be finite everywhere and the total probability must be finite.

The functions, $ce^{i\lambda x/\hbar}$, have been put forward as the eigenfunctions of the momentum operator with eigenvalues λ. To test this, equation (3.3) and equation (3.2) are substituted into equation (3.1), and the indicated operations are performed.

$$-i\hbar \frac{\partial}{\partial x}\left[ce^{i\lambda x/\hbar}\right] = -i\hbar\, i\,\lambda/\hbar\, \left[ce^{i\lambda x/\hbar}\right] = \lambda\left[ce^{i\lambda x/\hbar}\right]$$

The operator operating on the function gives back the identical function times a constant, that is,

$$\underline{P}\,|P\rangle = \lambda\,|P\rangle\,.$$

Therefore, the form given in equation (3.3) are the eigenfunctions of the free-particle momentum eigenvalue problem. The real numbers, λ, are the eigenvalues. In this nonrelativistic treatment, there are no restrictions on the values of λ. The momentum can take on any value, continuously, from ∞ to $-\infty$. This is the same result that is obtained from the nonrelativistic classical treatment, but it is arrived at in an entirely different manner. The eigenvalues, λ, are the values of the momenta of a free particle. So, $\lambda = p$, and the momentum eigenfunctions can be written as

$$|P\rangle = \Psi_p = ce^{ipx/\hbar} = ce^{ikx}, \tag{3.4}$$

where $p = \hbar k$, and k is called the wave vector. The momentum eigenfunctions are plane waves with wave vector k.

B. NORMALIZATION OF THE MOMENTUM EIGENFUNCTIONS

Before discussing the quantum mechanical description of a free particle further, the question of normalization of the momentum eigenfunctions will be taken up. It was proven in Chapter 2 that an eigenket of an operator can be multiplied by any nonzero constant, and it is still an eigenket with the same eigenvalue. It is the direction of the eigenvector, not its length, that defines the state of the system. Since the kets are interpreted as a type of probability function, which will be discussed in greater detail below, common usage takes probabilities to range between zero and one. In this sense, it may be convenient to take the probability of finding a particle, integrated over all space, to be unity, since this is just a statement that the particle exists somewhere. In the process of normalization, the Dirac delta function will be introduced. The Dirac delta function appears frequently in quantum theory.

A normalized ket obeys the condition

$$(\langle a|a\rangle)^{1/2} = 1.$$

Bras and kets are vectors, but a closed bracket is a number. $\langle a|b\rangle$ represents the scalar product of the vector $|b\rangle$ with the vector $|a\rangle$. $\langle a|a\rangle$ is the scalar product of ket $|a\rangle$ with itself. Furthermore, $\langle a| = \overline{|a\rangle}$; that is, the bra $\langle a|$ is the complex conjugate of the ket $|a\rangle$. Therefore, the length of a vector is determined by the scalar product of a vector with its complex conjugate. Operationally, the bracket is formed from the ket and its corresponding bra. When working with wavefunctions, which are vector functions, the scalar product takes on the form

$$\langle b|a\rangle = \int \psi_b^* \psi_a \, d\tau, \tag{3.5}$$

where the integration is over the configuration space of the wavefunctions. This is the standard form in linear algebra for the scalar product of two vector functions. Then the scalar product of a wavefunction with itself is

$$\langle a|a\rangle = \int \phi_{\mathrm{a}}^* \, \phi_{\mathrm{a}} \, d\tau. \tag{3.6}$$

Each momentum eigenket has a unique eigenvalue. The Orthogonality Theorem, proven in Chapter 2, states that eigenkets belonging to different eigenvalues are orthogonal. Therefore, the momentum eigenfunctions must be orthogonal. In the course of obtaining the normalization constant for the momentum eigenfunctions, it will also be shown that the momentum eigenkets are indeed orthogonal. To find the normalization constant and to demonstrate orthogonality, it is necessary to evaluate

$$\langle P'|P\rangle = \int_{-\infty}^{\infty} \psi_{P'}^* \, \psi_P \, dx = \begin{cases} 1 & \text{if } P' = P \\ 0 & \text{if } P' \neq P. \end{cases} \tag{3.7}$$

If $P' = P$, the integral is the scalar product of the eigenvector with itself and gives the length of the vector. If $P' \neq P$, the integral is the scalar product of a vector with a different vector, and if they are orthogonal, the integral should yield zero. Using equation (3.4), with $P = \hbar k$, the integral is

$$c^2 \int_{-\infty}^{\infty} e^{-ik'x} e^{ikx} \, dx = c^2 \int_{-\infty}^{\infty} e^{i(k-k')x} \, dx. \tag{3.8}$$

This is an integral that cannot be evaluated in a straightforward manner. To see this, write

$$\int_{-\infty}^{\infty} e^{i(k-k')x} \, dx = \int_{-\infty}^{\infty} \left[\cos(k-k')x + i \sin(k-k')x\right] dx.$$

The sin and cos terms oscillate with x, and the integral cannot be performed in this form. To avoid this problem, the Dirac delta function $\equiv \delta$, can be used. The delta function is defined by

$$\delta(x) = 0 \quad \text{if } x \neq 0. \tag{3.9a}$$

$$\int_{-\infty}^{\infty} \delta(x) \, dx = 1. \tag{3.9b}$$

The delta function is nonzero only at one point, but its integral is unity. An equivalent definition, for arbitrary function $f(x)$ continuous at $x = 0$, is

$$\int_{-\infty}^{\infty} f(x)\delta(x)\, dx = f(0), \tag{3.10}$$

or

$$\delta(x - a) = 0 \quad \text{if } x \neq a$$

$$\int_{-\infty}^{\infty} f(x)\delta(x - a)\, dx = f(a).$$

The δ function is a strange entity. To obtain insight into the nature of the δ function, consider a rectangular function centered about $x = 0$ with unit area shown in Figure 3.1a. If the height is doubled and the width is halved, as shown in Figure 3.1b, the area is still unity. If the height is repeatedly doubled and the width repeatedly halved, the result is a rectangle that is increasingly tall and increasingly narrow. In the limit that the width goes to zero and the height goes to infinity, the area remains equal to unity. This is like the δ function. It is infinitely narrow, only defined at a point, yet it has a finite area equal to one.

There are many mathematical representations of the δ function. Consider

$$\frac{(\sin gx)}{\pi x},$$

where g is any real number. This is J_0, the zeroth-order spherical Bessel function. It has a value of g/π at $x = 0$, since for small y, $\sin(y) = y$. It oscillates with decreasing amplitude as $|x|$ increases, and it has unit integral from $-\infty$ to ∞, independent of the choice of g. In the limit that $g \to \infty$, this becomes the δ function:

$$\delta(x) = \lim_{g\to\infty} \frac{\sin gx}{\pi x}. \tag{3.11}$$

The function has unit integral. The oscillations are infinitely fast, so over any finite interval, they average to zero.

This form of the δ function can be used to normalize the momentum eigenfunctions and to demonstrate their orthogonality. The integral in equation (3.8) can be written as

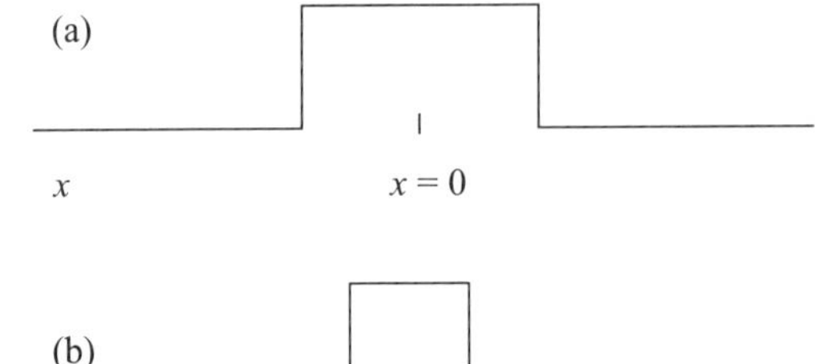

Figure 3.1 Illustration of the nature of the Dirac delta function. (a) Rectangle of unit area. (b) The height is doubled, the width is halved, and the area is still unity.

$$\int_{-\infty}^{\infty} e^{i(k-k')x}\,dx = \int_{-\infty}^{\infty} [\cos(k-k')x + i\sin(k-k')x]\,dx.$$

Rewriting the right-hand side in terms of a limit gives

$$\lim_{g\to\infty} \int_{-g}^{g} [\cos(k-k')x + i\sin(k-k')x]\,dx.$$

Performing the definite finite integral yields

$$\lim_{g\to\infty} \left\{ \frac{[\sin(k-k')x]}{(k-k')} - \frac{[i\cos(k-k')x]}{(k-k')} \right\} \Bigg|_{-g}^{g} = \lim_{g\to\infty} \frac{2\sin g(k-k')}{(k-k')}$$

and, using equation (3.11),

$$= 2\pi\delta(k-k'). \tag{3.12}$$

The result is 2π times the Dirac δ function, $\delta(k-k')$. Since $\delta(k-k') = 0$ for $k \neq k'$, equation (3.12) demonstrates the momentum eigenvectors are orthogonal.

To obtain the normalization constant, it is necessary to recognize that for a continuous range of vectors (Hilbert space) which are the variable of a vector function, the function is not defined at a point. It is necessary to do an integral over a small interval about the point. In this case,

$$\int_{k'=k-\varepsilon}^{k'=k+\varepsilon} \delta(k-k')\,dk' = 1 \quad \text{if } k = k'. \tag{3.13}$$

Therefore, with equations (3.7), (3.8), and (3.12),

$$\frac{1}{c^2}\langle P'|P\rangle = \begin{cases} 2\pi & \text{if } P' = P \\ 0 & \text{if } P' \neq P. \end{cases} \tag{3.14}$$

The kets $|P\rangle$ are orthogonal, and the momentum eigenfunction normalization constant is

$$c = \frac{1}{\sqrt{2\pi}}, \tag{3.15}$$

and the normalized free particle momentum eigenfunction is

$$\Psi_p(x) = \frac{1}{\sqrt{2\pi}} e^{ikx}, \tag{3.16}$$

with $\hbar k = p$.

C. WAVE PACKETS

A classical free particle is localized in space. Its position can be specified. Given the form of the quantum free-particle wavefunction in equation (3.16), what is the location in space of a particle with a well-defined momentum, $p = \hbar k$? The free-particle wavefunction is a plane wave. It can be written in terms of cos and sin as

$$\Psi_p = \frac{1}{\sqrt{2\pi}} e^{ikx} = \frac{1}{\sqrt{2\pi}} (\cos kx + i \ \sin \ kx).$$

Both the cos term and the sin term oscillate continuously from $x = -\infty$ to $x = \infty$. The wavefunction is related to the probability of finding a particle in a region of space. A plot of a portion of a momentum eigenfunction is shown in Figure 3.2. The cos and sin terms are 90° out of phase, so the function is "space filling." When cos is 0, sin is ± 1, and vice versa. The particle is not localized; it is uniformly spread out along the x axis. This doesn't go along very well with the classical notion of a particle. A plane wave of definite momentum is spread over all space.

The Superposition Principle says that the state of a system can be composed of the superposition of two or more other states. Consider the result of the superposition of a large number of momentum eigenstates. Since the state space is continuous, superposition involves an integral, that is,

$$\Psi_{\Delta p} = \int_p c(p) \Psi_p \, dp, \tag{3.17}$$

where $c(p)$ gives the amplitude of each momentum eigenstate in the superposition.

Two specific superpositions will be discussed. First consider a rectangular distribution of momentum eigenfunctions centered about a momentum $\hbar k_0$. The distribution is such that there is equal amplitude of each wavefunction in the superposition over a range of wavevectors. The range of the wavefunctions spans momenta $\hbar(k_0 - \Delta k)$ to $\hbar(k_0 + \Delta k)$. Outside of this range, the amplitude of any momentum state in the superposition is zero. Such a distribution can be implemented by setting the limits of integration in equation (3.17) with $c(p)$ equal to 1 for all p. Working with the k vectors, examine

$$\Psi_{\Delta k} = \int_{k_0 - \Delta k}^{k_0 + \Delta k} e^{ikx} \, dk, \tag{3.18}$$

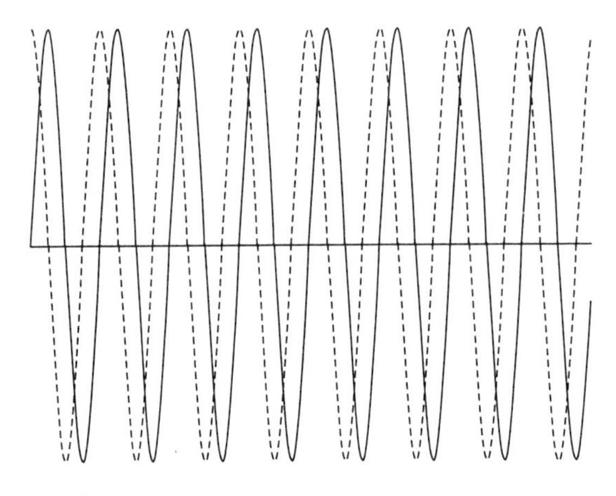

Figure 3.2 A plot of a portion of a momentum eigenfunction. The real component ($\cos(kx)$) is the solid line, and the imaginary component ($\sin(kx)$) is the dashed line.

with $\Delta k \ll k_0$ and neglecting normalization. This is like the integral, equation (3.8), performed above for normalization except then the integration was over x about $x = 0$. Here, the integration is over k about $k = k_0$. Neglecting normalization, the result is

$$\Psi_{\Delta k} = \frac{2 \sin \ \Delta k x}{x} e^{i k_0 x}, \tag{3.19}$$

or

$$\Psi_{\Delta k} = \frac{2 \sin(\Delta k x)}{x} [\cos k_0 x + i \ \sin \ k_0 x]. \tag{3.20}$$

This wavefunction is not spread out over all space uniformly. There is a high-frequency oscillation produced by the terms in k_0. The $\sin(\Delta k x)$ term oscillates more slowly because $\Delta k \ll k_0$, and these oscillations are damped by the x in the denominator. The low-frequency term divided by x is an "envelope" function that is filled in by the real and imaginary high-frequency oscillations. $\Psi_{\Delta k}$ is a wave packet.

Figure 3.3 shows a plot of the wave packet, that is, the filled-in envelope function. The wave packet is maximum at $x = 0$, and it dies out in a oscillatory manner as $|x|$ becomes large. The envelope is filled in by the high-frequency oscillations arising from terms containing k_0. As x becomes large, the amplitude of the function, $\Psi_{\Delta k}$, becomes rapidly small. The particle is now more or less localized in space. The integral in equation (3.18) is equivalent to taking the Fourier transform of a rectangle from momentum space into position space. The result is J_0, the zeroth-order spherical Bessel function. As Δk becomes larger, the peak of the packet is taller, and it dies out more rapidly with increasing $|x|$. The wave packet becomes more localized.

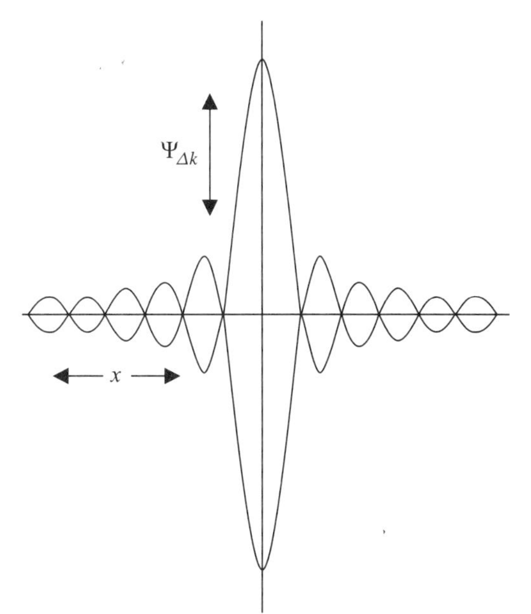

Figure 3.3 The outline of the filled in envelope of a free-particle wave packet formed by the superposition of free-particle momentum eigenstates. The packet is formed by an equal-amplitude superposition of eigenstates over an interval of $2\Delta k$ [see equation (3.18)].

Large Δk means large Δp; the greater the spread in the values of the momentum associated with the momentum eigenkets that make up the wave packet, the smaller the spread in the position of the wave packet. A large uncertainty in p yields a small uncertainty in x.

When Schrödinger introduced the concept of the quantum mechanical wavefunction as a solution to the Schrödinger equation (the energy eigenvalue problem in the Schrödinger representation discussed in Chapter 5), he proposed that the wavefunction represented matter waves and that these were real physical entities. However, Born put forward the correct interpretation of the wavefunction in terms of probabilities. This is known as the "Born interpretation of the wavefunction." The amplitude of a classical electromagnetic wavefunction, φ, is proportional to the electric field, E,

$$\varphi \propto E,$$

and

$$E^* E \propto I;$$

that is, the E field times its complex conjugate, or its absolute value squared, is proportional to I, the intensity. The Born interpretation of the quantum mechanical wavefunction states that the absolute value squared of the wavefunction is proportional to the probability of finding a particle in a region of space. The probability of finding a particle in the region between x and $x + \Delta x$ given by

$$P(x, x + \Delta x) = \int_{x}^{x+\Delta x} \Psi^*(x)\, \Psi(x)\, dx. \tag{3.21}$$

The wavefunction is the probability amplitude. Wavefunctions can be complex; probabilities are always real. In the limit that $\Delta x \rightarrow 0$, equation (3.21) is $P(x)$, the probability of finding the particle at the point x. Frequently, $P(x)$ is written as simply $\Psi^*(x)\Psi(x)$, but the probability of finding a particle at a point in space is zero because the volume element is zero. Therefore, an integral about the point x is implied. Thus the wavefunction in equation (3.20) describing the wave packet is the probability amplitude of finding the particle along the x axis. It is complex and has both positive and negative values. $\Psi^*\Psi$ is real and positive and gives the probability of finding the particle at various positions in space. The Born interpretation shows that the wave packet, composed of a superposition of momentum eigenkets, more or less localizes the particle in space. This is in contrast to a single momentum eigenket, which is delocalized over all space.

A wave packet is formed by regions of constructive interference and regions of destructive interference among probability amplitude waves. A wave packet is obtained by taking a superposition of the momentum eigenstates. A wave packet, centered at x_0, can be written in the form

$$\Psi_{\Delta k}(x) = \int_{-\infty}^{\infty} f(k - k_0)\, e^{ik(x-x_0)}\, dk, \tag{3.22}$$

where $f(k - k_0)$ is a weighting function that is nicely behaved; that is, it is continuous and dies out away from k_0. The weighting function causes the superposition to be composed only of k states near k_0. To see that equation (3.22) will yield a wave packet, it is necessary

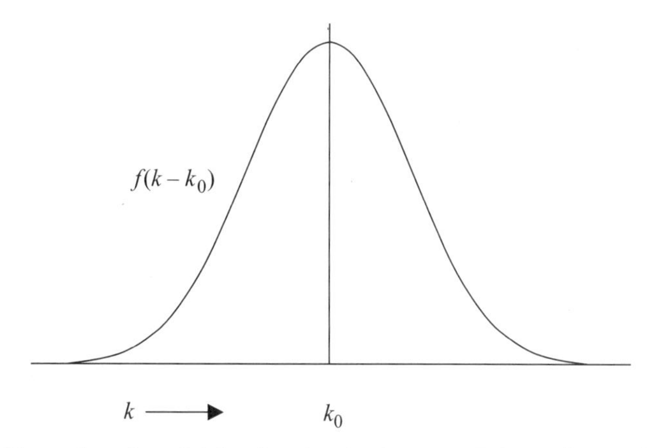

Figure 3.4 An illustration of a weighting function that is centered about a free-particle momentum eigenstate, k_0. The weighting function gives the amplitude of k states in the superposition that forms a wave packet.

to consider where the waves constructively and destructively interfere. At $x = x_0$, the argument of the exponential is zero, and

$$e^{ik(x-x_0)} = e^0 = 1 \quad \text{for all } k.$$

All of the k states in the superposition interfere constructively; all contributions to the integral add, and the value is maximum. For $x \neq 0$, the exponential can be written as

$$e^{ik(x-x_0)} = \cos\, k(x - x_0) + \sin\, k(x - x_0).$$

For $(x - x_0)$ large, the cos and sin terms oscillate wildly with changing k. Therefore, at a particular point, x, one k vector will make a positive contribution and another will make a negative contribution. The contributions from different k vectors cancel out; that is, there is destructive interference at large x. Thus for $|x| \gg x_0$, $\Psi_{\Delta k} \rightarrow 0$, and the probability of finding the particle goes to zero. The individual momentum eigenstates are delocalized over all space, but a superposition of the delocalized states has regions of constructive and destructive interference, which more or less localize the particle.

Another specific example of a wave packet is obtained by selecting the weighting function in equation (3.22) to be a Gaussian. A Gaussian function has the form

$$G(y) = \exp\left[-(y - y_0)^2/2\sigma^2\right], \tag{3.23}$$

where σ is the standard deviation. To form a Gaussian wave packet, the weighting function (not normalized) is

$$f(k) = \exp\left[-\frac{(k - k_0)^2}{2(\Delta k)^2}\right]. \tag{3.24}$$

This is a Gaussian in k, with the width of the spread in k given by the standard deviation of k; that is, $\Delta k = \sigma$. Then substituting equation (3.24) into equation (3.22) gives

$$\Psi_{\Delta k}(x) = \int_{-\infty}^{\infty} \exp\left[-\frac{(k-k_0)^2}{2(\Delta k)^2} + ik(x-x_0)\right] dk. \tag{3.25}$$

Performing the integral yields

$$\begin{aligned}\Psi_{\Delta k} &= \sqrt{2\pi}\,\Delta k\,\exp\left[ik_0(x-x_0) - \frac{1}{2}(x-x_0)^2(\Delta k)^2\right] \\ &= \sqrt{2\pi}\,\Delta k\,e^{-\frac{1}{2}(x-x_0)^2(\Delta k)^2}\,e^{ik_0(x-x_0)}.\end{aligned} \tag{3.26}$$

This looks like the momentum eigenket at $k = k_0$ multiplied by a Gaussian envelope function in position space. The standard deviation is $\sigma = 1/\Delta k$. The oscillatory term (the complex exponential) is multiplied by the rapidly decaying Gaussian function centered at $x = x_0$. The real and imaginary components of the high-frequency complex exponential fill in the Gaussian envelope. A Gaussian superposition of momentum eigenstates yields a position space Gaussian wave packet. Performing the integral in equation (3.25) is taking the Fourier transform of a momentum space Gaussian, and the Fourier transform of a Gaussian is another Gaussian—in this case in position space.

The width of the Gaussian is determined by the standard deviation,

$$\frac{1}{(\Delta k)^2} = \sigma^2.$$

A large Δk means a small σ. The larger the spread in momentum (wave vectors), the narrower the wave packet. Looking at this somewhat more quantitatively, the probability of finding the particle at a position x is proportional to the wavefunction times its complex conjugate,

$$\Psi_{\Delta k}^{*}\Psi_{\Delta k} \propto \exp\left[-(x-x_0)^2\,(\Delta k)^2\right]. \tag{3.27}$$

For values of x around x_0, the argument of the exponential is small; therefore, there is a significant probability of finding the particle. Where the argument is large, the probability of finding the particle is small. Deciding which values of the argument constitute large values is somewhat arbitrary. A quantitative discussion will be given in connection with the Uncertainty Principle in Chapter 4. For now, take the argument to be small when $(x-x_0)^2(\Delta k)^2 \geq 1$. With $(x-x_0) = \Delta x$, this condition can be written as

$$\Delta x\ \Delta k \geq 1.$$

Using $\hbar k = p$ gives

$$\Delta x\,\frac{\Delta p}{\hbar} \geq 1,$$

or

$$\Delta x\ \Delta p \geq \hbar. \tag{3.28}$$

The result given in equation (3.28) is within a factor of two of the Heisenberg uncertainty relation. Its origin can be seen clearly from the analysis of the wave packet problem. For a single momentum eigenstate, the momentum is perfectly defined, and Δp is zero. The position is spread out over all space, so Δx is infinite. A wave packet is composed of a superposition of momentum eigenkets. As the number of eigenkets in the superposition is increased, the possible values that will be observed in a measurement of the momentum is increased; that is, Δp grows. As Δp increases, the wave packet becomes more localized. When more waves are superimposed to make the wave packet, the spread in wavelengths is greater. This causes the onset of significant destructive interference to occur for smaller changes in position away from the center of the packet. The larger the value of Δp, the more the wave packet is localized. Large uncertainty in momentum results in small uncertainty in position, and conversely. This is a direct consequence of the superposition principle and the probability wave nature of the eigenkets.

D. WAVE PACKET MOTION AND GROUP VELOCITIES

So far the time-independent momentum eigenfunctions,

$$\Psi_p(x) = ce^{ikx},$$

have been discussed. The description of the wave packets given above did not include motion. The packets are time independent. The spatial extent of the packets was described, but how they propagate was not explained. Wave packets, which describe a particle as more or less localized in space, characterize particles such as photons or electrons. In the photoelectric effect, a photon of sufficiently high energy impinges on a metal surface, and an electron is ejected from the metal. Clearly both are moving, and a quantum mechanical description of a free particle must explain how they move in terms of wave packets.

In Chapter 2, in discussing the normalization of kets, it was shown that even when the direction and the normalization constant are known, a ket can be multiplied by a phase factor,

$$e^{i\phi},$$

where ϕ is a real number. In Chapter 5, it will be shown that the time-dependent Schrödinger equation for systems with time-independent energy gives the time dependence of the wavefunction as

$$e^{-iEt/\hbar} = e^{-i\omega t}, \tag{3.29}$$

where E is the energy and $\hbar\omega = E$. This is a time-dependent phase factor, that is,

$$e^{-i\omega t}\, e^{i\omega t} = 1,$$

which is independent of the time. Since the momentum eigenvalue problem was solved using the Schrödinger representation form of the momentum operator, equation (3.2), the time-dependent phase factor can be used to describe the time dependence of the free-particle momentum eigenfunctions. Thus, the time-dependent momentum eigenfunction is

$$\Psi_k(x, t) = e^{i[k(x-x_0)-\omega t]}. \tag{3.30}$$

The momentum eigenfunction given in equation (3.30) is still delocalized over all space. However, it is now a running wave. A plot of either the real or imaginary part of equation (3.30) as a function of time shows that the peaks of the wave move by a fixed point, x.

Using the time-dependent momentum eigenfunctions, a wave packet can be written in the same manner as in equation (3.22),

$$\Psi_{\Delta k}(x,t) = \int_{-\infty}^{\infty} f(k) \exp\left[i\left(k(x - x_0) - \omega t\right)\right] dk. \tag{3.31}$$

First consider a photon moving in a vacuum. For a photon in a vacuum, the relationship between the frequency, ω, and the wave vector, k, called the dispersion relation, is linear, that is,

$$\omega = ck, \tag{3.32}$$

where c is the velocity of light in vacuum. Using this dispersion relation, the photon wave packet is written as

$$\Psi_{\Delta k}(x,t) = \int_{\infty}^{\infty} f(k)\ \exp\left[ik(x - x_0 - ct)\right] dk, \tag{3.33}$$

where $f(k)$ is the weighting function that determines the shape of the wave packet. At $t = 0$, the form of equation (3.33) is the same as equation (3.22). The packet is peaked at $x = x_0$. This is the point where the argument of the exponential is zero for all values of k. Thus all of the waves have maximum constructive interference at $x = x_0$. At later times, the peak moves. The peak is located at

$$x = x_0 + ct,$$

because this is the point at which the argument of the exponential is zero independent of k, the point of maximum constructive interference. Therefore, the packet moves with the speed of light. Each k state (plane wave) that makes up the packet moves with the same velocity, so the packet maintains its shape as it moves. The motion is due to changing regions of constructive and destructive interference of the delocalized plane waves (momentum eigenstates) that make up the packet. Both localization and motion arise from the superposition of eigenstates.

If a photon enters a dispersive medium such as a piece of glass, the frequency ω is no longer linearly proportional the wave vector, k. The material has an index of refraction that depends on the wavelength, $n(\lambda)$. In a vacuum, the speed of light is equal to c for all wavelengths. In glass or in other materials, the velocity of light depends on the wavelength as follows:

$$V = c/n(\lambda). \tag{3.34}$$

The angular frequency is

$$\omega = 2\pi\ c/\lambda\, n(\lambda). \tag{3.35}$$

[The angular frequency, ω (in radians/s), is related to the frequency, ν (in Hz), by $\omega = 2\pi\ \nu$, and $\lambda = 2\pi/k$.] In general, for materials that are transparent for a range of wavelengths under consideration, $n(\lambda)$ is a complicated function that increases as the wavelength becomes shorter. Therefore, ω is a complicated function of k. In the middle of the visible region of the spectrum, the index of refraction of glass is ~ 1.5. (For wavelengths near an absorption, the index of refraction is a complex number with the imaginary component related to the strength of the absorption.)

The important feature of a dispersive material is that ω is no longer linearly dependent on the wave vector, k. Expressing λ in terms of k equation (3.35) becomes

$$\omega(k) = ck/n(k). \tag{3.36}$$

It is still possible to write an expression for a wave packet as in equation (3.31), but the simple substitution that gave equation (3.33) for a nondispersive medium cannot be made. The expression describing a wave packet in a dispersive medium is

$$\Psi_{\Delta k}(x, t) = \int_{-\infty}^{\infty} f(k) \exp\left[i(k(x - x_0)) - \omega(k)t\right] dk. \tag{3.37}$$

Equation (3.37) still looks like a wave packet. At $t = 0$, the expression is identical to equation 3.22, the equation for a wave packet before the time dependence was introduced. The packet still moves, but not with the velocity of light. Different states in the superposition move at different velocities. The velocity of a single state in the superposition is called the phase velocity.

To determine the motion of a wave packet in a dispersive medium, it is necessary to calculate the velocity of the maximum of the packet—that is, the center of the packet. At $t = 0$ and $x = x_0$, argument of exponential is zero for all k, and all k states add in phase. This is the point of maximum constructive interference, the "location" of the packet. The argument of exponential is

$$\phi = k(x - x_0) - \omega(k)t. \tag{3.38}$$

Since $\omega(k)$ is not linearly proportional to k, for $t > 0$, the argument will never again be zero simultaneously for all k in the wave packet. However, the greatest degree of constructive interference will occur in the spatial region where the argument changes as slowly as possible with k since this produces slow oscillations. When ϕ changes slowly within a range of k's determined by $f(k)$, constructive interference occurs. The location of the minimum change in ϕ with k is the peak of the wave packet. The point x where this occurs can be found.

$$\frac{\partial \phi}{\partial k} = 0 = (x - x_0) - t\frac{\partial \omega}{\partial k} \tag{3.39}$$

Thus, the location of the peak of the packet at time t is

$$x = x_0 + t\,\frac{\partial \omega}{\partial k}. \tag{3.40}$$

The distance the packet has moved in time t is

$$d = (x - x_0) = \left(\frac{\partial \omega}{\partial k}\right) t. \tag{3.41}$$

This has the form $d = Vt$; that is, distance equals velocity multiplied by the time. Therefore, the maximum of the packet moves with a velocity called the group velocity, V_g, with

$$V_g = \left(\frac{\partial \omega}{\partial k}\right)_{k=k_0}. \tag{3.42}$$

The group velocity is the speed of a wave packet in a dispersive medium. The phase velocity is

$$V_p = \lambda \nu = \omega / k. \tag{3.43}$$

The group velocity and the phase velocity are only the same for a photon when

$$\omega(k) = ck,$$

that is, in a nondispersive medium, a vacuum.

The wave packet group velocity given in equation (3.42) was obtained in the context of a photon in a dispersive medium. However, nothing limits the discussion to zero rest mass particles. The momentum eigenstates used to form the wave packet are the same for a photon, an electron, or a rock. To apply equation (3.42) to a material particle, such as an electron wave packet, it is necessary to know the dispersion relation, $\omega(k)$. A wavelength associated with a particle unifies the theories of matter and light. De Broglie introduced the relationship between a particle's momentum and its wavelength:

$$p = \hbar k = h/\lambda, \tag{3.44}$$

where λ is the de Broglie wavelength.

The necessary dispersion relationship for a material particle is found by using the relationships, $E = \hbar\omega$ and $E = p^2/2m$, which is the nonrelativistic free-particle energy. Then,

$$\omega(k) = \frac{E}{\hbar} = \frac{p^2}{2m\hbar} = \frac{\hbar k^2}{2m}. \tag{3.45}$$

Using $\omega(k)$, the group velocity of a nonzero rest mass free particle is

$$V_g = \frac{\partial \omega(k)}{\partial k} = \frac{\hbar}{2m}\frac{\partial k^2}{\partial k} = \frac{\hbar k}{m} = \frac{p}{m}. \tag{3.46}$$

This is an important result. The group velocity of the wave packet moves with classical velocity, p/m. However, the quantum mechanical description of a free particle is fundamentally different from the classical description. Quantum mechanics describes a particle as a superposition of momentum eigenstates. It is partially localized with $\Delta x \Delta p \approx \hbar$. The particle moves because of time-dependent changing regions of constructive and destructive interference. Nonetheless, the quantum description of a material particle yields the same velocity as classical mechanics. This is in accord with the Correspondence Principle. In a realm in which classical mechanics gives a proper result, for example, the velocity of a material particle, quantum theory will produce the same result.

Although the quantum and classical velocities of a nonzero rest mass free particle are the same, the quantum description provides an avenue for understanding phenomena that the classical picture lacks. The quantum description of a photon and an electron are identical. Both are wave packets composed of a superposition of momentum eigenstates. Both are partially localized. The differences in their group velocities arises from the differences in their dispersion relations. Because their descriptions are fundamentally the same, it is clear that a photon and an electron can exhibit both wave and particle characteristics. Whether a wave or particle characteristic is observed depends only on the context of the experiment. When a single photon ejects a single electron in the photoelectric effect, light is displaying its particle-like nature, its localization. When light diffracts from a grating, it is exhibiting its wave-like character. When an electron is accelerated by the electron "gun" in a colored television set, it can be aimed by deflection plates to hit a small spot on the screen that is stimulated to emit a particular color. The electron is acting as a localized particle. When an electron is diffracted from the surface of a crystal in a low-energy electron diffraction (LEED) surface science experiment, it is exhibiting its wave-like character. Wave-particle duality applies to both light and material particles. Particles of all types are wave packets. The nature of the observable determines which aspect of the wave packet will be manifested. The quantum theory of wave packets provides a unified description of the wave and particle aspects of light and matter.

CHAPTER 4

Commutators, Dirac's Quantum Condition, and the Uncertainty Principle

A. Dirac's Quantum Condition

The formal connection between classical mechanics and quantum mechanics is made by the correspondence between the classical Poisson bracket of the functions f and g and the quantum mechanical commutator of the operators, $\underline{f}$ and $\underline{g}$.

The commutator of two linear operators is

$$\left[\underline{A}, \underline{B}\right] = \underline{A}\,\underline{B} - \underline{B}\,\underline{A}. \tag{4.1}$$

If A and B are numbers, then $A\ B - B\ A = 0$. Numbers commute under the operation of multiplication. However, linear operators don't necessarily commute, that is,

$$\begin{aligned} \underline{A}\,\underline{B}\,|C\rangle &= \underline{A}\left[\underline{B}\,|C\rangle\right] \\ &= \underline{A}\,|Q\rangle \\ &= |Z\rangle\,.. \end{aligned}$$

Operating $\underline{B}$ on the ket $|C\rangle$ will, in general, yield a new ket, $|Q\rangle$, and then operating $\underline{A}$ on $|Q\rangle$ will yield another ket, $|Z\rangle$. While

$$\begin{aligned} \underline{B}\,\underline{A}\,|C\rangle &= \underline{B}\left[\underline{A}\,|C\rangle\right] \\ &= \underline{B}\,|S\rangle \\ &= |T\rangle\,, \end{aligned}$$

$|Z\rangle$ and $|T\rangle$ are not necessarily equal. Therefore, the operators $\underline{A}$ and $\underline{B}$ do not commute because applying the commutator, $[\underline{A}, \underline{B}]$, to an arbitrary ket gives a nonzero result. Frequently $[\underline{A}, \underline{B}] \neq 0$ is written as a statement that the operators do not commute. This equation is not taken in the algebraic sense, but, rather, it means that operating the commutator on an arbitrary ket gives a nonzero result.

The classical Poisson bracket is

$$\{f, g\} = \frac{\partial f}{\partial x}\frac{\partial g}{\partial p} - \frac{\partial g}{\partial x}\frac{\partial f}{\partial p}. \tag{4.2}$$

Here, $f = f(x, p)$ and $g = g(x, p)$ are functions of classical position, x, and momentum, p. f and g represent classical dynamical variables. They are not quantum mechanical operators.

The Poisson bracket for the classical dynamical variables position and momentum is

$$\{x, p\} = \frac{\partial x}{\partial x}\frac{\partial p}{\partial p} - \frac{\partial x}{\partial p}\frac{\partial p}{\partial x} \tag{4.3}$$

Therefore,

$$\{x, p\} = 1 \tag{4.4}$$

because the first term on the right-hand side of equation (4.3) is 1 and the second term is 0.

Dirac put forward a relationship between the classical Poisson bracket of two classical dynamical variables and the commutator of the corresponding quantum mechanical operators that defines acceptable sets of quantum operators. The relationship is called Dirac's Quantum Condition. Dirac stated it as follows: "The quantum mechanical operators $\underline{f}$ and $\underline{g}$, which in quantum theory replace the classically defined functions f and g, must always be such that the commutator of $\underline{f}$ and $\underline{g}$ corresponds to the Poisson bracket of f and g according to

$$i\hbar\{f, g\} \rightarrow \left[\underline{f}, \underline{g}\right]." \tag{4.5}$$

Operators that comply with Dirac's Quantum Condition form an internally consistent set that can be used to calculate observables. Sets of operators that obey the Quantum Condition are not unique. Each such set forms the basis for a particular representation of quantum theory.

The quantum mechanical commutator of the operators $\underline{x}$ and $\underline{P}$ must obey the relation

$$\left[\underline{x}, \underline{P}\right] = i\hbar\{x, p\}; \tag{4.6}$$

and since the Poisson bracket is equal to 1,

$$\left[\underline{x}, \underline{P}\right] = i\hbar. \tag{4.7}$$

As stated previously, an equation like equation (4.7) means that operating the commutator on an arbitrary ket is the same as operating $i\hbar$ on the ket. A number, such as $i\hbar$, can always be considered to be an operator because it is implicitly multiplied by the identity operator. So $i\hbar$ actually means $i\hbar\,\underline{1}$. $\underline{1}$ is the identity operator. When it operates on any ket, it yields the identical ket; that is, the ket's direction and length are unchanged.

In Chapter 3, the Schrödinger representation form of the momentum operator was used. For this form to obey Dirac's Quantum Condition, the classical functions for momentum and position are replaced by the operators such that

$$p \rightarrow \underline{P} = -i\hbar\frac{\partial}{\partial x} \tag{4.8}$$

$$x \rightarrow \underline{x}. \tag{4.9}$$

To see that this choice obeys the Quantum Condition, the commutator of the $\underline{x}$ and $\underline{P}$ operators is applied to an arbitrary ket, $|S\rangle$.

$$[\underline{x}, \underline{P}]\,|S\rangle =$$

$$(\underline{x}\,\underline{P} - \underline{P}\,\underline{x})\,|S\rangle =$$

$$\underline{x}\left(-i\hbar\frac{\partial}{\partial x}\right)|S\rangle + i\hbar\frac{\partial}{\partial x}\,x\,|S\rangle\,.$$

Applying the chain rule for differentiation to the second term yields

$$i\hbar\left(-\underline{x}\frac{\partial}{\partial x}\,|S\rangle + |S\rangle + \underline{x}\frac{\partial}{\partial x}\,|S\rangle\right).$$

The first term and third terms cancel, and the result is

$$= i\hbar\,|S\rangle\,.$$

Therefore,

$$\left[\underline{x}, \underline{P}\right]|S\rangle = i\hbar\,|S\rangle$$

and

$$\left[\underline{x}, \underline{P}\right] = i\hbar, \tag{4.10}$$

since these are equivalent when operating on an arbitrary ket.

Dirac's Quantum Condition is the direct connection between classical and quantum mechanics.

B. COMMUTATORS AND SIMULTANEOUS EIGENFUNCTIONS

Commutators play a fundamental role in quantum theory. Dirac's Quantum Condition uses commutator relationships to determine allowable sets of linear operators that are used to represent dynamical variables (observable quantities). In the last section it was seen that the position and momentum operators do not commute. In connection with the wave packet problem discussed in Chapter 3, Section C, it was shown that the position and momentum of a particle cannot be measured exactly simultaneously. A qualitative relationship for the product in the uncertainties in position and momentum was given in equation (3.28). However, there are some observables that can be measured exactly, simultaneously, and some operators do commute.

Consider the eigenvalue equations for two dynamical variables represented by the linear operators $\underline{A}$ and $\underline{B}$:

$$\underline{A}\,|S\rangle = \alpha\,|S\rangle \tag{4.11a}$$

$$\underline{B}\,|S\rangle = \beta\,|S\rangle\,. \tag{4.11b}$$

The $|S\rangle$ are said to be the simultaneous eigenvectors of the operators $\underline{A}$ and $\underline{B}$, with sets of eigenvalues α and β. Since the eigenvalues of linear operators are the observables, if the $|S\rangle$ are the simultaneous eigenvectors of two or more linear operators representing observables, then these observables can be measured exactly, simultaneously.

Left-multiplying equation (4.11a) by $\underline{B}$ and left-multiplying equation (4.11b) by $\underline{A}$ gives

$$\underline{B}\,\underline{A}\,|S\rangle = \underline{B}\,\alpha\,|S\rangle \qquad \underline{A}\,\underline{B}\,|S\rangle = \underline{A}\,\beta\,|S\rangle\,.$$

Since a number always commutes with an operator, the right-hand sides of the two equations are

$$= \alpha\,\underline{B}\,|S\rangle \qquad\qquad = \beta\,\underline{A}\,|S\rangle\,.$$

Operating on $|S\rangle$ with $\underline{A}$ and $\underline{B}$,

$$= \alpha\,\beta\,|S\rangle \qquad\qquad = \beta\,\alpha\,|S\rangle\,.$$

These are equal because α and β are numbers and multiplication of numbers is commutative. Therefore,

$$\underline{A}\,\underline{B}\,|S\rangle = \underline{B}\,\underline{A}\,|S\rangle$$

and

$$(\underline{A}\,\underline{B} - \underline{B}\,\underline{A})\,|S\rangle = 0.$$

Since $(\underline{A}\,\underline{B} - \underline{B}\,\underline{A})$ is the commutator of $\underline{A}$ and $\underline{B}$, then

$$[\underline{A}, \underline{B}] = 0.$$

If $[\underline{A}, \underline{B}] = 0$, the operator $\underline{A}$ and $\underline{B}$ are said to commute. Operators having all of their eigenvectors as simultaneous eigenvectors commute. It can be proven that the eigenvectors of commuting operators can always be constructed in such a way that they are simultaneous eigenvectors.

As a very simple example of commuting operators, consider the momentum and energy operators for the free particle. The energy operator is obtained from the classical Hamiltonian, which is the sum of the kinetic energy and the potential energy, by replacing the classical functions with the appropriate quantum mechanical operators. A free particle only has kinetic energy; the potential energy is zero everywhere. Therefore the Hamiltonian operator for a free particle is

$$\underline{H} = \frac{\underline{P}^2}{2m}. \tag{4.12}$$

This is the square of the momentum operator divided by twice the mass. An operator raised to a power n means that the operator is applied successively n times. So $\underline{P}^2$

acts as $\underline{P}\,\underline{P}$. The commutator of the Hamiltonian and momentum operators for the free particle is

$$[\underline{H}, \underline{P}] = \frac{1}{2m}\left[\underline{P}^2, \underline{P}\right]. \tag{4.13}$$

Applying the commutator to an arbitrary ket gives

$$\frac{1}{2m}[\underline{P}^2, \underline{P}]\,|S\rangle = \underline{P}^2\underline{P}\,|S\rangle - \underline{P}\,\underline{P}^2\,|S\rangle = \underline{P}\,\underline{P}\,\underline{P}\,|S\rangle - \underline{P}\,\underline{P}\,\underline{P}\,|S\rangle = 0. \tag{4.14}$$

Therefore,

$$[\underline{H}\,, \underline{P}] = 0, \tag{4.15}$$

since this is the result of applying the commutator to an arbitrary ket. $\underline{H}$ and $\underline{P}$ commute for the free particle, the energy and momentum operators have simultaneous eigenvectors, and the energy and momentum can be measured exactly simultaneously.

Commuting operators are important because they indicate which observables can be simultaneously measured. A state of a system is characterized by its observable properties. Frequently, a single observable is insufficient to completely define the state of a system. For example, the states of the hydrogen atom, which will be discussed in detail in Chapter 7, cannot be distinguished solely by their energy. The second hydrogen atom energy level is composed of four orbitals, namely, the $2s$ and three different $2p$ orbitals. All four of these states have the same energy, so measuring the energy does not determine the state of the system. However, the $2s$ state and the three $2p$ states have different total orbital angular momenta. The orbital angular momentum operator and the Hamiltonian (energy) operator commute. Therefore, the energy and the total orbital angular momentum can be measured simultaneously. This allows the $2s$ and $2p$ orbitals to be distinguished. The three different $2p$ orbitals differ by their projections of the angular momentum vector on an axis (see Chapter 15). There is an operator for the projection on an axis, and it commutes with both the Hamiltonian and the total angular momentum operator. Therefore, all three of these properties can be measured, and the states can be distinguished. The states are defined in terms of the eigenvalues of a set of commuting operators. There is always a complete set of commuting operators (i.e., enough commuting operators) to completely define the state of a system.

In working with commutators, there are some useful relations for their manipulation:

$$[\underline{A}\,, \underline{B}] = -[\underline{B}\,, \underline{A}] \tag{4.16a}$$

$$[\underline{A}\,, \underline{BC}] = [\underline{A}\,, \underline{B}]\underline{C} + \underline{B}[\underline{A}\,, \underline{C}] \tag{4.16b}$$

$$[\underline{AB}\,, \underline{C}] = [\underline{A}\,, \underline{C}]\underline{B} + \underline{A}[\underline{B}\,, \underline{C}] \tag{4.16c}$$

$$[\underline{A}\,, [\underline{B}\,, \underline{C}]] + [\underline{B}\,, [\underline{C}\,, \underline{A}]] + [\underline{C}\,, [\underline{A}\,, \underline{B}]] = 0 \tag{4.16d}$$

$$[\underline{A}, \underline{B} + \underline{C}] = [\underline{A}, \underline{B}] + [\underline{A}, \underline{C}]. \tag{4.16e}$$

C. EXPECTATION VALUES AND AVERAGES

Consider the eigenvalue equation,

$$\underline{A}\,|a\rangle = \alpha\,|a\rangle\,. \tag{4.17}$$

$\underline{A}$ is the operator representing a dynamical variable. $|a\rangle$ is an eigenvector of $\underline{A}$ with eigenvalue α. α is the quantity that will be observed when a measurement of the dynamical variable represented by $\underline{A}$ is made on the system in state $|a\rangle$. Take $|a\rangle$ to be normalized. Left-multiplying equation (4.17) by the bra $\langle a|$ gives

$$\begin{aligned}\langle a|\,\underline{A}\,|a\rangle &= \langle a|\,\alpha\,|a\rangle \\ &= \alpha\,\langle a|a\rangle \\ &= \alpha.\end{aligned}$$

For an eigenvector of an operator, a closed bracket like $\langle a|\,\underline{A}\,|a\rangle$ is the eigenvalue, a real number.

The state of a system may not be in an eigenstate of a particular operator. Therefore, it is necessary to evaluate

$$\underline{A}\,|b\rangle\,,$$

to determine the outcome of a measurement of an observable, represented by the operator $\underline{A}$, made on a state $|b\rangle$ that is not an eigenvector of $\underline{A}$. The eigenkets of $\underline{A}$, the $|a_i\rangle$, form a complete set because there is one for each state of the system. The $|a_i\rangle$ can be taken to be normalized and orthogonal. In general, the kets can be normalized. It was proven in Chapter 2, Section C, that nondegenerate eigenkets are orthogonal. If some of the eigenkets are degenerate, orthogonal superpositions can always be formed that are still eigenkets. Therefore, the $|a_i\rangle$ form a complete orthonormal basis set.

The ket $|b\rangle$ can be expanded in the complete set of states $|a_i\rangle$ because the Superposition Principle says that a state can always described in terms of a sum (superposition) of other states. Then,

$$|b\rangle = c_1|a_1\rangle + c_2|a_2\rangle + c_3|a_3\rangle + \cdots. \tag{4.18}$$

If the eigenvectors $|a_i\rangle$ are continuous (e.g., the momentum eigenkets of a free particle), then the sum is replaced by an integral. If there is a discrete range and a continuous range of eigenkets, then there will be a sum plus an integral. The following analysis will be done only with discrete eigenkets. However, the results are general.

$$|b\rangle = \sum_i c_i|a_i\rangle\,. \tag{4.19}$$

The simplest case is an expansion that involves only two states, that is,

$$|b\rangle = c_1|a_1\rangle + c_2|a_2\rangle\,. \tag{4.20}$$

Now applying the operator $\underline{A}$ to the ket $|b\rangle$ yields

$$\underline{A}\,|b\rangle = \underline{A}(c_1|a_1\rangle + c_2|a_2\rangle) \tag{4.21a}$$

$$= c_1\underline{A}|a_1\rangle + c_2\underline{A}|a_2\rangle \tag{4.21b}$$

$$= \alpha_1 c_1|a_1\rangle + \alpha_2 c_2|a_2\rangle\,. \tag{4.21c}$$

Since $|b\rangle$ is not an eigenket of $\underline{A}$, operating $\underline{A}$ on $|b\rangle$ will not give $|b\rangle$ back. By expanding $|b\rangle$ in the eigenkets of $\underline{A}$ it is possible to determine the result of applying $\underline{A}$ to $|b\rangle$ because the result of operating $\underline{A}$ on the eigenkets of $\underline{A}$ is known. The result is the eigenkets multiplied by their eigenvalues. This is shown in equation (4.21c).

Equation (4.21c) can now be left-multiplied by the bra $\langle b|$:

$$\langle b|\,\underline{A}\,|b\rangle = (c_1^*\,\langle a_1| + c_2^*\,\langle a_2|)(\alpha_1 c_1|a_1\rangle + \alpha_2 c_2|a_2\rangle) \tag{4.22a}$$

$$= \alpha_1 c_1^* c_1 + \alpha_2 c_2^* c_2 \tag{4.22b}$$

$$= \alpha_1\,|c_1|^2 + \alpha_2\,|c_2|^2\,. \tag{4.22c}$$

The α_i are real because they are the eigenvalues of a Hermitian operator. While the coefficients c_i may be complex, the coefficient times its complex conjugate is a real number. Therefore, the bracket $\langle b|\,\underline{A}\,|b\rangle$ is a real number. Each term in equation (4.22c) is an eigenvalue times the absolute value squared of the coefficient of the ket in the expansion, equation (4.20), corresponding to the eigenvalue. If there are more than two states in the expansion, then

$$|b\rangle = \sum_i c_i\,|a_i\rangle \tag{4.23}$$

and

$$\langle b|\,\underline{A}\,|b\rangle = \sum_i \alpha_i\,|c_i|^2\,. \tag{4.24}$$

$|c_1|^2$, the absolute value squared of the coefficient c_i in the expansion of $|b\rangle$ in terms of the eigenvectors $|a_i\rangle$ of the operator (observable) $\underline{A}$, is the probability that a measurement of observable property represented by $\underline{A}$ on the state $|b\rangle$ will yield the eigenvalue α_i.

When a single measurement of a dynamical variable represented by an operator is made on a system, an eigenvalue of the operator will always be measured. If the system is in an eigenstate of the operator, then the particular, associated eigenvalue will certainly be measured. If the system is prepared again in an identical manner, so that it is in the same eigenstate, then the same eigenvalue will be measured. However, a system is not necessarily in an eigenstate of the operator for the dynamical variable of interest. In Chapter 1, the problem of polarization of photons was discussed qualitatively. The Superposition Principle was used to explain why a photon initially in a state of polarization α, when observed with a polarizer, would always yield either polarization parallel or perpendicular. The Superposition Principle says that the state of a system can always be expressed as a superposition of two or more other states. When a state of a system is not an eigenstate of the operator representing the observable of interest, it can be expressed as a superposition of the eigenstates of the operator. When a single measurement is made, one of the eigenstates will be observed. If the system is prepared again identically, in general, a measurement will yield a different observable value—that is a different eigenvalue. The probability that a particular eigenvalue will be measured in a single measurement is given (for normalized kets) by the absolute value squared of the coefficient of the eigenstate which belongs to that eigenvalue when the state under observation is expressed as an expansion in terms of the eigenkets of the operator (observable). When a system is not in an eigenstate, repeated measurements on the system, prepared identically each time, will yield different values. By making repeated measurements on an identically prepared system, the probability distribution of the eigenvalues is obtained. This probability distribution provides information on the nature of the state as described as a superposition of the eigenstates of the operator representing the dynamical variable of interest.

❖ **Definition:** The average is the value of a particular outcome times its probability, summed over all possible outcomes.

From the definition of the average, $\langle b|\, \underline{A}\, |b\rangle$ [equation (4.24)] is the average value of the observable $\underline{A}$ when many measurements are made on a system that is prepared each time in an identical manner so it is in the state $|b\rangle$. Take a single system, make a measurement, and record the value. Prepare the system again in the same manner, and make a second measurement. Do this repeatedly. $\langle b|\, \underline{A}\, |b\rangle$ gives the average of these repeated measurements.

In many experimental situations, a measurement is made on many identical systems rather than making many measurements on a single system. For example, rather than making many measurements on a single atom, a measurement is made on large number of identical atoms. It should be noted that it is now possible to make measurements on single atoms. However, in practice, most measurements are made on a macroscopic quantity of a substance.

❖ **Assume:** One measurement on a large number of identically prepared noninteracting systems is the same as the average of many repeated measurements on one such system prepared each time in an identical manner.

$\langle b|\, \underline{A}\, |b\rangle$ is called the expectation value of the operator $\underline{A}$. It is the quantum mechanical average value of the observable. Since in many experimental situations, a system, or more frequently a large number of identical systems, is not in a single eigenstate, measurements will yield the expectation value of the observable.

In a particular representation of quantum theory, the kets and bras are written as vector functions or wavefunctions. In terms of particular wavefunctions, the expectation value is written as

$$\langle b|\, \underline{A}\, |b\rangle = \int \psi_b^* \underline{A}\, \psi_b d\tau, \tag{4.25}$$

where $d\tau$ means to integrate over all spatial coordinates.

D. THE UNCERTAINTY PRINCIPLE

Commuting Hermitian operators have simultaneous eigenvectors. As discussed in Section B, this means that the two observables associated with the two operators can be measured exactly, simultaneously. Observable quantities corresponding to operators can be arbitrarily well-defined at the same time. In Section A, it was shown that $[\underline{x}\,, \underline{P}] \neq 0$. For the Gaussian wave packet, discussed in Chapter 3, Section C, it was found that position and momentum could not be arbitrarily well-defined simultaneously and that $\Delta x\, \Delta P \approx \hbar$. This is referred to as an uncertainty relationship. For any pair of noncommuting operators, there is an uncertainty relation that quantifies to what extent the two observables can be specified simultaneously.

In general, for two Hermitian operators, $\underline{A}$ and $\underline{B}$, with the commutator relation

$$[\underline{A}, \underline{B}] = i\, \underline{C}, \tag{4.26}$$

where $\underline{C}$ is another operator or a number (a special case of a linear operator),

$$\Delta \underline{A}\, \Delta \underline{B} \geq \frac{1}{2} \left| \langle \underline{C} \rangle \right| \tag{4.27}$$

$$\langle \underline{C} \rangle = \langle S |\, \underline{C}\, | S \rangle\,. \tag{4.28}$$

$\langle \underline{C} \rangle$ is the expectation value of $\underline{C}$, and $\langle S|$ *and* $|S\rangle$ are an arbitrary, normalized bra and ket. A Hermitian operator is an operator for which $\langle S|\, \overline{\underline{A}}\, |T\rangle = \langle S|\, \underline{A}\, |T\rangle$.

To develop the relationship in equation (4.27), consider the Hermitian operator $\underline{D}$,

$$\underline{D} = \underline{A} + \alpha \underline{B} + i \beta \underline{B}. \tag{4.29}$$

α and β are real numbers. Then

$$\underline{D}\, |S\rangle = |Q\rangle\,, \tag{4.30}$$

since the application of an operator to an arbitrary ket, in general, yields a different ket.

$$\langle Q | Q \rangle = \langle S |\, \overline{\underline{D}}\underline{D}\, | S \rangle \geq 0 \tag{4.31}$$

since $\langle Q|Q\rangle$ is the length of a vector. Using the definition in equation (4.29), along with equation (4.31), it is straightforward to show that

$$\langle \underline{A}^2 \rangle + (\alpha^2 + \beta^2) \langle \underline{B}^2 \rangle + \alpha \langle \underline{C}' \rangle - \beta \langle \underline{C} \rangle \geq 0, \tag{4.32}$$

where $\underline{C}' = \underline{A}\underline{B} + \underline{B}\underline{A}$. The operator $\underline{C}'$ is called the anticommutator of the operators $\underline{A}$ and $\underline{B}$. It is symbolized as $\underline{A}\underline{B} + \underline{B}\underline{A} = [\underline{A}, \underline{B}]_+$. The $+$ denotes that this is the anticommutator rather than the commutator of the operators $\underline{A}$ and $\underline{B}$.

For an arbitrary ket $|S\rangle$, $\underline{B}\,|S\rangle \neq 0$. Then equation (4.32) can be rearranged to give

$$\langle \underline{A}^2 \rangle + \langle \underline{B}^2 \rangle \left(\alpha + \frac{1}{2} \frac{\langle \underline{C}' \rangle}{\langle \underline{B}^2 \rangle} \right)^2 + \langle \underline{B}^2 \rangle \left(\beta - \frac{1}{2} \frac{\langle \underline{C} \rangle}{\langle \underline{B}^2 \rangle} \right)^2 - \frac{1}{4} \frac{\langle \underline{C}' \rangle^2}{\langle \underline{B}^2 \rangle} - \frac{1}{4} \frac{\langle \underline{C} \rangle^2}{\langle \underline{B}^2 \rangle} \geq 0. \tag{4.33}$$

Since this holds for any values of α and β, they are selected such that the terms in parentheses are zero. Then,

$$\langle \underline{A}^2 \rangle \langle \underline{B}^2 \rangle \geq \frac{1}{4} \left(\langle \underline{C} \rangle^2 + \langle \underline{C}' \rangle^2 \right) \geq \frac{1}{4} \langle \underline{C} \rangle^2 . \tag{4.34}$$

The expression after the first $\geq$ is obtained by setting the terms in parentheses equal to zero and rearranging. $\langle \underline{C} \rangle$ and $\langle \underline{C}' \rangle$ are both real numbers because they are expectation values of Hermitian operators, and Hermitian operators have real eigenvalues. Then $\langle \underline{C} \rangle^2$ and $\langle \underline{C}' \rangle^2$ are both positive numbers because they are the square of real numbers. This leads to the term after the second $\geq$ in equation (4.34) because a positive number is greater than or equal to the sum of itself and another positive number; the equality holds for $\langle \underline{C}' \rangle = 0$. Therefore,

$$\langle \underline{A}^2 \rangle \langle \underline{B}^2 \rangle \geq \frac{1}{4} \langle \underline{C} \rangle^2 . \tag{4.35}$$

Defining,

$$(\Delta A)^2 = \langle \underline{A}^2 \rangle - \langle \underline{A} \rangle^2 . \tag{4.36}$$

ΔA is the spread or uncertainty in the observable A which is associated with the operator $\underline{A}$. $(\Delta A)^2$ is the second moment of the distribution of values obtained for the observable A. For a Gaussian distribution, it is the square of the standard deviation. Similarly,

$$(\Delta B)^2 = \langle \underline{B}^2 \rangle - \langle \underline{B} \rangle^2 . \tag{4.37}$$

For the frequently encountered special case of

$$\langle \underline{A} \rangle = \langle \underline{B} \rangle = 0, \tag{4.38}$$

combining equations (4.35)–(4.37) and writing $\left(\langle \underline{C} \rangle^2 \right)^{1/2} = |\langle \underline{C} \rangle|$, yields the fundamental uncertainty relation, or the Heisenberg Uncertainty Principle:

$$\Delta A \, \Delta B \geq \frac{1}{2} |\langle \underline{C} \rangle| . \tag{4.39}$$

As an example, consider position and momentum. The commutator of the position and momentum operators, as given in equation (4.7), is

$$[\underline{x}, \underline{P}] = i\hbar, \tag{4.40}$$

which is in the form of equation (4.26), and

$$\langle \underline{x} \rangle = \langle \underline{P} \rangle = 0. \tag{4.41}$$

Since $\hbar$ is a number, it is implicitly multiplied by the identity operator, and its expectation value is simply $\hbar$. Therefore, the position–momentum uncertainty relation is

$$\Delta x \, \Delta P \geq \hbar/2. \tag{4.42}$$

This rigorous relationship is similar to the one that was derived qualitatively for the free-particle wave packet in Chapter 3. The wave packet problem provides physical insight into the origin of the uncertainty relation.

For the case in which $\langle \underline{A} \rangle \neq 0$ and $\langle \underline{B} \rangle \neq 0$, two new Hermitian operators can be formed:

$$\underline{A}' = \underline{A} - \langle \underline{A} \rangle \, \underline{1} \tag{4.43}$$

and

$$\underline{B}' = \underline{B} - \langle \underline{B} \rangle \, \underline{1} \tag{4.44}$$

The expectation values are real numbers, and $\underline{1}$ is the identity operator. The new operators have the same commutator relationship as the original operators:

$$\left[\underline{A}', \underline{B}'\right] = i\,\underline{C}. \tag{4.45}$$

However, the expectation value of A' is

$$\begin{aligned}
\langle \underline{A}' \rangle &= \langle \underline{A} - \langle \underline{A} \rangle \, \underline{1} \rangle \\
&= \langle \underline{A} \rangle - \langle \langle \underline{A} \rangle \, \underline{1} \rangle \\
&= \langle \underline{A} \rangle - \langle \underline{A} \rangle \, \langle \underline{1} \rangle \\
&= \langle \underline{A} \rangle - \langle \underline{A} \rangle \\
&= 0
\end{aligned}$$

since the expectation value of the identity operator is unity. The same result is obtained for $\langle \underline{B}' \rangle$. Therefore,

$$\langle \underline{A}' \rangle = \langle \underline{B}' \rangle = 0 \tag{4.46}$$

Since the expectation values of the operators are zero, the procedure used above can be employed. The result is the general form of the uncertainty relation,

$$(\Delta A)^2 (\Delta B)^2 \geq \left(\frac{\langle \underline{C} \rangle}{2} \right)^2 + \left(\frac{\langle \underline{C}' \rangle}{2} - \langle \underline{A} \rangle \langle \underline{B} \rangle \right)^2. \tag{4.47}$$

CHAPTER 5

THE SCHRÖDINGER EQUATION, TIME-DEPENDENT AND TIME-INDEPENDENT

A. THE SCHRÖDINGER EQUATION

The Schrödinger equation is the basis for one of the widely used representations of quantum theory. The time-dependent Schrödinger equation is

$$i\hbar \frac{\partial \Phi(x, y, z, t)}{\partial t} = \underline{H}(x, y, z, t)\Phi(x, y, z, t), \tag{5.1}$$

where the position- and time-dependent function, $\Phi(x, y, z, t)$, is called a wavefunction. $\underline{H}(x, y, z, t)$ is the Hamiltonian operator; that is, it is the energy operator in the Schrödinger representation. The energy depends on the coordinates of the particles and on time since there may be time-dependent forces acting on the system. Initially the Hamiltonian will be considered for situations in which it is time-independent. This means that the energy of the system is constant. For a time-independent Hamiltonian, properties of a system may still evolve in time, but the energy is constant. An explicit example of this type is discussed in Chapter 8, and situations in which the Hamiltonian is time-dependent are treated in Chapters 11, 12, and 14.

The classical Hamiltonian is the sum of the kinetic energy and the potential energy,

$$H_{\text{classical}} = \frac{P^2}{2m} + V, \tag{5.2}$$

where P is the classical momentum and V is the potential energy. The quantum mechanical Hamiltonian operator is obtained by replacing the classical dynamical variables with the corresponding quantum operators. The operators in the Schrödinger representation are discussed in Chapter 4. In one dimension, the Hamiltonian operator is

$$\underline{H} = \frac{-\hbar^2}{2m}\frac{\partial^2}{\partial x^2} + V(x). \tag{5.3}$$

In three dimensions, the Hamiltonian is

$$\underline{H} = \frac{-\hbar^2}{2m}\nabla^2 + V(x, y, z), \tag{5.4}$$

where ∇^2, the Laplacian operator, is

$$\nabla^2 = \frac{\partial^2}{\partial x^2} + \frac{\partial^2}{\partial y^2} + \frac{\partial^2}{\partial z^2}. \tag{5.5}$$

When the Hamiltonian is time-independent—that is, $\underline{H}(x, y, z,)$—the time-dependent Schrödinger equation can be separated into time-dependent and time-independent parts by making the substitution

$$\Phi(x, y, z, t) = \phi(x, y, z)F(t). \tag{5.6}$$

The wavefunction is taken as a product of a time-independent function, $\phi(x, y, z)$, and time-dependent function, $F(t)$. Substituting equation (5.6) into equation (5.1) gives

$$i\hbar\frac{\partial}{\partial t}\phi(x, y, z)F(t) = \underline{H}(x, y, z)\phi(x, y, z)F(t). \tag{5.7}$$

On the left-hand side, ϕ is not a function of time, so it can be pulled outside of the derivative. On the right-hand side, $\underline{H}$ does not depend on t, so it will not operate on $F(t)$. $F(t)$ can be moved to the left of $\underline{H}$. Then the equation (5.7) can be written as

$$i\hbar\,\phi(x, y, z)\frac{\partial}{\partial t}F(t) = F(t)\,\underline{H}(x, y, z)\phi(x, y, z). \tag{5.8}$$

Dividing through by $\Phi = \phi\ F$ yields

$$\frac{i\hbar\dfrac{dF(t)}{dt}}{F(t)} = \frac{\underline{H}(x, y, z)\,\phi(x, y, z)}{\phi(x, y, z)}. \tag{5.9}$$

The left-hand side of equation (5.9) depends only on t and right-hand side depends only on spatial coordinates. Thus, no matter how x, y, and z are changed, the right-hand side must remain unchanged because there is no change in the value of the left-hand side; and no matter how t is changed, the left-hand side must remain unchanged because there is no change in the right-hand side. Therefore, each side of equation (5.9) is equal to a constant, E:

$$\frac{i\hbar\dfrac{dF}{dt}}{F} = E = \frac{\underline{H}\phi}{\phi}. \tag{5.10}$$

The right-hand side of equation (5.10) yields

$$\underline{H}(x, y, z)\phi(x, y, z) = E\,\phi(x, y, z). \tag{5.11}$$

This is the time-independent Schrödinger equation. It is the energy eigenvalue problem in the Schrödinger representation. Operating the Hamiltonian operator, which is the energy operator, on the time-independent wavefunction, $\phi(x, y, z)$ yields $\phi(x, y, z)$ times the constant E. Therefore, E is the energy of the state of the system ϕ because it is the eigenvalue of the energy operator. This form of the energy eigenvalue problem is obtained when the Hamiltonian is time-independent, and therefore the energy is time-independent.

The left-hand side of equation (5.10) yields an equation for the time-dependent part of the wavefunction, $F(t)$:

$$\frac{i\hbar \dfrac{dF(t)}{dt}}{F(t)} = E. \tag{5.12}$$

Multiplying both sides by $F(t)$ gives a differential equation for F, that is,

$$i\hbar \frac{dF(t)}{dt} = E\,F(t). \tag{5.13}$$

Rearranging gives

$$\frac{dF(t)}{F(t)} = -\frac{i}{\hbar} E\,dt, \tag{5.14}$$

and integrating both sides yields

$$\ln F = -\frac{i\,E\,t}{\hbar} + C, \tag{5.15}$$

where C is the constant of integration. Taking the antilog of the equation and finding the value of C that makes the time-dependent part of the wavefunction normalized results in

$$F(t) = e^{-i\,E\,t/\hbar}. \tag{5.16}$$

$F(t)$ is the time-dependent part of the wavefunction for a system with a time-independent Hamiltonian. It is a time-dependent phase factor.

The total wavefunction for the state of energy E is

$$\Phi_E(x, y, z, t) = \phi_E(x, y, z)e^{-i\,E\,t/\hbar}. \tag{5.17}$$

E is the energy eigenvalue. The observable E is used to label the state. Normalization of Φ depends only on the spatial part of the wavefunction:

$$\langle \Phi_E | \Phi_E \rangle = \int \Phi_E^* \, \Phi_E \, d\tau = \int \phi_E^* \, e^{+i\,E\,t/\hbar} \, \phi_E \, e^{-i\,E\,t/\hbar} \, d\tau = \int \phi_E^* \, \phi_E \, d\tau. \tag{5.18}$$

The total wavefunction is normalized if the time-independent part is normalized. Consider the expectation value of an operator $\underline{S}$ where $\underline{S}$ does not depend on time:

$$\langle \underline{S} \rangle = \langle \Phi | \, \underline{S} \, | \Phi \rangle = \int \Phi_E^* \, \underline{S} \, \Phi_E \, d\tau = \int \phi_E^* \, e^{i\,E\,t/\hbar} \, \underline{S} \, \phi_E \, e^{-i\,E\,t/\hbar} \, d\tau. \tag{5.19}$$

Since $\underline{S}$ does not depend on time, it commutes with $e^{-i\,E\,t/\hbar}$. Then the expression for the expectation value is

$$\langle \Phi | \, \underline{S} \, | \Phi \rangle = \int \phi_E^* \, \underline{S} \, \phi_E \, d\tau. \tag{5.20}$$

The expectation value is time-independent, and it depends only on the time-independent part of the wavefunction.

B. THE EQUATION OF MOTION OF THE EXPECTATION VALUE

Operators in the Schrödinger representation, such as the position and momentum operators, do not change in time. The time dependence is contained in the wavefunction. For a state $|S\rangle$, the expectation value of an operator $\underline{A}$ representing an observable can be calculated:

$$\langle \underline{A} \rangle = \langle S | \underline{A} | S \rangle .$$

For a system described by classical mechanics, in principle a classical dynamical variable (e.g., an observable such as P, the momentum), can be calculated for any state of the system. For a quantum system, the expectation value of the corresponding operator can be calculated. For momentum, the expectation value is $\langle S | \underline{P} | S \rangle$, which is the average value of the observable. In classical mechanics, it is possible to calculate the time derivative of a dynamical variable by taking the time derivative of the classical function. For classical momentum the time derivative is

$$\dot{P}(t) = \frac{\partial P(t)}{\partial t}. \tag{5.21}$$

Although the classical momentum goes over to the quantum mechanical operator, $P \rightarrow \underline{P}$, since the quantum operators are not time-dependent, the equivalent of equation (5.21) cannot be obtained by taking the time derivative of the quantum operator. It is necessary to define the quantum equivalent of equation (5.21) and to determine how to calculate it.

❖ **Definition:** The time derivative of an operator $\underline{A}$ (i.e., $\dot{\underline{A}}$) is an operator whose expectation value in any state $|S\rangle$ is the time derivative of the expectation value of the operator $\underline{A}$.

Then, an operator is needed that has an expectation value that is equivalent to

$$\frac{d \langle \underline{A} \rangle}{dt} = \frac{\partial}{\partial t} \langle S | \underline{A} | S \rangle . \tag{5.22}$$

The necessary operator can be obtained using the Schrödinger equation,

$$\underline{H} | S \rangle = i \hbar \frac{\partial}{\partial t} | S \rangle . \tag{5.23}$$

The right-hand side of equation (5.22) can be written as

$$\frac{\partial}{\partial t} \langle S | \underline{A} | S \rangle = \left(\frac{\partial}{\partial t} \langle S | \right) \underline{A} | S \rangle + \langle S | \underline{A} \frac{\partial}{\partial t} | S \rangle . \tag{5.24}$$

The derivative is taken using the chain rule. $\underline{A}$ is time-independent, so its derivative with respect to t is zero. The complex conjugate of the Schrödinger equation is

$$\langle S|\,\underline{H} = -i\hbar\,\frac{\partial}{\partial t}\,\langle S|\,, \tag{5.25}$$

where $\underline{H}$ operates to the left on the bra. Using equation (5.25), the term in brackets in equation (5.24) is

$$\left(\frac{\partial}{\partial t}\,\langle S|\right) = \frac{i}{\hbar}\,\langle S|\,\underline{H}. \tag{5.26}$$

Using equation (5.23) gives

$$\frac{\partial}{\partial t}\,|S\rangle = \frac{-i}{\hbar}\underline{H}\,|S\rangle\,. \tag{5.27}$$

Substituting equations (5.26) and (5.27) into equation (5.24) gives

$$\frac{\partial}{\partial t}\,\langle S|\,\underline{A}\,|S\rangle = \frac{i}{\hbar}[\langle S|\,\underline{H}\,\underline{A}\,|S\rangle - \langle S|\,\underline{A}\,\underline{H}\,|S\rangle]. \tag{5.28}$$

Therefore,

$$\frac{d\left\langle \underline{A}\right\rangle}{dt} = \frac{i}{\hbar}\,\langle S|\,\underline{H}\,\underline{A} - \underline{A}\,\underline{H}\,|S\rangle\,. \tag{5.29}$$

The right-hand side of equation (5.29) is $i/\hbar$ times the expectation value of the commutator of $\underline{H}$ with $\underline{A}$. The operator that represents the time derivative of a quantum mechanical dynamical variable according to the definition given below equation (5.21) is

$$\dot{\underline{A}} = \frac{i}{\hbar}[\underline{H}, \underline{A}] \tag{5.30a}$$

since

$$\frac{d\left\langle \underline{A}\right\rangle}{dt} = \frac{i}{\hbar}\left\langle [\underline{H}, \underline{A}]\right\rangle. \tag{5.30b}$$

An equation similar to equation (5.30) forms the basis for density matrix representation of quantum mechanics. The density operator and the density matrix are discussed in Chapter 14. The density operator is time-dependent. The operator $\underline{A}$ was taken to be time-independent. In general, if $\underline{A}$ is time-dependent, then

$$\frac{d\left\langle \underline{A}\right\rangle}{dt} = \frac{i}{\hbar}\left\langle [\underline{H}, \underline{A}]\right\rangle + \left\langle \frac{\partial \underline{A}}{\partial t}\right\rangle. \tag{5.31}$$

C. THE FREE-PARTICLE ENERGY EIGENVALUE PROBLEM

The time-independent Schrödinger equation is a form of the energy eigenvalue problem. As a first example of solving the time-independent Schrödinger equation, consider the free-particle problem. The momentum eigenstates and the wave packet nature of free particles were discussed in detail in Chapter 3. The momentum eigenvalue equation is

$$\underline{P}\,|P\rangle = p\;|P\rangle\,.$$

In the Schrödinger representation, it was determined that the momentum eigenfunctions for the free particle are

$$\psi_p(x) = \frac{1}{\sqrt{2\pi}}e^{i\,k\,x}, \tag{5.32}$$

with the momentum eigenvalues, $p = \hbar k$, and $k = p/\hbar$. For a free particle, the Hamiltonian is

$$\underline{H} = \frac{\underline{P}^2}{2m}, \tag{5.33}$$

since the potential energy for a free particle is zero; the total energy is the kinetic energy. The commutator of momentum and free-particle Hamiltonian operators is

$$[\underline{P}, \underline{H}] = \frac{\underline{P}^3}{2m} - \frac{\underline{P}^3}{2m} = 0. \tag{5.34}$$

The operators commute. Therefore, the $\underline{P}$ and $\underline{H}$ will have simultaneous eigenfunctions; that is, the same wavefunctions will be eigenfunctions of both operators.

The energy eigenvalue problem is

$$\underline{H}\,|P\rangle = E\,|P\rangle\,, \tag{5.35}$$

where the kets are the momentum eigenkets. Using the Schrödinger form of the Hamiltonian for the free particle (the kinetic energy operator),

$$\underline{H} = -\frac{\hbar^2}{2m}\frac{\partial^2}{\partial x^2}, \tag{5.36}$$

equation (5.35) can be evaluated.

$$\underline{H}\,|P\rangle = -\frac{\hbar^2}{2m}\frac{\partial^2}{\partial x^2}\left(\frac{1}{\sqrt{2\pi}}e^{i\,k\,x}\right) \tag{5.37}$$

$$= -\frac{\hbar^2}{\sqrt{2\pi}\,2m}(ik)^2 e^{i\,k\,x} \tag{5.38}$$

$$= \frac{\hbar^2 k^2}{2m}\left(\frac{1}{\sqrt{2\pi}}e^{i\,k\,x}\right) \tag{5.39}$$

$$= \frac{p^2}{2m}\,|P\rangle\,. \tag{5.40}$$

The momentum kets are eigenkets of the free-particle Hamiltonian, and the energy eigenvalues, the values of the observable energies for a nonrelativistic quantum mechanical free particle, are

$$E = \frac{p^2}{2m}. \tag{5.41}$$

The observable energies of a quantum free particle are the same as a classical free particle. Although the quantum and classical energies of a free particle are the same, as has been discussed in detail in Chapter 3, the description of a quantum free particle as a wave packet is fundamentally different from the classical description. The quantum description is necessary to describe physical processes such as electron diffraction and the photoelectric effect.

D. THE PARTICLE IN A BOX ENERGY EIGENVALUE PROBLEM

The free-particle Hamiltonian contains only a kinetic energy term. The potential energy is zero. Other problems, such as electrons in atoms or molecules, have both kinetic energy and potential energy terms in the Hamiltonian as in equations (5.3) and (5.4). While it is not physically realizable, the simplest quantum mechanical problem with a potential that gives bound states is the one-dimensional particle in a box. A real system that has some characteristics like the particle in a box is an electron in a quantum well. The one-dimensional box is defined by a potential $V(x)$ that is infinite everywhere except inside the box, where $V(x)$ is zero (see Figure 5.1). Qualitatively, this can be thought of as a box with walls that are infinitely massive and impenetrable. A particle placed inside the box cannot escape because the potential is infinite. Classically, this can be pictured as a one-dimensional racquet ball court with infinitely massive walls and a perfect ball in the sense that the collisions of the ball with the walls are perfectly elastic. There is no air resistance in this hypothetical one-dimensional racquet ball court. Under these conditions, if a player hits the ball against the wall, it will bounce off with no loss of kinetic energy. Therefore, it will continue to bounce between the walls indefinitely. The energy of the ball is its kinetic energy. The ball can be hit hard or soft to give it more or less kinetic energy. The energy for this classical particle in a box is continuous. It can take on any value (neglecting relativistic effects). It can have zero energy, in which case it will be located at some position x and have zero momentum.

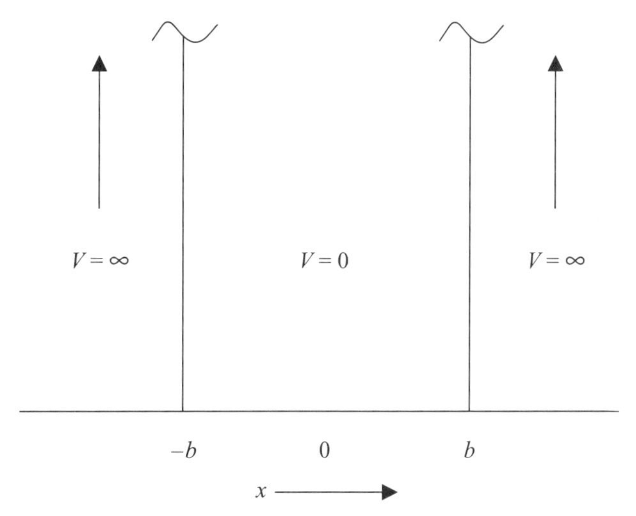

Figure 5.1 The potential energy for a particle in an infinite box. The potential is zero inside the box ($-b$ to b) and infinite outside the box.

The quantum mechanical particle in a box consists of a particle, such as an electron, in a box that is small in an absolute sense; that is, the size of the box is on the order of tens of nanometers. The box, the region where $V = 0$, extends from $-b$ to b. It is possible to determine immediately that the quantum particle in a box will have characteristics that are different from its classical equivalent. If the lowest energy of a quantum particle in a box could be zero, then the particle would be stationary and located at some point in the box. Since both the particle's position and momentum would be known, that is $\Delta x \Delta p = 0$, in violation of the Uncertainty Principle. Therefore, the lowest energy state of the quantum mechanical particle in a box problem cannot be zero, in contrast to the classical problem.

To find the allowed energy states of the quantum particle in a box, the energy eigenvalue problem is solved. The energy eigenvalue problem can take the form of the time-independent Schrödinger equation in one dimension:

$$-\frac{\hbar^2}{2m}\frac{d^2\varphi(x)}{dx^2} + V(x)\varphi(x) = E\,\varphi(x), \tag{5.42}$$

where the φ are the eigenfunctions, and the E are the energy eigenvalues.

$$V(x) = 0, \qquad |x| < b \tag{5.43a}$$

$$V(x) = \infty, \qquad |x| \geq b. \tag{5.43b}$$

For $|x| < b$, $V(x) = 0$, and the Hamiltonian has only the kinetic energy term:

$$-\frac{\hbar^2}{2m}\frac{d^2\varphi(x)}{dx^2} = E\,\varphi(x). \tag{5.44}$$

Equation (5.44) is a differential equation. Solving it will yield functions φ. However, these functions are interpreted as probability amplitudes as discussed in Chapter 3, Section C. For φ to solve the differential equation and also be a good wavefunction with physical meaning, it is necessary to impose auxiliary conditions on the nature of the acceptable solutions. The Born interpretation of the wavefunction as a probability amplitude function leads to the Born conditions that a solution to the Schrödinger equation must meet to be an acceptable, physically meaningful solution to the quantum mechanical eigenvalue problem and not merely a mathematical function that solves a differential equation. The four Born conditions for an acceptable wavefunction are:

1. The wavefunction must be finite everywhere.
2. The wavefunction must be single-valued.
3. The wavefunction must be continuous.
4. The first derivative of wavefunction must be continuous.

The four conditions are put forward to make the probability amplitude functions, which are solutions to the Schrödinger equation, consistent with a reasonable physical picture of nature. Condition 1 states the probability of finding a particle in some region of space can't be infinite. Condition 2 states that there can be only one value for the probability of finding a particle in some region of space. Conditions 3 and 4 arise from the view that nature is not discontinuous. While the properties of a system may change very rapidly for a small change in position, the change is not discontinuous. The Born conditions act as boundary conditions on the solution of the Schrödinger differential equation.

Equation (5.44) must be solved with a function that meets the four conditions. Rearranging gives

$$\frac{d^2\varphi(x)}{dx^2} = -\frac{2mE}{\hbar^2}\varphi(x); \tag{5.45}$$

that is, the second derivative of a function equals the function times a negative constant. Functions with this property are sin and cos:

$$\frac{d^2 \sin(ax)}{dx^2} = -a^2 \sin(ax) \tag{5.46a}$$

$$\frac{d^2 \cos(ax)}{dx} = -a^2 \cos(ax). \tag{5.46b}$$

Either the sin or cos functions or any combination of sin and cos solve the differential equation, provided that

$$a^2 = \frac{2mE}{\hbar^2}. \tag{5.47}$$

Although equations (5.46) with equation (5.47) solve the differential equation, the result is not a good wavefunction. Since the potential is infinite for $|x| \geq b$, the wavefunction must vanish at the wall of the box. The probability of finding the particle outside of the box is zero. Therefore, φ must be zero outside of the box. In addition, the wavefunction must be continuous. The function drawn in Figure 5.2 is a solution to the differential equation, but it is not continuous at the walls of the box. It has a finite value at $|x| = b$, but for $|x| \geq b$, φ must equal zero.

The acceptable sin and cos wavefunctions must be zero at $|x| = b$. φ will vanish at $|x| = b$ and be an acceptable wavefunction, provided that

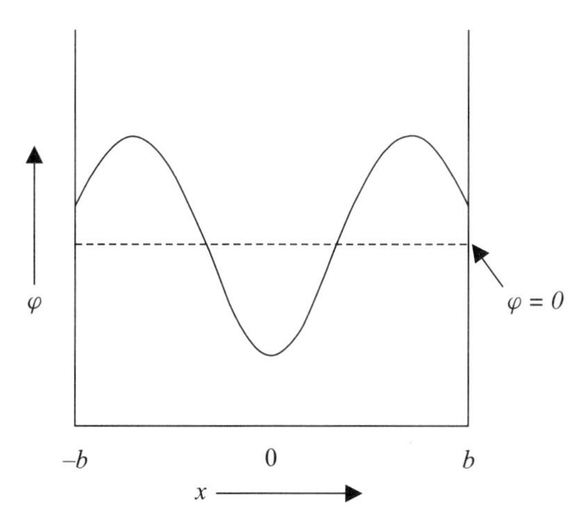

Figure 5.2 A solution to the particle in an infinite box Schrödinger equation that solves the differential equation but is not an acceptable wavefunction. The function shown is not continuous at the walls because the wavefunction is zero outside the box.

$$a = \frac{n\pi}{2b} \equiv a_n, \tag{5.48}$$

where n is an integer. For

$$\cos a_n x, \qquad n = 1, 3, 5, \ldots, \tag{5.49a}$$

and for

$$\sin a_n x, \qquad n = 2, 4, 6, \ldots. \tag{5.49b}$$

The relation, equation (5.48), means that there must be an integral number of half wavelengths in the box with zero amplitude at the walls. Any sin or cos function with an integral number of half wavelengths is an acceptable wavefunction. There are now two conditions for a: the one given in equation (5.47) and the one given in equation (5.48). Combining these yields

$$\frac{n^2\pi^2}{4b^2} = \frac{2mE}{\hbar^2}. \tag{5.50}$$

Equation (5.50) can be solved for the energy eigenvalues, E_n:

$$E_n = \frac{n^2\pi^2\hbar^2}{8mb^2} = \frac{n^2h^2}{8mL^2}, \tag{5.51}$$

where the length of the box, L, equals $2b$. The energy eigenvalues are labeled by the integer n, which is referred to as the quantum number. The energies for the quantum particle in a box take on discrete values. The energy levels of the quantum particle in a box are in contrast to the continuous energy values that a particle can have in a classical box. The lowest energy level has $n = 1$, and E is nonzero. Thus, even the lowest energy state has kinetic energy. As argued qualitatively above, the particle is never perfectly localized and stationary, and the Uncertainty Principle is not violated.

A very important feature of the solution to the particle in the box problem is the use of the Born conditions as boundary conditions to obtain an acceptable wavefunction. The solution to the differential equation did not yield quantized energy levels. The quantization occurred when the Born conditions were imposed to make the mathematical solution to equation (5.45) into an acceptable wavefunction. Imposition of physical considerations on the mathematical problem led to the solutions that are physically meaningful and resulted in the quantization of the energy levels.

The particle in the box is a highly artificial problem. In nature, potentials do not become infinite discontinuously. While the first three Born conditions are met in the solutions to the particle in the box problem, the fourth condition, that the first derivative of wavefunction must be continuous, is not met. The derivative is not continuous at $|x| = b$. Outside the box the slope is zero, while inside the box the slope is finite at $|x| = b$. This failure to obey the all of the Born conditions arises from the discontinuous nature of the potential, a feature that does not occur in nature.

The particle in the box wavefunctions are

$$\varphi_n(x) = \left(\frac{1}{b}\right)^{1/2} \cos\frac{n\pi x}{2b}, \qquad |x| \le b \qquad n = 1, 3, 5, \ldots \tag{5.52a}$$

and

$$\varphi_n(x) = \left(\frac{1}{b}\right)^{1/2} \sin\frac{n\pi x}{2b}, \qquad |x| \le b, \qquad n = 2, 4, 6, \ldots, \tag{5.52b}$$

where $2b$ is the length of the box, and the factors preceding cos and sin are the normalization constant. Outside of the box, the wavefunctions are zero. The first few wavefunctions inside the box are displayed in Figure 5.3. The wavefunctions show the spatial dependence of the probability amplitude of finding the particle at some location in the box. The wavefunctions can be positive or negative (or complex). The absolute value squared of the function (i.e., $\varphi^*\varphi$) is a real, positive function that gives the probability of finding the particle in some region of space. As discussed in Chapter 3, Section C, the probability of finding the particle in a region of space is given by the integral of $\varphi^*\varphi$ over the region [see equation (3.21)]. The wavefunctions with $n > 1$ have nodes inside the box at locations other than at the walls of the box. The probability of finding the particle at a node is zero. The question frequently arises of how the particle can move from one side of the box to the other and pass through the nodes. Implicit in this question is the view that the particle travels in a well-defined trajectory. This is a classical picture, and it is not applicable.

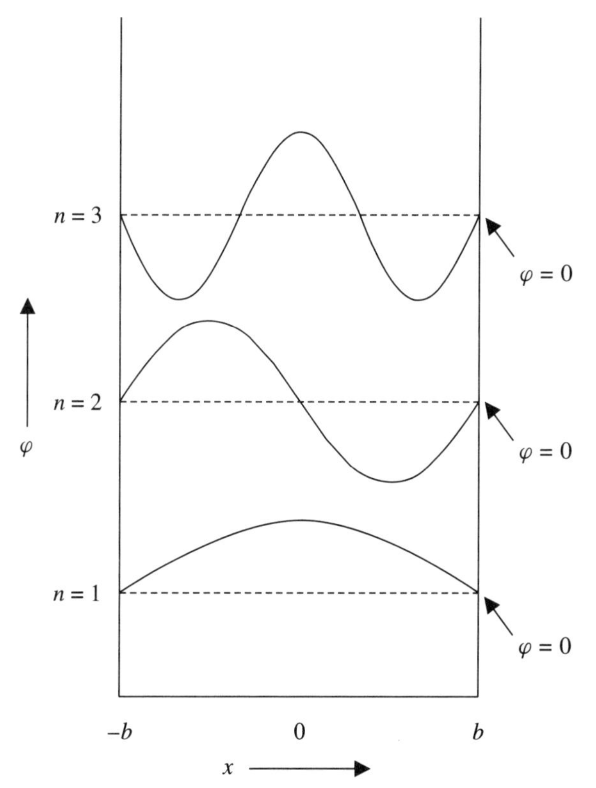

Figure 5.3 Particle in an infinite box wavefunctions for the three lowest energy eigenstates.

Waves, even classical waves, can have nodes. The wavefunction is a probability amplitude wave. It does not have a description of the particle in terms of a trajectory associated with it.

The quantization of the energy levels and the form of the wavefunctions, both brought about by the Born boundary conditions, are analogous to the classical problem of the modes of a string fixed at both ends. The fixed end points of the string result in a boundary condition that requires the modes of the string to be integral multiples of half wavelengths. The fact that the ends of the string are fixed forces the amplitude of the oscillation to be zero at the ends. The lowest mode, the fundamental, looks like the $n = 1$ function plotted in Figure 5.3. This is the mode that occurs when a guitar string is plucked. It is also possible to produce higher harmonics. These correspond to modes of the string in which there are one or more nodes in the wave. These modes look like the $n > 1$ modes of the particle in the box wavefunctions. However, the analogy cannot be carried too far. The energy of the classical string system is continuous. If the string is plucked harder, the string will vibrate with greater amplitude, corresponding to higher energy. Also, the string can be at rest, which is the state of zero energy.

Although the particle in the box is not a physically realistic problem, the results obtained from it can be used to get an estimate of the magnitudes of the energies of an electron in an atom or molecule. As the simplest possible model, consider an electron in a molecule to be a particle in a one-dimensional box of molecular size. This ignores the three-dimensional nature of molecules and the Coulomb interactions among the charged particles that make up molecules. Nevertheless, it is interesting to see what the effect of the size of a molecule has on the energy states. Consider the molecule anthracene (see Figure 5.4). The π electrons are delocalized over the three-ring system. The lowest allowed electronic transition is the promotion of one of the π electrons from the highest filled π bonding molecular orbital to the lowest empty π antibonding molecular orbital. This transition can be induced by the absorption of a photon. The photon energy is the difference in energy between the ground electronic state and the first excited electronic state. The ground state is referred to as S_0, and the first singlet excited state is S_1. The S designation refers to the state of the electrons' intrinsic angular momentum, called spin. The states are singlets if the electrons spins are paired. Electron spin, singlet states, and states with unpaired electron spins (e.g., triplet states) are discussed in Chapter 16.

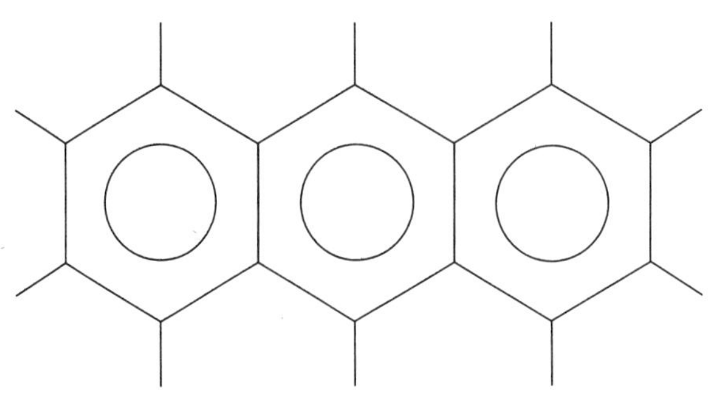

Figure 5.4 Anthracene.

As a very crude model of anthracene, take it to be a one-dimensional box. Then, the S_0 to S_1 transition energy is the difference in energy between the $n = 1$ and $n = 2$ particle in a box states. Using equation (5.51), the energy difference is

$$\Delta E = E_2 - E_1 = \frac{3h^2}{8mL^2}, \tag{5.53}$$

where m is the electron mass and L is the length of the box, which for anthracene is ~6 Å. This yields a wavelength of 393 nm for light that will be absorbed. The experimentally observed first strong anthracene absorption is at ~400 nm. (The wavelength of absorption varies, depending on the solvent, temperature, and pressure.) The very close agreement is fortuitous for anthracene. Other molecules will not agree as well. 393 nm is just into the ultraviolet. Larger molecules will absorb in the visible; smaller molecules will absorb further into the ultraviolet. The important point is that an electron confined to a molecular-sized box will have energy level spacings that correspond to wavelengths in the visible and ultraviolet, independent of the details of the molecular structure and Coulomb interactions. The size of the system (a molecule) and the mass of the particle (an electron) set the energy scale.

E. PARTICLE IN A FINITE BOX, TUNNELING AND IONIZATION

A particle in a one-dimensional box in which the potential does not go to infinity at the walls, but rather becomes a constant but finite value, has important properties that are fundamentally different from a particle in an infinite box. Inside the box the potential is $V(x) = 0$, and outside of the box it is $V(x) = V$. This is illustrated in Figure 5.5.

The time-independent Schrödinger equation is

$$-\frac{\hbar^2}{2m}\frac{d^2\varphi(x)}{dx^2} + V(x)\,\varphi(x) = E\,\varphi(x), \tag{5.54a}$$

with

$$V(x) = 0, \qquad |x| < b \tag{5.54b}$$

$$V(x) = V, \qquad |x| \geq b. \tag{5.54c}$$

Since the potential is not infinite outside the box, there are wavefunction solutions to the Schrödinger equation that are nonzero both inside and outside the box.

Inside the box, the Schrödinger equation is identical to the infinite box problem:

$$\frac{d^2\varphi(x)}{dx^2} = -\frac{2mE}{\hbar^2}\varphi(x). \tag{5.55}$$

The solutions are

$$\varphi(x) = q_1 \sin\sqrt{\frac{2mE}{\hbar^2}}x \tag{5.56a}$$

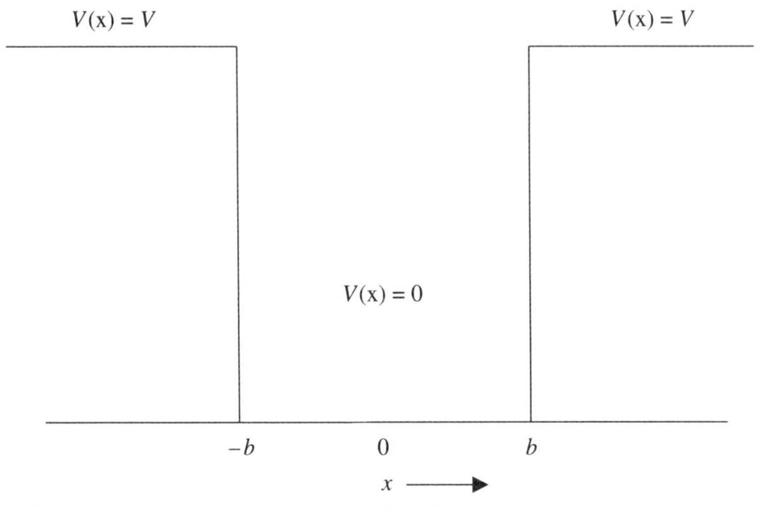

Figure 5.5 The potential energy for a particle in a finite box The potential is zero inside the box ($-b$ to b) and a constant value of V outside the box. The walls have a finite height but are infinitely thick.

or

$$\varphi(x) = q_2 \cos \sqrt{\frac{2mE}{\hbar^2}}x. \tag{5.56b}$$

Outside the box, the Schrödinger equation has the form

$$\frac{d^2\varphi(x)}{dx^2} = -\frac{2m(E-V)}{\hbar^2}\varphi(x). \tag{5.57}$$

There are two cases. The energy of the particle is less than potential energy outside of the box, that is, $E < V$. This is the situation in which the particle is bound. In the other case, $E > V$. The particle has enough energy to be above the potential wall. The particle is not bound. It is the equivalent to an electron that has been removed from an atom by ionization.

1. Bound States and Tunneling

The bound states are the states for which $E < V$. Then $E - V$ in equation (5.57) is negative, and the equation that describes the system outside the box has the form

$$\frac{d^2\varphi(x)}{dx^2} = \frac{2m(V-E)}{\hbar^2}\varphi(x), \tag{5.58}$$

where the factor multiplying the wavefunction on the right-hand side is positive. Unlike equation (5.55), the solutions to this equation are not oscillatory, but have the form $\exp(\pm ax)$ since

$$\frac{d^2 e^{\pm ax}}{dx^2} = a^2 e^{\pm a x}.$$

Therefore, the solution to the differential equation [equation (5.58)] for $|x| \geq b$ and $E < V$ is

$$\varphi(x) = e^{\pm\left[\frac{2m(V-E)}{\hbar^2}\right]^{1/2} x}. \tag{5.59}$$

However, for this solution to be an acceptable wavefunction, it must obey the Born condition that $\varphi(x)$ must be finite everywhere. To prevent $\varphi(x)$ from blowing up as $|x| \to \infty$, the wavefunction becomes

$$\varphi(x) = r_1 e^{-\left[\frac{2m(V-E)}{\hbar^2}\right]^{1/2} x}, \qquad x \geq b \tag{5.60a}$$

$$\varphi(x) = r_2 e^{+\left[\frac{2m(V-E)}{\hbar^2}\right]^{1/2} x}, \qquad x \leq -b. \tag{5.60b}$$

Equations (5.60) illustrate an important feature of quantum theory. The wavefunction, and therefore the probability of finding the particle, is not zero in spatial regions where $E < V$. The particle can be found where the potential energy is greater than the energy of the particle. This cannot happen in a classical system. The region where $E < V$ is referred to as the classically forbidden region. A classical particle can never enter such a region. The fact that a quantum particle can enter the classically forbidden region is responsible for important physical effects that are discussed below.

For the bound states, equations (5.56) give the solutions inside the box, and equations (5.60) give the solutions outside the box. To meet the Born conditions, the total wavefunctions must be continuous and their first derivatives must be continuous at the walls of the box. The functions are oscillatory inside the box and exponentially decaying outside the box. The oscillatory and exponential functions must meet smoothly at the walls of the box. This is illustrated schematically for the $n = 3$ state in Figure 5.6. Unlike the particle in the infinite box, the oscillatory part of the wavefunction does not go to zero at the wall. Rather, it merges smoothly with the exponential component in the classically forbidden region. For the finite box, all four of the Born conditions are obeyed; that is, unlike the infinite box, the wavefunction and the first derivative of the wavefunction are continuous at the walls. Using both continuity conditions and equations

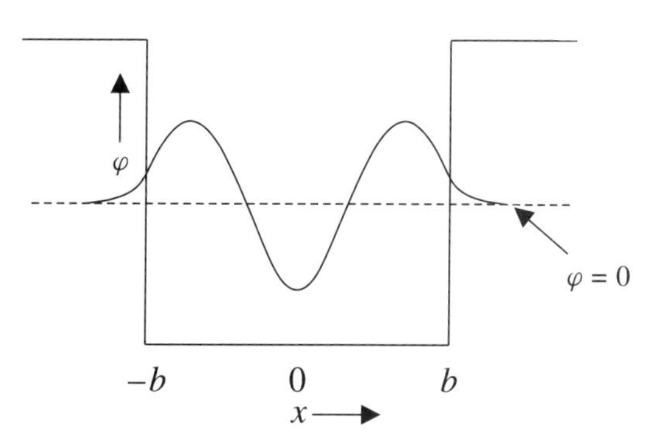

Figure 5.6 A wavefunction for a particle in a finite box. The wavefunction and its first derivative are continuous at the walls of the box. The wavefunction extends into the classically forbidden region.

(5.56) and (5.60), it is possible to numerically solve for the four constants q_1, q_2, r_1, and r_2 and to obtain the quantized energy eigenvalues. This is not done here. Rather, the results of the penetration of the wavefunction into the classically forbidden region are described.

Figure 5.7 shows a blowup of the region where the wavefunction penetrates into the classically forbidden region. Since the function decays exponentially outside the box, it approaches zero asymptotically. The rate of the decay of the probability amplitude depends on the difference in the potential energy (the height of the wall) and the energy of the particle, $V - E$. If the difference is not too great, the wavefunction will have substantial amplitude for a significant distance into the wall. For a wall with a finite thickness, the exponentially decaying tail of the wavefunction will be nonzero at the far side of the wall. If the potential on the far side of the wall is zero (or less than the particle energy), the wavefunction is oscillatory, with finite probability amplitude for finding the particle on the far side of the wall—that is, outside of the box. The particle is said to have tunneled through the wall.

Figure 5.8 shows a schematic illustration of the behavior of the wavefunction for a particle in a box with an infinite wall on the left-hand side of the box and a wall of finite height and finite thickness on the right-hand side of the box. The wavefunction penetrates through the wall; that is, it has a finite amplitude on the far side of the wall. On the far side of the right-hand wall, the potential is zero. Therefore, the Hamiltonian only has a kinetic energy term, and the wavefunction is oscillatory. A particle in the box can leak out by tunneling through the wall of the box. Tunneling has no classical counterpart. A classical particle inside the box with insufficient energy to go over the potential barrier will remain inside the box. A quantum particle can pass through the barrier even though it has insufficient energy to go over the barrier. The probability of finding the particle outside the box depends on the extent of the exponential decay of the wavefunction in the classically forbidden region.

Classically, a particle moving inside the box with energy less than that of the potential barrier will remain inside the box. For an idealized situation, in which the collisions of the particle with the walls of the box are perfectly elastic and there are no other damping

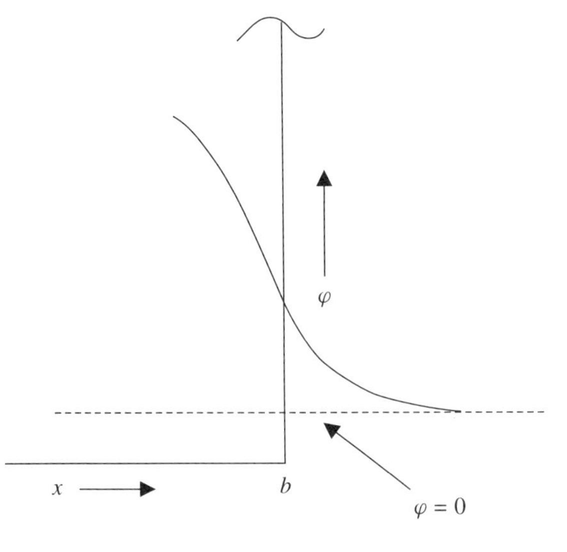

Figure 5.7 A blowup of the wall region of a particle in a finite box showing the penetration of the wavefunction into the classically forbidden region.

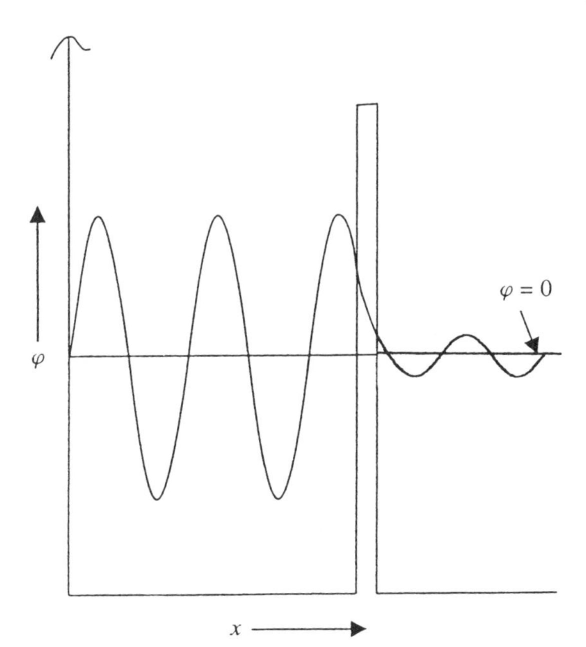

Figure 5.8 The potential energy and a wavefunction for a particle in a box with one wall having finite height and thickness. The wavefunction penetrates the classically forbidden region and has finite amplitude outside of the box. The particle tunnels through the classically forbidden region.

mechanisms, such as air resistance, the classical particle will bounce back and forth indefinitely. The quantum equivalent of the classical particle in the box is a wave packet analogous to the wave packets discussed in Chapter 3. However, the wave packet would be composed of a superposition of eigenstates of the Hamiltonian for the particle in the box with a finite wall rather than a superposition of momentum free-particle eigenstates. The particle in the box wave packet is more or less localized and moving back and forth between the walls of the box. It will move back and forth with a group velocity (Chapter 3, Section D) determined by the superposition of eigenstates that comprise the packet. For a packet relatively well localized in space near the left-hand wall of the box (Figure 5.8), the probability amplitude of finding the particle at the finite right-hand wall can be very small. Therefore, the probability of tunneling is negligible. However, when the particle is in the vicinity of the right-hand wall, the probability amplitude of the wavefunction at the right-hand wall is large, and the probability of tunneling is maximum. If the box is wide compared to the width of the wave packet, tunneling will mainly occur when the particle "hits" the right-hand wall. The collision frequency, ν_c, with the right-hand wall is $\nu_c = V_g/2L$, where V_g is the group velocity of the packet and L is the width of the box. The probability per unit time of finding the particle outside of the box is on the order of $\nu_c P_c$, where P_c is the probability of finding that the particle has tunneled out of the box on a single collision. P_c is determined by the superposition of states that comprise the packet and the height and width of the barrier. Given a large number of identical particle in finite box systems prepared identically at the same time, an observer would see bursts of particles appear outside the box with frequency ν_c.

An expression can be derived that gives insight into the probability of tunneling through the rectangular potential displayed in Figure 5.8. The barrier begins at the point $x = b$ and ends at the point $x = c$. Using equation (5.60a), the ratio of the probabilities at points b and c can be found. The ratio of the probabilities is given by

$$\frac{P_c}{P_b} = \frac{\int_c^{c+\varepsilon} \psi^*(x)\psi(x)\,dx}{\int_b^{b+\varepsilon} \psi^*(x)\psi(x)\,dx}. \tag{5.61}$$

Since the probability is not rigorously defined at a point, an integral must be performed over an infinitesimal interval of width ε. It does not matter which side of the point c the interval is chosen since the function is continuous and its first derivative is continuous. Substituting the wavefunction, equation (5.60a), into the top and bottom of equation (5.61) and integrating yields

$$\frac{P_c}{P_b} = \frac{\psi^*(c)\,\psi(c)}{\psi^*(b)\,\psi(b)}. \tag{5.62}$$

This is the same result that is obtained if the less rigorous expression for the probability at a point, $\psi^*\psi$, is used. The specific expression is

$$\frac{P_c}{P_b} = e^{-2[2m(V-E)/\hbar^2]^{1/2}(c-b)}; \tag{5.63}$$

or with $c - b = d$, the width of the rectangular barrier, the expression is

$$\frac{P_c}{P_b} = e^{-2d[2m(V-E)/\hbar^2]^{1/2}}. \tag{5.64}$$

Here, V is the barrier height and E is the energy of the particle. For a particle with the mass of an electron, and $E = 1000\ \text{cm}^{-1}(2 \times 10^{-20}\ \text{J})$ and $V = 2000\ \text{cm}^{-1}\ (4 \times 10^{-20}\ \text{J})$, the ratios of probabilities for $d = 1$ Å, 10 Å, and 100 Å, are 0.68, 0.02, and 3×10^{-17}, respectively. Thus, an electron can readily tunnel through a moderately high barrier with distances on the order of the size of an atom or molecule. However, once the distance becomes significantly greater than atomic dimensions, the tunneling probability becomes very small.

In real systems, a barrier will not be rectangular. A frequently used model for the shape of a potential barrier is an inverted parabola. For a parabolic potential barrier, the exponential decay constant, λ, called the tunneling parameter, is given by

$$\lambda = d(2mV/\hbar^2)^{1/2}, \tag{5.65}$$

where d is the distance through which the particle tunnels, m is the particle mass, and V is the height of the barrier; that is, it is the difference in energy between the top of the barrier and the energy of the particle. This expression is almost identical to that for the rectangular barrier. From the expression for the tunneling parameter, it is clear that tunneling will not occur over large distances and that light particles will have a greater probability of tunneling than heavy particles. Tunneling is very important for electrons and is responsible for many effects in chemistry and in semiconductor electronics. In chemistry, hydrogen tunneling is also important, but as the mass of the atom or group of atoms increases, the importance of tunneling in chemical processes decreases.

Figure 5.9 shows a schematic of a one-dimensional potential surface for a chemical reaction. The "reaction coordinate" is q. Classically, the reaction can only proceed from reactants to products via thermal activation, which takes the system over the potential barrier. The temperature dependence of such a classical reaction would have an activation energy corresponding to the top of the barrier. However, it is observed that chemical reactions can have activation energies which appear to be less than the barrier height. As shown in Figure 5.9, the reactants can be thermally promoted to a point near the top of the barrier, and tunneling can take the reactants to products by passage through the barrier. For the process shown in Figure 5.9, it is possible for products to be formed directly, without thermal activation. However, near the bottom of the reactant potential well, the barrier is very wide. The tunneling parameter [equation (5.65)] depends on d, the barrier width. Therefore, the probability of passing through the barrier will fall off exponentially with its width, and thermal activation to bring the system near the top of the barrier is required to allow the reaction to have a significant rate.

2. Unbound States and Ionization

When the particle energy E is greater than the potential energy V, the particle is no longer bound; that is, it no longer has an exponentially decaying probability outside of the box. Return to equation (5.57). When $E > V$, $E - V$ is a positive number. Inside the box, $V = 0$, and the solutions are given by equations (5.56). Outside the box, equation (5.57) has the form discussed previously; the second derivative of the wavefunction is equal to a negative

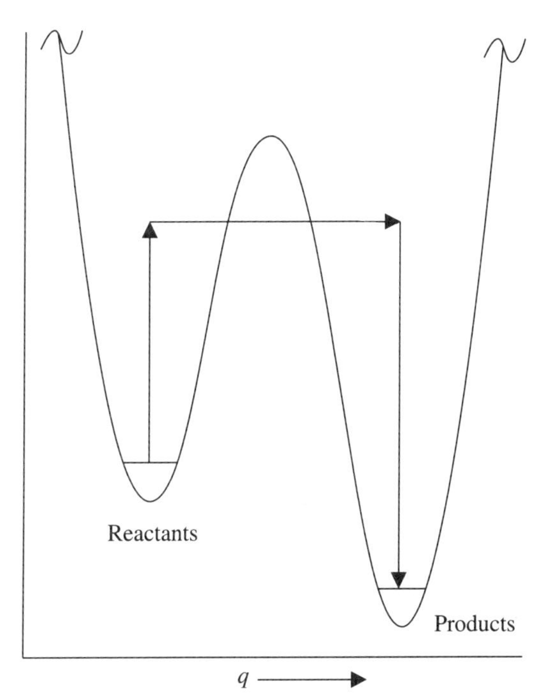

Figure 5.9 A schematic illustration of a double well potential surface for a chemical reaction. q is the reaction coordinate. The reaction proceeds by tunneling through the potential barrier.

constant times the same function. Then the solutions to equation (5.57) for $E > V$ are sin and cos:

$$\varphi(x) = s_1 \sin \sqrt{\frac{2m(E-V)}{\hbar^2}}x \tag{5.66a}$$

or

$$\varphi(x) = s_2 \cos \sqrt{\frac{2m(E-V)}{\hbar^2}}x. \tag{5.66b}$$

The solutions are oscillatory both inside and outside the box. To obtain the wavefunctions and the energy eigenvalues, the continuity conditions that require the wavefunctions and their first derivatives to be continuous at the walls of the box ($x = b$ and $x = -b$) are used with the functions given in equations (5.56) and (5.66). For example, at $x = b$ combining equation (5.56a) and equation (5.66a) gives

$$q_1 \sin \sqrt{\frac{2mE}{\hbar^2}}b = s_1 \sin \sqrt{\frac{2m(E-V)}{\hbar^2}}b. \tag{5.67}$$

Using the relations obtained from the continuity conditions, numerical solutions can be obtained for the constants q_1, q_2, s_1, and s_2 and the energies.

Figure 5.10 shows a schematic of a wavefunction with $E > V$. Although the energy is greater than the potential energy, the eigenstates are influenced by the presence of the square well potential. The wavefunction is oscillatory everywhere, and it extends to infinity. However, the square well acts as an attractive potential and increases the probability of finding the particle in the vicinity of the well.

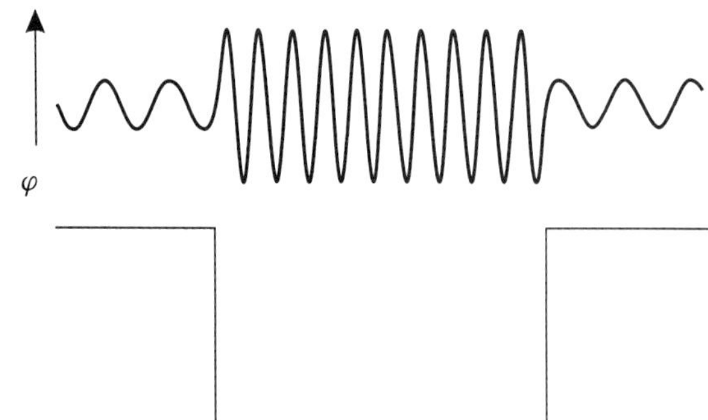

Figure 5.10 The potential energy and wavefunction for a particle in a box with finite height walls and with the particle energy greater than the potential energy outside the box.

In the limit that $E \gg V$, $(E - V) \cong E$. In this limit, $q_1 = s_1$ in equation (5.67); the wavefunction has equal amplitude everywhere. The wavefunction for $E \gg V$ is just

$$\varphi(x) = \sin \sqrt{\frac{2mE}{\hbar^2}} x. \tag{5.68}$$

The wavefunction takes on a form as if there are no walls. There is a continuous range of energies because there are no longer continuity boundary conditions that force the quantization of the energy levels. For sufficiently high energy, the particle is not influenced by the potential well. The particle behaves as a free particle. It has been "ionized" in the sense that it has gone from being bound to being free. For a free particle,

$$\sqrt{\frac{2mE}{\hbar^2}} = \sqrt{\frac{2mp^2}{2m\hbar^2}} = \sqrt{\frac{\hbar^2 k^2}{\hbar^2}} = k. \tag{5.69}$$

Therefore, the wavefunction is

$$\varphi(x) = \sin kx. \tag{5.70}$$

This is a free-particle wavefunction.

A particle in a box with infinite walls is always bound. The wavefunctions cannot extend beyond the walls into the classically forbidden region, nor can the addition of sufficient energy ionize the particle. Real potentials do not have these features. The particle in a finite box exhibits properties characteristic of real systems. The wavefunctions of the bound states penetrate the walls into the classically forbidden region. If a wall has both a finite height and finite thickness, the exponentially decaying part of the wavefunction in the classically forbidden region (inside the wall) will have nonzero amplitude at the far side of the wall. This gives rise to tunneling. If the particle is given sufficient energy, it will be able to access a continuum of unbound states. Thus finite potentials give rise to tunneling and to ionization.

CHAPTER 6 THE HARMONIC OSCILLATOR IN THE SCHRÖDINGER AND DIRAC REPRESENTATIONS

The one-dimensional harmonic oscillator problem is important in both classical and quantum mechanics because it is the basis for the analysis of many other more complex problems. In chemistry, it is the starting point for understanding the vibrations and infrared spectroscopy of molecules. In physics, it is important in the quantum theory of solids and the quantum theory of radiation. An example of a classical harmonic oscillator is a mass attached to an ideal spring. When the spring is stretched or compressed and released, the mass will oscillate back and forth around its equilibrium position. For a harmonic oscillator, the spring obeys Hooke's law; that is, the force is proportional to the displacement,

$$F = -kx, \tag{6.1}$$

where the displacement of the mass is along the x axis. Force is the negative of the derivative of the potential,

$$F = -\frac{\partial V(x)}{\partial x}. \tag{6.2}$$

Therefore, the potential energy for a harmonic oscillator is

$$V(x) = \int kx\,dx = \frac{1}{2}kx^2, \tag{6.3}$$

where the force constant, k, is

$$k = 4\pi^2 m\nu^2 = m\omega^2. \tag{6.4}$$

m is the mass of the oscillating object, ν is the oscillator frequency, and ω is the angular frequency.

The potential for the harmonic oscillator is parabolic as shown in Figure 6.1. For a classical oscillator, the spring is stretched and released. At the point of release, the total energy is potential energy. The mass accelerates as it moves back toward the center of the parabolic potential. At the center, the total energy is kinetic energy; the potential energy is zero. As the mass continues to move, the spring is compressed. It reaches a point where the total energy is again potential energy. The mass stops, reverses direction, and again accelerates toward the center. The points where the mass stops because the total energy is potential energy are called the classical turning points. A trajectory is shown schematically in Figure 6.1, and the classical turning points are indicated by the dashed lines. A classical harmonic oscillator can have any energy. The range of energy values is continuous. The energy is determined by how much the spring is stretched—that is, the value of the potential energy at the turning points. The motion of the mass is oscillatory. The position varies sinusoidally with time. The oscillator is moving fastest in the center, and therefore it spends the least time near the center. It is possible for the oscillator to have zero energy. The particle is not moving and is located at the center of the parabolic potential well. This is shown in Figure 6.1.

A quantum harmonic oscillator behaves very differently from its classical counterpart. An initial indication of the differences is the fact that the lowest energy state of a quantum harmonic oscillator cannot be zero. The state of zero energy corresponds to the position being well-defined at the center of the potential and the momentum being well defined as zero. Therefore, $\Delta x \, \Delta p = 0$ in violation of the Uncertainty Principle, $\Delta x \, \Delta p \geq \hbar/2$.

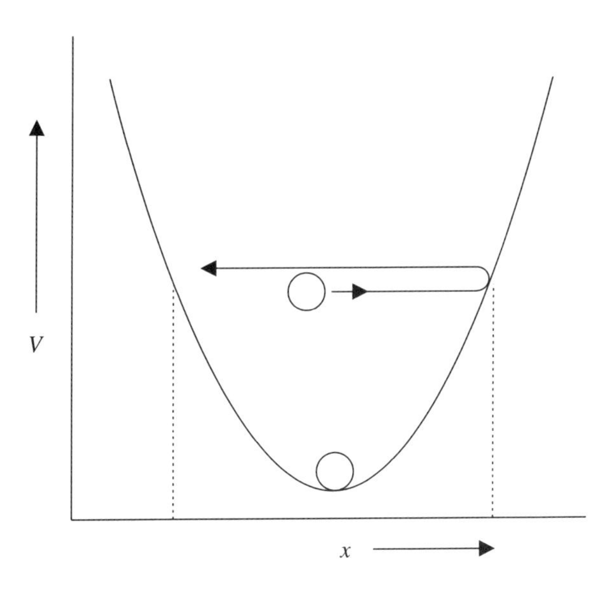

Figure 6.1 The parabolic potential for a harmonic oscillator. The circles represent two possible situations of a classical particle. The dashed lines are the classical turning points.

A. THE QUANTUM HARMONIC OSCILLATOR IN THE SCHRÖDINGER REPRESENTATION

First, the harmonic oscillator energy eigenvalue problem will be solved in the Schrödinger representation. In Chapter 6, Section B, the problem will be solved again using the Dirac raising and lowering operator approach. The Hamiltonian operator for the one-dimensional harmonic oscillator is

$$\underline{H} = -\frac{\hbar^2}{2m}\frac{d^2}{dx^2} + \frac{1}{2}kx^2. \tag{6.5}$$

The energy eigenvalue equation,

$$\underline{H}\,|\psi\rangle = E|\psi\rangle$$

can be written as

$$(\underline{H} - E)|\psi\rangle = 0.$$

In the Schrödinger representation, this takes the form

$$\frac{d^2\psi\,(x)}{dx^2} + \frac{2m}{\hbar^2}[E - 2\pi^2 m\nu^2 x^2]\,\psi\,(x) = 0. \tag{6.6}$$

With the substitutions

$$\lambda = \frac{2mE}{\hbar^2} \tag{6.7a}$$

and

$$\alpha = \frac{2\pi m\nu}{\hbar} \tag{6.7b}$$

the differential equation is

$$\frac{d^2\psi\,(x)}{dx^2} + (\lambda - \alpha^2 x^2)\,\psi\,(x) = 0. \tag{6.8}$$

To solve the eigenvalue problem, it is necessary to find a set of functions, $\psi(x)$, that satisfy this equation from $-\infty$ to $+\infty$. The functions must also obey the four Born conditions given in Chapter 5, Section D.

To solve equation (6.8), first a solution is found in the limit that x becomes very large. Once the $x \to \infty$ solution is found, a power series is introduced to make the large x solution valid for all x. This is referred to as the polynomial method, and it will be used again in the solution of the hydrogen atom problem in Chapter 7. For very large x

$$\alpha^2 x^2 \gg \lambda,$$

and therefore the constant λ can be dropped. The resulting equation is

$$\frac{d^2\psi}{dx^2} = \alpha^2 x^2 \psi. \tag{6.9}$$

The solution to this equation for very large x is

$$\psi = e^{\pm\frac{\alpha}{2}x^2}, \tag{6.10}$$

since

$$\frac{d^2\psi}{dx^2} = \alpha^2 x^2 e^{\pm\frac{\alpha}{2}x^2} \pm \alpha\, e^{\pm\frac{\alpha}{2}x^2}, \tag{6.11}$$

and the second term on the right-hand side is negligible compared to the first term as x approaches infinity.

Of the two solutions given in equation (6.10),

$$e^{-\frac{\alpha}{2}x^2} \quad \text{and} \quad e^{+\frac{\alpha}{2}x^2},$$

only the first is an acceptable wavefunction. The second solution becomes infinite as $x \to \pm\infty$, in violation of one of the Born conditions. Therefore, the large x solution is

$$\psi(x) = e^{-\frac{\alpha}{2}x^2}. \tag{6.12}$$

To find a solution for all x, the function

$$\psi(x) = e^{-\frac{\alpha}{2}x^2} f(x) \tag{6.13}$$

is used, where $f(x)$ is an unknown function to be determined. Equation (6.13) is substituted into the original equation, equation (6.8). To proceed, it is necessary to take the second derivative,

$$\frac{d^2\,\psi(x)}{dx^2} = e^{-\frac{\alpha}{2}x^2}(\alpha^2 x^2 f - \alpha f - 2\alpha\, x f' + f''),$$

with

$$f' = \frac{df}{dx} \quad \text{and} \quad f'' = \frac{d^2 f}{dx^2}.$$

Substituting these into equation (6.8) and dividing by

$$e^{-\frac{\alpha}{2}x^2}$$

yields a differential equation for $f(x)$,

$$f'' - 2\alpha x f' + (\lambda - \alpha) f = 0. \tag{6.14}$$

Dividing by α gives

$$\frac{1}{\alpha} f'' - 2x f' + \left(\frac{\lambda}{\alpha} - 1\right) f = 0. \tag{6.15}$$

Making the substitution

$$y = \sqrt{\alpha}\, x \tag{6.16}$$

rescales the x axis and gives

$$f(x) = H(y). \tag{6.17}$$

Using equations (6.16) and (6.17) with equation (6.15) yields a standard form

$$\frac{d^2H(\gamma)}{d\gamma^2} - 2\gamma\frac{dH(\gamma)}{d\gamma} + \left(\frac{\lambda}{\alpha} - 1\right)H(\gamma) = 0. \tag{6.18}$$

Equation (6.18) is called Hermite's equation. The solutions to differential equations such as this were investigated in the 1880s. The substitution, equation (6.16), was made to obtain this standard form since the method for its solution and the properties of the solutions are well known. Finding $H(\gamma)$ gives $f(x)$, and multiplying $f(x)$ by the large x solution [equation (6.13)] yields $\psi(x)$.

To solve equation (6.18) and obtain $H(\gamma)$, a series expansion is used:

$$H(\gamma) = \sum_{\nu} a_\nu \gamma^\nu = a_0 + a_1\gamma + a_2\gamma^2 + a_3\gamma^3 + \cdots. \tag{6.19a}$$

The derivatives are obtained by term-by-term differentiation:

$$\frac{dH(\gamma)}{d\gamma} = \sum_{\nu} \nu\, a_\nu \gamma^{\nu-1} = a_1 + 2a_2\gamma + 3a_3\gamma^2 + \cdots \tag{6.19b}$$

$$\frac{d^2H}{d\gamma^2} = \sum_{\nu} \nu(\nu-1)\, a_\nu \gamma^{\nu-2} = 2a_2 + 6a_3\gamma + \cdots. \tag{6.19c}$$

Substituting the three series [equation (6.19)] into the differential equation (6.18) gives

$$\begin{aligned}
&2a_2 + 6a_3\gamma + 12a_4\gamma^2 + 20a_5\gamma^3 + \cdots \\
&- 2a_1 - 4a_2\gamma^1 - 6a_3\gamma^2 - \cdots \\
&+ \left(\frac{\lambda}{\alpha} - 1\right)a_0 + \left(\frac{\lambda}{\alpha} - 1\right)a_1\gamma + \left(\frac{\lambda}{\alpha} - 1\right)a_2\gamma^2 + \left(\frac{\lambda}{\alpha} - 1\right)a_3\gamma^3 + \cdots = 0.
\end{aligned} \tag{6.20}$$

Equation (6.20) contains an infinite number of terms that sum to zero regardless of the choice of the position variable, γ. The series contains all powers of γ—that is, $\gamma^0, \gamma^1, \gamma^2, \gamma^3 \ldots$. Since γ can take on any value and the series still sums to zero, the coefficients of the individual powers of γ all must be identically equal to zero. By collecting terms in equation (6.20), the coefficient of each power of γ can be found and set equal to zero:

$$2a_2 + \left(\frac{\lambda}{\alpha} - 1\right)a_0 = 0 \qquad [\gamma^0]$$

$$6a_3 + \left(\frac{\lambda}{\alpha} - 3\right)a_1 = 0 \qquad [\gamma^1]$$

$$12a_4 + \left(\frac{\lambda}{\alpha} - 5\right)a_2 = 0 \qquad [\gamma^2]$$

$$20a_5 + \left(\frac{\lambda}{\alpha} - 7\right)a_3 = 0. \qquad [\gamma^3]$$

The expression in the bracket following each equation indicates the power of γ associated with each coefficient. In general, the coefficient of the νth power of γ, which is equal to zero, is

$$(\nu + 1)(\nu + 2)a_{\nu+2} + \left(\frac{\lambda}{\alpha} - 1 - 2\,\nu\right) a_\nu = 0. \tag{6.21}$$

Then, solving for $a_{\nu+2}$ yields

$$a_{\nu+2} = -\frac{\left(\dfrac{\lambda}{\alpha} - 2\,\nu - 1\right)}{(\nu + 1)(\,\nu + 2)}\, a_\nu. \tag{6.22}$$

Equation (6.22) is called a recursion formula. If a_0 and a_1 are given (which at this point are arbitrary), then $a_2,\ a_3,\ a_4,\ \ldots$ can be calculated. If $a_0 = 0$, then only odd terms are obtained. If $a_1 = 0$, then only even terms are obtained. Thus, there is an odd series and an even series. Using equation (6.22) with equation (6.19a) gives a series expansion solution to the differential equation. Combining these with equations (6.17) and (6.13) gives solutions to the original differential equation.

The solution to the differential equation involves a power series with an infinite number of terms. While the power series substituted for $f(x)$ in equation (6.13) solves the differential equation, it is not a good wavefunction because it becomes infinite as x goes to infinity. This is so even though the power series is multiplied by the solution for large γ,

$$e^{-\frac{\gamma^2}{2}}.$$

To see this, consider the expansion

$$e^{\gamma^2} = 1 + \gamma^2 + \frac{\gamma^4}{2!} + \frac{\gamma^6}{3!} + \cdots + \frac{\gamma^\nu}{(\nu/2)!} + \frac{\gamma^{\nu+2}}{((\nu/2) + 1)!} + \cdots. \tag{6.23}$$

For large enough values of γ, the first terms in this series will be unimportant. Call the coefficient in this expansion

$$b_\nu = \frac{1}{(\nu/2)!}.$$

Then

$$b_{\nu+2} = \frac{1}{((\nu/2) + 1)!} = \frac{1}{(\nu/2)!((\nu/2) + 1)},$$

and

$$b_{\nu+2} = \frac{1}{(\nu/2 + 1)} b_\nu.$$

For large values of ν, such that $(\nu/2) \gg 1$, the equation reduces to

$$b_{\nu+2} = \frac{2}{\nu} b_\nu. \tag{6.24}$$

For very large values of ν, the recursion relation, equation (6.22), becomes

$$a_{\nu+2} = -\frac{(-2\nu)a_\nu}{\nu^2}$$

or

$$a_{\nu+2} = \frac{2}{\nu} a_\nu. \tag{6.25}$$

Taking the ratio of higher-order terms in the two series gives

$$\frac{a_{\nu+2}}{b_{\nu+2}} = \frac{a_\nu}{b_\nu} = C; \tag{6.26}$$

that is, the ratios of the higher-order terms of the two series is a constant that is independent of the index, ν. The higher-order terms of the series for $H(\gamma)$ differ from the series for e^{γ^2} only by a multiplicative constant. For large values of $|\gamma|$, for which the lower-order terms in the expansion of H are unimportant, H behaves like e^{γ^2}.

The solution to the differential equation [equation (6.8)] is

$$\psi(\gamma) = e^{-\frac{\gamma^2}{2}} H(\gamma), \tag{6.27}$$

which for large values of $|\gamma|$ is

$$= e^{-\gamma^2/2} e^{\gamma^2} = e^{\gamma^2/2}. \tag{6.28}$$

Therefore, if the series for $H(\gamma)$ has an infinite number of terms, $\psi(\gamma)$ blows up as $\gamma \to \infty$ and is unacceptable as a wavefunction. This is an important point. Equation (6.27) with $H(\gamma)$ having an infinite number of terms is a solution to the differential equation, but it is not a good wavefunction because it fails to meet the mathematical constraints imposed by the Born probability interpretation of the wavefunction solutions to the Schrödinger equation.

To obtain a solution to the differential equation that is also a valid wavefunction, it is necessary for the series to have a finite number of terms. For any finite number of terms, as $\gamma \to \infty$ the Gaussian factor in equation (6.27), which multiplies the polynomial H, will go to zero faster than any power of γ in the polynomial goes to infinity. Therefore, the total function will not blow up, but rather go to zero, as $\gamma \to \infty$. If the parameter λ, which contains the energy, is restricted to values

$$\lambda = (2n + 1)\alpha, \tag{6.29}$$

then the recursion relation [equation (6.22)]

$$a_{\nu+2} = -\frac{\left(\frac{\lambda}{\alpha} - 2\nu - 1\right)}{(\nu+1)(\nu+2)} a_\nu$$

will yield zero for all coefficients with $\nu > n$. It is then possible to make the odd series or the even series terminate after a finite number of terms. If a_1 is set equal to zero and a_0 is finite, the even series is obtained with a finite number of terms determined by the choice of n in equation (6.29). If a_0 is set equal to zero and a_1 is finite, the odd series is obtained with a finite number of terms again determined by the choice of n in equation (6.29). Selecting λ according to equation (6.29) yields a polynomial with a finite number of terms and therefore gives an acceptable wavefunction. Using the definitions of λ and α given in equations (6.7),

$$\lambda = \frac{2mE}{\hbar^2} = \frac{(2n+1)\,2\pi m\nu}{\hbar}. \tag{6.30}$$

Solving this equation for E yields

$$E_n = \left(n + \frac{1}{2}\right) h\nu, \tag{6.31}$$

the quantized energy eigenvalues of the quantum harmonic oscillator. The quantization of the energy levels arose from the solution of the differential equation and the imposition of the Born conditions on the wavefunction.

The lowest energy of a quantum harmonic oscillator occurs for $n = 0$, and

$$E_0 = \frac{1}{2} h\nu. \tag{6.32}$$

This is in contrast to a classical harmonic oscillator for which the lowest allowed energy is zero. The $n = 0$ energy of the quantum harmonic oscillator is referred to as the zero point energy. The spacing between the levels is $h\nu$. The levels are equally spaced. This is very different from a classical harmonic oscillator for which the energy can vary continuously.

The harmonic oscillator wavefunctions are

$$\psi_n(x) = N_n e^{-\gamma^2/2} H_n(\gamma), \tag{6.33}$$

where N_n is the normalization constant, and H_n are called the Hermite polynomials. To this point, a_0 and a_1, which begin the even and odd series, respectively, have arbitrary values. Normalization determines these values, and the normalization constant is

$$N_n = \left\{\left(\frac{\alpha}{\pi}\right)^{1/2} \frac{1}{2^n n!}\right\}^{1/2}. \tag{6.34}$$

(Note that $0! = 1$.) N_n is derived subsequently. The first few Hermite polynomials are

$$H_0(\gamma) = 1 \tag{6.34a}$$

$$H_1(\gamma) = 2\gamma \tag{6.34b}$$

$$H_2(\gamma) = 4\gamma^2 - 2 \tag{6.34c}$$

$$H_3(\gamma) = 8\gamma^3 - 12\gamma \tag{6.34d}$$

$$H_4(\gamma) = 16\gamma^4 - 48\gamma^2 + 12 \tag{6.34e}$$

$$H_5(\gamma) = 32\gamma^5 - 160\gamma^3 + 120\gamma \tag{6.34f}$$

$$H_6(\gamma) = 64\gamma^6 - 480\gamma^4 + 720\gamma^2 - 120. \tag{6.34g}$$

The wavefunction for the lowest-energy eigenstate of the quantum harmonic oscillator is

$$\psi_0(x) = \left(\frac{\alpha}{\pi}\right)^{1/4} e^{-\frac{\alpha}{2}x^2} = \left(\frac{\alpha}{\pi}\right)^{1/4} e^{-\gamma^2/2}. \tag{6.35}$$

ψ_0 is a Gaussian. It is plotted in Figure 6.2 with γ as the distance parameter. The wavefunction is the probability amplitude. The probability of finding the particle at some point in space is $\psi^*\psi$ or, since the functions are real, $(\psi)^2$. [As discussed in Chapter 3, Section C, formally the probability is defined in a small interval about the point of interest and is found by performing an integral of $(\psi)^2$ over the interval.] $(\psi)^2$ is plotted in Figure 6.3. Both the wavefunction and the probability are peaked at the center—that is, where the potential is a minimum (see Figure 6.1). This is in contrast to a classical harmonic oscillation. In a classical harmonic oscillator, the velocity is maximum at the center of the parabolic potential well and goes to zero at the classical turning points. While the classical oscillator is not described in terms of a probability function, the oscillator spends the least time in the center and the most time at the turning points. Therefore, the classical oscillator is more likely to be observed at the turning points and less likely to be found in the center.

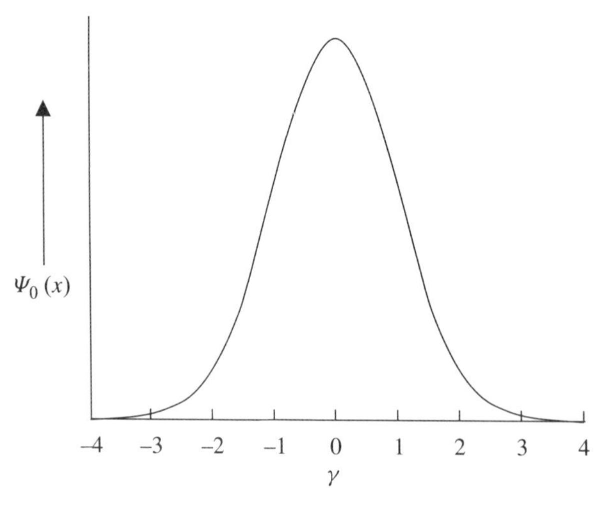

Figure 6.2 The lowest-energy eigenfunction ($n = 0$) of a quantum harmonic oscillator plotted against the distance parameter, γ. The probability amplitude function is peaked in the center, in contrast to a classical harmonic oscillator.

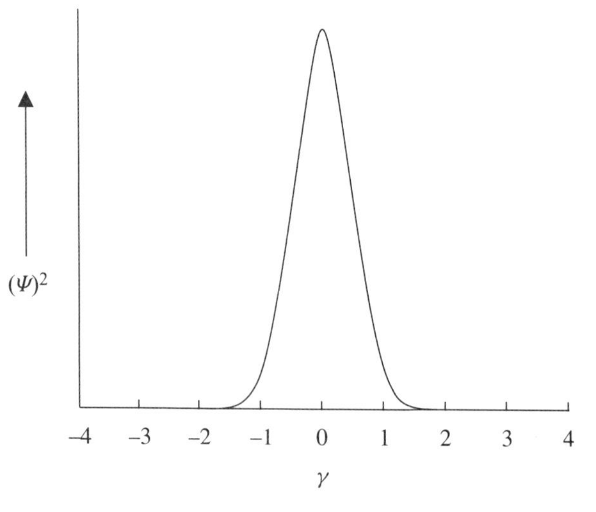

Figure 6.3 The probability distribution function for the $n = 0$ level of a quantum harmonic oscillator plotted against the dimensionless distance parameter, γ. The classical turning points are at $\gamma = \pm 1$. The probability distribution extends beyond the classical turning points.

The classical turning points occur when the total energy is equal to the potential energy. To locate the classical turning points on the probability distribution for the ground state of the quantum harmonic oscillator, the $n = 0$ oscillator energy is set equal to the potential energy,

$$\frac{1}{2}h\nu = \frac{1}{2}kx^2. \tag{6.36}$$

Substituting the definitions of k, α, and γ [equations (6.4), (6.7b), and (6.16), respectively], and solving for γ yields the dimensionless classical turning points,

$$\gamma = \pm 1. \tag{6.37}$$

In terms of the dimensionless distance parameter, γ, the classical turning points are independent of the mass, frequency, and spring constant of the oscillator. From the plot of the probability in Figure 6.3, it is seen that the probability distribution extends past the classical turning points. There is finite probability of finding the quantum particle in the classically forbidden region of space. Some of the consequences of this nonclassical behavior are discussed in Chapter 5, Section E in connection with the particle in the finite box. Integrating $\psi^*\psi$ between the classical turning points gives the fraction of the probability that is within the classically allowed region. Since the wavefunctions are normalized, subtracting this fraction from 1 gives the probability of finding the particle in the classically forbidden region. This probability is 0.16, and it is independent of the oscillator parameters. Therefore, a significant fraction of the probability is in the classically forbidden region.

The $n = 1$ to 6 wavefunctions are shown in Figure 6.4. The horizontal axes are in units of γ, and the horizontal line in each graph is the line of zero probability amplitude. The wavefunctions have an increasing number of nodes, points of zero probability amplitude,

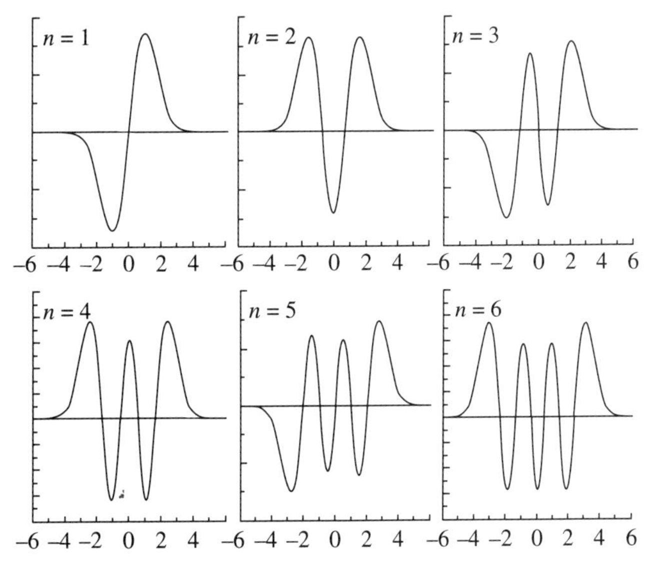

Figure 6.4 Harmonic oscillator wavefunctions for the states $n = 1$ to 6. The horizontal axes are in units of dimensionless distance, γ.

as the quantum number increases. The probability amplitude changes sign when a node is crossed. While the wavefunctions can oscillate from positive to negative, the probabilities, $\psi^*\psi$, are always positive. Figure 6.5 shows a plot of the probability distribution for the state $n = 10$.

The $n = 0$ state probability distribution function (Figure 6.2) is peaked in the center, in contrast to the $n = 10$ state. As the quantum number increases, the nodes become more closely spaced. For large quantum numbers, the probability distribution begins to look like that of a classical harmonic oscillator; that is, it is peaked at the ends of the travel with the minimum in the center. The general features of the probability distribution approach that of a classical oscillator as n increases. The quantum oscillator always has an oscillatory probability distribution with nodes (i.e., points of zero probability), which is not a property of a classical oscillator. However, the quantum description of a macroscopic oscillator is equivalent to the classical description. To see this, consider a 1-g mass connected to a spring that is initially displaced 1 cm, and the oscillator period is $\nu = 1$ Hz. From equation (6.4), the spring constant is $k = 4\pi^2$ g/s^2. At the classical turning point, $x = 1$ cm, the energy of the oscillator is $\frac{1}{2}kx^2 = 2\pi^2 \; g\,cm^2/s^2$. This is the total energy because, at the turning point, all of the energy is potential energy. Setting this energy equal to the energy of a quantum harmonic oscillator, $\left(n + \frac{1}{2}\right) h\nu = 2\pi^2$, yields a quantum number of $n \approx 3 \times 10^{27}$. The macroscopic oscillator has a quantum number that is exceedingly large. There are approximately 10^{27} nodes in the 2-cm travel of the oscillatory. The separation between nodes is on the order of 10^{-27} cm, or the distance between nodes is thirteen orders of magnitude less than the size of a single atomic nucleus. Since the probability of finding a particle is only defined for a small interval, any reasonable, finite-size interval will contain a vast number of nodes. The nodes will be averaged out in the interval, and a plot of the probability versus position will yield a smooth curve with no nodes, identical to the classically calculated distribution.

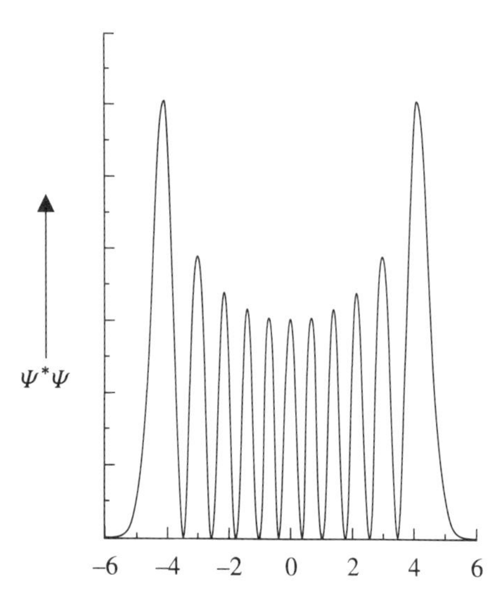

Figure 6.5 The probability function, $\psi^*\psi$, for the harmonic oscillator state, $n = 10$. The horizontal axis is in units of γ.

There are an infinite number of harmonic oscillator wavefunctions. As the quantum number n increases, the wavefunction involves a polynomial with an increasing number of terms. Frequently it is desirable to obtain relationships that apply to all of the states—for example, the normalization constant. As will be discussed in the next section when the Dirac approach to the quantum harmonic oscillator is introduced, these can be rather simple to derive using the Dirac formalism. However, in the Schrödinger representation, it is necessary to deduce relationships that apply to all of the wavefunctions in spite of the fact that each has associated with it a different polynomial. Such relationships can be obtained using a variety of generating function techniques. One useful generating function involving the Hermite polynomials is

$$S(\gamma, s) = \sum_n \frac{H_n(\gamma)s^n}{n!} = e^{\gamma^2-(s-\gamma)^2}. \tag{6.38}$$

As an example of the use of the generating functions, the normalization constant, given in equation (6.34), will be derived, and the wavefunctions will be shown to be orthogonal. The wavefunctions given in equation (6.33) with the normalization constant defined in equation (6.34) obey the relation

$$\int_{\infty}^{\infty} \psi_n^*(x)\, \psi_m(x) dx = \delta_{nm}, \tag{6.39}$$

where the Kronecker delta is defined as

$$\delta_{mn} \equiv \begin{cases} 1, & m = n \\ 0, & m \neq n. \end{cases} \tag{6.40}$$

Since each eigenfunction belongs to a different eigenvalue, the eigenfunctions are orthogonal [see Chapter 2, Section C]; that is, the integral in equation (6.39) equals zero for $m \neq n$. This will be shown explicitly as part of the derivation of the normalization constant.

Using the generating function [equation (6.38)] and a second generating function,

$$T(\gamma, t) = \sum_m \frac{H_m(\gamma)t^m}{m!} = e^{\gamma^2-(t-\gamma)^2}, \tag{6.41}$$

the following integral can be expressed in terms of the generating functions,

$$\int_{-\infty}^{\infty} STe^{-\gamma^2} d\gamma = \sum_n \sum_m s^n t^m \int_{-\infty}^{+\infty} \frac{H_n(\gamma) H_m(\gamma)}{n!m!} e^{-\gamma^2} d\gamma. \tag{6.42}$$

The left-hand side of equation (6.42) can be written in terms of the far right-hand sides of equations (6.38) and (6.41):

$$= \int_{-\infty}^{\infty} e^{-s^2-t^2+2s\gamma+2t\gamma-\gamma^2} d\gamma = e^{2st} \int_{-\infty}^{\infty} e^{-(\gamma-s-t)^2} d(\gamma - s - t). \tag{6.43}$$

Making the substitution, $y = \gamma - s - t$, and integrating gives

$$= e^{2st} \int_{-\infty}^{\infty} e^{-y^2} dy = \pi^{\frac{1}{2}} e^{2st}. \tag{6.44}$$

Expanding the exponential yields

$$= \pi^{\frac{1}{2}} \left(1 + \frac{2st}{1!} + \frac{2^2 s^2 t^2}{2!} + \cdots + \frac{2^n s^n t^n}{n!} + \cdots\right). \tag{6.45}$$

The left-hand side of equation (6.42) has been expressed as a series. The right-hand side of equation (6.42) is a series times an integral. For each choice of m and n, the integral is a constant that is

$$\int_{-\infty}^{\infty} \frac{H_n(\gamma) H_m(\gamma)}{n!m!} e^{-\gamma^2}\, d\gamma \equiv \theta_{mn}. \tag{6.46}$$

Rewriting equation (6.42) in terms of the two series gives

$$\sum_n \sum_m s^n t^m \theta_{mn} = \pi^{1/2} \left(1 + \frac{2st}{1!} + \frac{2^2 s^2 t^2}{2!} + \cdots + \frac{2^n s^n t^n}{n!} + \cdots\right). \tag{6.47}$$

For $n = m$, the series on the left-hand side of equation (6.47) is

$$\sum_n s^n t^n \theta_{nn}. \tag{6.48}$$

The series with $n = m$ [expression (6.48)] is equal to the right-hand side series in equation (6.47), provided that

$$\theta_{nn} = \frac{2^n \pi^{1/2}}{n!}. \tag{6.49}$$

Since the two series are equal with $n = m$, the other terms in the left-hand series in equation (6.47) $(n \neq m)$ must be equal to zero, that is,

$$\theta_{nm} = 0, \qquad m \neq n. \tag{6.50}$$

For $n = m$,

$$\theta_{nn} = \int_{-\infty}^{\infty} \frac{H_n H_n}{(n!)^2} e^{-\gamma^2 d\gamma} = \frac{2^n \pi^{1/2}}{n!}. \tag{6.51}$$

Multiplying by $(n!)^2$ gives the desired integral for the normalization constant:

$$\int H_n H_n e^{-\gamma^2} d\gamma = \pi^{1/2} 2^n n!. \tag{6.52}$$

Furthermore, equation (6.50) demonstrates the orthogonality of the wavefunctions since $\theta_{nm} = 0$ for $n \neq m$. The unnormalized wavefunction is

$$\psi_n(x) = e^{-\gamma^2/2} H_n(\gamma).$$

To normalize the wavefunction, the unnormalized wavefunction must be multiplied by the inverse of the square root of the right-hand side of equation (6.52). Using x as the position variable instead of γ requires changing the differential operator in equation (6.52), $d\gamma = \sqrt{\alpha}\, dx$. The final result is given in equation (6.34).

The generating function approach used to obtain the normalization constant for all states of the harmonic oscillator can also be used to prove other relationships among the harmonic oscillator states. One of these, the selection rule for absorption of light, will be discussed below in the context of the Dirac treatment of the quantum harmonic oscillator.

B. THE QUANTUM HARMONIC OSCILLATOR IN THE DIRAC REPRESENTATION

In this section, the quantum harmonic oscillator problem will be solved again using the Dirac representation. The eigenvalues and eigenvectors will be found. The eigenvalues, which are observables, are independent of the representation used to solve the problem. However, the eigenvectors and the method of solving the problem are fundamentally different. The Dirac representation provides a mathematically appealing method for the calculation of properties associated with the harmonic oscillator. In addition, the Dirac representation of the harmonic oscillator problem serves as a basis for problems in solid-state physics and the quantum theory of radiation.

The one-dimensional harmonic oscillator Hamiltonian is the sum of the kinetic energy and the potential energy:

$$\underline{H} = \frac{\underline{P}^2}{2m} + \frac{1}{2}k\underline{x}^2, \tag{6.53}$$

where $\underline{P}$ is the momentum operator and $\underline{x}$ is the position operator. The eigenvalue equation is

$$\underline{H}\,|E\rangle = E\,|E\rangle\,. \tag{6.54}$$

The eigenkets, $|E\rangle$, are taken to be orthonormal. They can be normalized, and it was proven in Chapter 2, Section C that they will be orthogonal because each eigenvector belongs to a distinct eigenvalue.

The position-momentum commutator is

$$[\underline{x},\ \underline{P}] = i\hbar\,\underline{1}.$$

The $\underline{1}$ is the identity operator. Introducing units of mass, length, and time such that

$$m = k = \hbar = 1$$

and using these units to write the Hamiltonian and the commutator gives

$$\underline{H} = \frac{1}{2}\left(\underline{P}^2 + \underline{x}^2\right) \tag{6.55}$$

$$[\underline{x},\ \underline{P}] = i\,\underline{1}. \tag{6.56}$$

The operators $\underline{a}$ and $\underline{a}^+$ are defined as

$$\underline{a} = \frac{i}{\sqrt{2}}\left(\underline{P} - i\underline{x}\right) \tag{6.57}$$

$$\underline{a}^+ = \frac{1}{i\sqrt{2}}\left(\underline{P} + i\underline{x}\right). \tag{6.58}$$

$\underline{a}^+$ is the Hermitian conjugate of $\underline{a}$ because $\underline{P}$ and $\underline{x}$ are Hermitian, that is,

$$\underline{a}^+ = \overline{\underline{a}}. \tag{6.59}$$

The Hamiltonian operator can be written in terms of $\underline{a}^+$ and $\underline{a}$:

$$\underline{a}\,\underline{a}^+ = \frac{1}{2}\left[(\underline{P} - i\underline{x})(\underline{P} + i\underline{x})\right]$$

$$= \frac{1}{2}\left[\underline{P}^2 - i\underline{x}\,\underline{P} + i\underline{P}\,\underline{x} + \underline{x}^2\right]$$

$$= \frac{1}{2}\left[\underline{P}^2 - i(\underline{x}\,\underline{P} - \underline{P}\,\underline{x}) + \underline{x}^2\right]$$

$$= \frac{1}{2}\left[\underline{P}^2 + \underline{x}^2\right] - \frac{i}{2}\left[\underline{x}, \underline{P}\right]$$

$$\underline{a}\,\underline{a}^+ = \underline{H} + \frac{1}{2}\,\underline{I}. \tag{6.60}$$

Similarly,

$$\underline{a}^+\underline{a} = \frac{1}{2}\left[\underline{P}^2 + i(\underline{x}\,\underline{P} - \underline{P}\,\underline{x}) + \underline{x}^2\right]$$

$$\underline{a}^+\underline{a} = \underline{H} - \frac{1}{2}\,\underline{I}. \tag{6.61}$$

Therefore, in terms of $\underline{a}$ and $\underline{a}^+$, the Hamiltonian is

$$\underline{H} = \frac{1}{2}(\underline{a}\,\underline{a}^+ + \underline{a}^+\underline{a}). \tag{6.62}$$

In an analogous manner, the following commutator relationships can be established:

$$\left[\underline{a}, \underline{a}^+\right] = \underline{I} \tag{6.63a}$$

$$\left[\underline{a}, \underline{H}\right] = \underline{a} \tag{6.63b}$$

$$\left[\underline{a}^+, \underline{H}\right] = -\underline{a}^+. \tag{6.63c}$$

The Hamiltonian can be written in terms of the operators $\underline{a}$ and $\underline{a}^+$. However, the characteristics of $\underline{a}$ and $\underline{a}^+$ have not been defined. The nature of $\underline{a}$ and $\underline{a}^+$ can be determined and, in the process, the eigenvalue problem is solved. Consider

$$\underline{a}\,|E\rangle = |Q\rangle\,, \tag{6.64}$$

where $|E\rangle$ is an eigenket of $\underline{H}$ and $|Q\rangle$ is another ket, as yet undefined. The complex conjugate of equation (6.64) is

$$\langle Q| = \langle E|\,\overline{\underline{a}} = \langle E|\,\underline{a}^+,$$

and

$$\langle Q \,|Q\rangle \geq 0$$

since this is the scalar product of a vector with itself. $\langle Q \,|Q\rangle = 0$ only if $|Q\rangle = 0$. Therefore,

$$\langle Q \,|Q\rangle = \langle E|\, a^{+} a \,|E\rangle \geq 0. \tag{6.65}$$

Using equation (6.61),

$$\langle E|\, \underline{a}^{+} \underline{a} \,|E\rangle = \langle E|\, \underline{H} - \tfrac{1}{2} 1 \,|E\rangle = \left(E - \tfrac{1}{2}\right) \langle E \,|E\rangle \geq 0. \tag{6.66}$$

Since the eigenkets, $|E\rangle$, are normalized, the right hand side of equation (6.66) yields

$$E \geq \tfrac{1}{2}. \tag{6.67}$$

This is the first important result. It gives information on the possible eigenvalues of the harmonic oscillator. If conventional units had been used, equation (6.67) would read $E \geq 1/2h\nu$, a result that is obtained from the Schrödinger treatment.

Now consider

$$\underline{a}\, \underline{H} \,|E\rangle = E\, \underline{a} \,|E\rangle\,. \tag{6.68}$$

$|E\rangle$ is an eigenket of $\underline{H}$ with eigenvalue E; and E, being a number, commutes with the operator $\underline{a}$, giving the right-hand side of equation (6.68). From the commutator relation, equation (6.63b),

$$\begin{aligned} [\underline{a},\ \underline{H}] &= \underline{a}\,\underline{H} - \underline{H}\,\underline{a} = \underline{a} \\ \underline{a}\,\underline{H} &= \underline{H}\,\underline{a} + \underline{a}. \end{aligned} \tag{6.69}$$

Combining equations (6.68) and (6.69) gives

$$(\underline{H}\,\underline{a} + \underline{a}) \,|E\rangle = E\, \underline{a} \,|E\rangle\,.$$

Then

$$\underline{H}\,\underline{a} \,|E\rangle + \underline{a} \,|E\rangle = E\, \underline{a} \,|E\rangle$$

and

$$\underline{H}\,\underline{a} \,|E\rangle = E\, \underline{a} \,|E\rangle - \underline{a} \,|E\rangle\,.$$

Therefore,

$$\underline{H}[\underline{a} \,|E\rangle] = (E - 1)[\underline{a} \,|E\rangle]. \tag{6.70}$$

Equation (6.70) states that $[\underline{a} \,|E\rangle]$ is an eigenket of $\underline{H}$ with eigenvalue $(E - 1)$ because operating $\underline{H}$ on $[\underline{a} \,|E\rangle]$ gives the same ket back multiplied by the number, $(E - 1)$. $(E - 1)$ is the eigenvalue of the ket, $[\underline{a} \,|E\rangle]$. This defines the operator $\underline{a}$:

$$\underline{a} \,|E\rangle = |E - 1\rangle\,. \tag{6.71}$$

$\underline{a} \,|E\rangle$ is an eigenket of $\underline{H}$ with eigenvalue $(E - 1)$; $\underline{a} \,|E\rangle$ must be the ket $|E - 1\rangle$, since

$$\underline{H}\,|E-1\rangle = (E-1)\,|E-1\rangle\,.$$

$\underline{a}$ is called a lowering operator. When $\underline{a}$ operates on $|E\rangle$ it gives a new eigenket, $|E-1\rangle$, one unit lower in energy. As will be discussed below, when $\underline{a}$ operates on $|E\rangle$ it actually yields a constant times $|E-1\rangle$. However, the equality in equation (6.71) still holds because the state of a system is defined by the direction of the ket vector, not its length. The state of the system is represented by the ket regardless of any multiplicative factor other than zero. A ket multiplied by zero does not represent a state of the system because a vector with zero length is not a vector. So the equality in equation (6.71) should be read as the ket $\underline{a}\,|E\rangle$ is the same ket as the ket $|E-1\rangle$, not that $\underline{a}$ operating on $|E\rangle$ gives one times $|E-1\rangle$.

Repeated applications of the lowering operator gives

$$\begin{aligned} \underline{a}\,|E\rangle &= |E-1\rangle \\ \underline{a}^2\,|E\rangle &= |E-2\rangle \\ \underline{a}^3\,|E\rangle &= |E-3\rangle \\ &\vdots \end{aligned} \tag{6.72}$$

However, the lowering cannot continue indefinitely because equation (6.67) states

$$E \geq 1/2.$$

At some point there will be a value of E, call it E_0, such that if 1 is subtracted from it, the inequality will be violated, that is,

$$E_0 - 1 < 1/2.$$

Therefore, for the eigenket, $|E_0\rangle$, corresponding to the energy eigenvalue E_0,

$$\underline{a}\,|E_0\rangle = 0. \tag{6.73}$$

Then, for the eigenvector $|E_0\rangle$, using equation (6.61),

$$\underline{a}^+\left[\underline{a}\,|E_0\rangle\right] = \left(\underline{H} - 1/2\,\underline{1}\right)\,|E_0\rangle \tag{6.74}$$

$$= \left(E_0 - 1/2\right)\,|E_0\rangle = 0. \tag{6.75}$$

Since $|E_0\rangle$ is not zero, its coefficient is zero, that is,

$$\left(E_0 - 1/2\right) = 0,$$

and therefore

$$E_0 = 1/2. \tag{6.76}$$

Equation (6.76) is the second important result. There is a lowest-energy eigenvalue, $E_0 = 1/2$. In terms of conventional units, the result is

$$E_0 = {}^1\!/_2 h\nu. \tag{6.77}$$

The effect of the operator $\underline{a}^+$ can be determined following a procedure analogous to that given in equation (6.68) through equation (6.71). The commutator relation, equation (6.63a), is used:

$$\underline{a}^+ \left[\underline{H}\,|E\rangle\right] = E\underline{a}^+|E\rangle \tag{6.78}$$

$$\underline{a}^+ \left[\underline{H}\,|E\rangle\right] = (\underline{H}\,\underline{a}^+ - \underline{a}^+)\,|E\rangle\,. \tag{6.79}$$

Combining equations (6.78) and (6.79) gives

$$\underline{H}\,\left[\underline{a}^+|E\rangle\right] = (E+1)\,\left[\underline{a}^+|E\rangle\right]. \tag{6.80}$$

Therefore, $\left[\underline{a}^+\,|E\rangle\right]$ is an eigenket of $\underline{H}$ having eigenvalue $(E+1)$:

$$\underline{a}^+|E\rangle = |E+1\rangle\,; \tag{6.81}$$

$\underline{a}^+$ is a raising operator. It takes the eigenket $|E\rangle$ into a new eigenket, $|E+1\rangle$, having one unit higher energy.

The energy eigenvalues of the harmonic oscillator can be found by applying $\underline{a}^+$ repeatedly to the lowest eigenket, $|E_0\rangle$, which has the eigenvalue, $E_0 = {}^1\!/_2$:

$$\begin{aligned} \underline{H}[\underline{a}^+\,|E_0\rangle] &= {}^3\!/_2\,|E_0+1\rangle \\ \underline{H}[\underline{a}^{+2}\,|E_0\rangle] &= {}^5\!/_2\,|E_0+2\rangle \\ \underline{H}[\underline{a}^{+3}\,|E_0\rangle] &= {}^7\!/_2\,|E_0+3\rangle \\ &\vdots \end{aligned} \tag{6.82}$$

Therefore, the energy eigenvalues are

$$E = {}^1\!/_2,\ {}^3\!/_2,\ {}^5\!/_2,\ {}^7\!/_2, \cdots$$

or

$$E_n = \left(n + {}^1\!/_2\right). \tag{6.83}$$

Using conventional units, the eigenvalues are

$$E_n = \left(n + {}^1\!/_2\right) h\nu. \tag{6.84}$$

The eigenvalues in equation (6.84) are identical to the eigenvalues found using the Schrödinger equation [equation (6.31)]. However, in the Dirac representation it is not necessary to solve a differential equation and then impose the Born boundary conditions to obtain acceptable wavefunctions and the quantized energy eigenvalues. All of the necessary information is contained in the Hamiltonian and the commutator relationship given in equations (6.53) and (4.7). In many respects, the Dirac approach is also mathematically simpler, and as will be shown below, the resulting eigenkets frequently provide an easier approach to calculation of the various quantum mechanical properties of the harmonic oscillator.

The eigenkets have been labeled by the energy of the state, $E = (n + {}^1\!/_2)$, where n can take on values, 0, 1, 2, 3, ... It is equivalent to label the eigenkets with the quantum number n rather than $n + {}^1\!/_2$, that is,

$$\left|E = n + {}^1\!/_2\right\rangle = |n\rangle \, . \tag{6.85}$$

The eigenkets, $|n\rangle$, are taken to be normalized. The raising operator increases n by one, and the lowering operator decreases n by one, that is,

$$\underline{a}^{+}\,|n\rangle = \beta_n\,|n+1\rangle$$
$$\underline{a}\,|n\rangle = \alpha_n\,|n-1\rangle \, .$$

β_n and α_n can be evaluated to give

$$\beta_n = \sqrt{n+1}$$
$$\alpha_n = \sqrt{n}.$$

Therefore, the result of operating a raising or lowering operator on a ket $|n\rangle$ is

$$\underline{a}^{+}\,|n\rangle = \sqrt{(n+1)}\,|n+1\rangle \tag{6.86a}$$
$$\underline{a}\,|n\rangle = \sqrt{n}\,|n-1\rangle \, . \tag{6.86b}$$

As will be demonstrated below, these equations are important because they are used in calculating properties associated with the harmonic oscillator.

Consider the operator, $\underline{a}^{+}\underline{a}$, operating on a ket, $|n\rangle$:

$$\underline{a}^{+}\underline{a}\,|n\rangle = \underline{a}^{+}\sqrt{n}\,|n-1\rangle$$
$$= n\,|n\rangle \, .$$

Therefore,

$$\underline{a}^{+}\underline{a}\,|n\rangle = n\,|n\rangle \, . \tag{6.87}$$

The kets $|n\rangle$ are eigenkets of the operator $\underline{a}^{+}\underline{a}$ with eigenvalues n, the quantum number or number of excitations of the harmonic oscillator. $\underline{a}^{+}\underline{a}$ is called the number operator. It is important in the quantum theory of radiation and the quantum theory of solids. These problems map onto the quantum harmonic oscillator problem. In the quantum theory of radiation, the number operator gives the number of photons in the radiation field. In the quantum theory of solids, the number operator gives the number of phonons, which are the quantized excitations of a lattice. In these applications, the raising and lowering operators are frequently called creation and annihilation operators, respectively. In the theory of radiation, the creation operator "creates" a photon, increasing the number of photons in the field by one. The annihilation operator "annihilates" a photon, decreasing the number of photons in the field by one. Absorption of light by a molecule takes the molecule from a lower-energy state to a higher-energy state. To conserve energy, in a fully quantum mechanical theory, an annihilation operator reduces the number of photons in the field by one. Fluorescent emission of a photon by a molecule involves a creation operator, increasing the field by one photon. In the semiclassical theory of absorption and emission of light by molecules, the molecules are treated quantum mechanically, but the radiation field is treated classically. There is no explicit consideration of changes in the amplitude of the radiation field. The semiclassical theory, using time-dependent perturbation theory, will be presented in Chapter 12. The semiclassical theory of radiation cannot account for spontaneous emission. The creation and annihilation operators will be qualitatively discussed in connection with spontaneous emission in Chapter 12, Section E.

In the derivation given above, special units were used to simplify the expressions. Using conventional units, the Hamiltonian operator is

$$\underline{H} = {}^{1}\!/_{2}\hbar\,\omega(\underline{a}\,\underline{a}^{+} + \underline{a}^{+}\underline{a}), \tag{6.88}$$

with

$$\omega = 2\pi\nu = (k/m)^{1/2}. \tag{6.89}$$

To see explicitly that the Dirac approach gives the same results as the Schrödinger treatment of the quantum harmonic oscillator, the eigenvalues of the Hamiltonian operator given in equation (6.88) can be found:

$$\begin{aligned}
\underline{H}\,|n\rangle &= {}^{1}\!/_{2}\hbar\omega\left(\underline{a}\,\underline{a}^{+}\,|n\rangle + \underline{a}^{+}\underline{a}\,|n\rangle\right)\\
&= {}^{1}\!/_{2}\hbar\omega\left(\underline{a}(n+1)^{1/2}\,|n+1\rangle + \underline{a}^{+}n^{1/2}\,|n-1\rangle\right)\\
&= {}^{1}\!/_{2}\hbar\omega\left((n+1)^{1/2}(n+1)^{1/2}\,|n\rangle + n^{1/2}\,n^{1/2}\,|n\rangle\right)\\
&= {}^{1}\!/_{2}\hbar\omega(2n+1)\,|n\rangle\\
&= \hbar\omega\left(n + {}^{1}\!/_{2}\right)|n\rangle\\
&= \left(n + {}^{1}\!/_{2}\right)h\nu\,|n\rangle\,.
\end{aligned}$$

Therefore, $|n\rangle$ are the eigenvectors of $\underline{H}$ with eigenvalues $(n + {}^{1}\!/_{2})h\nu$.

To use the raising and lowering operators to calculate properties of the harmonic oscillator other than the eigenvalues, it is necessary to write them with conventional units:

$$\underline{a} = \frac{i}{(2\hbar\,\omega)^{1/2}}\left(\frac{1}{m^{1/2}}\underline{P} - ik^{1/2}\underline{x}\right) \tag{6.90}$$

$$\underline{a}^{+} = \frac{1}{i(2\hbar\,\omega)^{1/2}}\left(\frac{1}{m^{1/2}}\underline{P} + ik^{1/2}\underline{x}\right). \tag{6.91}$$

The position and momentum operators can be written in terms of the raising and lower operator. Subtracting $\underline{a}^{+}$ from a and rearranging gives the position operator:

$$(\underline{a}+\underline{a}^{+}) = \frac{1}{(2\,\hbar\omega)^{1/2}}(2k^{1/2}\underline{x})$$

$$= \left(\frac{2k}{\hbar\omega}\right)^{1/2} \underline{x}$$

$$\underline{x} = \left(\frac{\hbar\omega}{2k}\right)^{1/2} (\underline{a}+\underline{a}^{+}). \tag{6.92}$$

Similarly,

$$\underline{P} = -i\left(\frac{\hbar\, m\,\omega}{2}\right)^{1/2} (\underline{a}-\underline{a}^{+}). \tag{6.93}$$

Using these operators, properties of the quantum harmonic oscillator can be calculated. Notice the differences in the mathematical expressions for these operators compared to the expressions for the momentum and position operators in the Schrödinger representation; that is, $\underline{x} = x$ and $\underline{P} = -i\,\hbar(d/dx)$. For many calculations involving the quantum harmonic oscillator, the operators in the Dirac representation are easier to work with.

As a first example of applications of the Dirac treatment of the harmonic oscillator, consider the absorption and emission of light. In Chapter 12, the problem of the interaction of a radiation field with molecules will be treated using time-dependent perturbation theory. The results for weak radiation fields show that the probability per unit time of absorption or emission, $P(t)$, depends on the absolute value squared of a bracket:

$$P(t) \propto \left|\langle F|\,\underline{x}\,|I\rangle\right|^{2}. \tag{6.94}$$

$|I\rangle$ is the initial state of the system, and $|F\rangle$ is the final state. The absolute value squared of a bracket is the bracket times its complex conjugate. The operator, $\underline{x}$, represents the radiation field with x polarization; that is, the electric field of the light is oscillating along the x axis. In Chapter 12, the full derivation of equation (6.94) is presented and the other factors that come into play, such as the intensity of the light, are included. For the harmonic oscillator,

$$P(t) \propto \left|\langle n|\,\underline{x}\,|m\rangle\right|^{2}, \tag{6.95}$$

where $|m\rangle$ and $|n\rangle$ are eigenkets of the harmonic oscillator Hamiltonian. Using equation (6.95), an important selection rule for the harmonic oscillator can be obtained. Substituting equation (6.92) for $\underline{x}$ gives

$$P(t) \propto \frac{\hbar\omega}{2k}\left|\langle n|\,\underline{a}+\underline{a}^{+}\,|m\rangle\right|^{2}.$$

Expressing the absolute value squared of the bracket as the bracket times its complex conjugate, along with using equation (6.59) gives

$$\begin{aligned}
\left|\langle n|\,\underline{a}+\underline{a}^{+}\,|m\rangle\right|^{2} &= \langle n|\,\underline{a}+\underline{a}^{+}\,|m\rangle\,\langle m|\,\underline{a}^{+}+\underline{a}\,|n\rangle \\
&= (\langle n|\,\underline{a}\,|m\rangle + \langle n|\,\underline{a}^{+}\,|m\rangle)(\langle m|\,\underline{a}^{+}\,|n\rangle + \langle m|\,\underline{a}\,|n\rangle) \\
&= \langle n|\,\underline{a}\,|m\rangle\,\langle m|\,\underline{a}^{+}\,|n\rangle + \langle n|\,\underline{a}^{+}\,|m\rangle\,\langle m|\,\underline{a}^{+}\,|n\rangle \\
&\quad + \langle n|\,\underline{a}\,|m\rangle\,\langle m|\,\underline{a}^{+}\,|n\rangle + \langle n|\,\underline{a}^{+}\,|m\rangle\,\langle m|\,\underline{a}\,|n\rangle\,.
\end{aligned}$$

The first and second terms in the last equation are zero. Consider the right-hand bracket of the second term. $\underline{a}^+$ operating on $|n\rangle$ gives $|n+1\rangle$. For this bracket to be nonzero, $\langle m|$ must be $\langle n+1|$. Then for the left-hand bracket of this pair, $\underline{a}^+$ operating on $|m\rangle = |n+1\rangle$ gives $|n+2\rangle$. Closing this bracket with $\langle n|$ gives zero by orthogonality of the eigenkets of the harmonic oscillator. In a similar manner, the first term is also zero. The third and last terms are products of brackets times their complex conjugates, and each can be written as the absolute value squared of their first brackets. Therefore,

$$P(t) \propto \frac{\hbar\omega}{2k}\left(\left|\langle n|\,\underline{a}\,|m\rangle\right|^2 + \left|\langle n|\,\underline{a}^+\,|m\rangle\right|^2\right)$$

Application of the lowering operator to $|m\rangle$ gives $|m-1\rangle$, and application of the raising operator to $|m\rangle$ gives $|m+1\rangle$. Then,

$$= \frac{\hbar\omega}{2k}\left(|\sqrt{m}\,\langle n|m-1\rangle\,|^2 + |\sqrt{m+1}\,\langle n|m+1\rangle\,|^2\right)$$

$$= \frac{\hbar\omega}{2k}\left(m|\,\langle n|m-1\rangle\,|^2 + (m+1)|\,\langle n|m+1\rangle\,|^2\right).$$

To obtain a nonzero result, $n = m - 1$ or $n = m + 1$. If neither of these conditions is met, both brackets are zero because of the orthogonality of the harmonic oscillator eigenkets, and no absorption or emission of radiation will occur. The selection rule is

$$n = m \pm 1. \tag{6.96}$$

Under normal spectroscopic conditions (weak fields), light can only change the quantum number of the harmonic oscillator by ± 1. For $n = m + 1$, light is absorbed. If $n = m - 1$, light is emitted. Note that for $m = 0$, light can be absorbed because the factor in front of the absolute value squared of the bracket is $m + 1$, but light cannot be emitted because the factor in front of $m - 1$ term is m. The $n = m \pm 1$ selection rule is an important starting point for the analysis of the vibrational spectra of molecules. Because real molecular vibrations are anharmonic oscillators, there are deviations from the selection rule. Nonetheless, the strongest vibrational transitions are those in which the vibrational quantum number changes by one. For a high-frequency vibration ($\hbar\omega \gg kT$), initially the vibrational quantum number will be 0. Therefore, the strong peaks observed in a molecular infrared vibrational spectrum are the $m = 0$ to $m = 1$ transitions.

At the outset of the discussion of the quantum harmonic oscillator, it was pointed out that the lowest energy state could not be $E = 0$ because this would violate the position–momentum uncertainty relation, $\Delta x \Delta p \geq \hbar/2$. Solving the eigenvalue problem showed that the lowest energy eigenvalue is not zero but $E = {}^1\!/_2 h\nu$. It is informative to calculate the actual values of the uncertainty for the energy eigenstates of the quantum harmonic oscillator. To calculate $\Delta x \Delta p$, it is necessary to obtain expressions for $(\Delta x)^2$ and $(\Delta p)^2$, take the product, and then take the square root:

$$(\Delta x)^2 = \langle \underline{x}^2 \rangle - \langle \underline{x} \rangle^2 \tag{6.97a}$$

$$(\Delta p)^2 = \langle \underline{P}^2 \rangle - \langle \underline{P} \rangle^2. \tag{6.97b}$$

Using equation (6.92), the expectation value of $\underline{x}$ is given by

$$\begin{aligned}\langle n|\,\underline{x}\,|n\rangle &\propto \langle n|\,\underline{a}+\underline{a}^{+}\,|n\rangle\\ &= \langle n|\,\underline{a}\,|n\rangle + \langle n|\,\underline{a}^{+}\,|n\rangle\\ &= \sqrt{n}\Big\langle n\,|n-1\rangle + \sqrt{n+1}\,\langle n|\,n+1\Big\rangle\\ &= 0\end{aligned}$$

because eigenkets belonging to different eigenvalues are orthogonal. The average value of x is zero. This can be seen immediately by inspection because the $\underline{x}$ operator contains one raising and one lowering operator. Therefore, operating on $|n\rangle$ cannot give $|n\rangle$ back. In the same manner

$$\langle n|\,\underline{P}\,|n\rangle = 0$$

because the $\underline{P}$ operator also contains one raising and one lowering operator. The average value of the momentum is zero.

Returning to equation (6.97a), it is necessary to calculate the average value of the square of the position:

$$\begin{aligned}\langle n|\,\underline{x}^2\,|n\rangle &= \left(\frac{\hbar\omega}{2k}\right)\langle n|\,(\underline{a}+\underline{a}^{+})^2\,|n\rangle\\ &= \left(\frac{\hbar\omega}{2k}\right)\langle n|\,(\underline{a}^2+\underline{a}\,\underline{a}^{+}+\underline{a}^{+}\underline{a}+\underline{a}^{+2})\,|n\rangle\,.\end{aligned}$$

The terms involving the squares of the lowering and raising operators will give zero because operating them on $|n\rangle$ will not return $|n\rangle$. Thus,

$$\begin{aligned}&= \left(\frac{\hbar\omega}{2k}\right)\langle n|\,(\underline{a}\,\underline{a}^{+}+\ \underline{a}^{+}\underline{a})\,|n\rangle\\ &= \left(\frac{\hbar\omega}{2k}\right)\left[\langle n|\,\underline{a}\,\underline{a}^{+}\,|n\rangle + \langle n|\,\underline{a}^{+}\underline{a}\,|n\rangle\right]\\ &= \left(\frac{\hbar\omega}{2k}\right)\left[\sqrt{n+1}\,\langle n|\,\underline{a}\,|n+1\rangle + \sqrt{n}\,\langle n|\,\underline{a}^{+}\,|n-1\rangle\right]\\ &= \left(\frac{\hbar\omega}{2k}\right)\left[(n+1)\,\langle n|\,n\rangle + n\,\langle n\,|n\rangle\right].\end{aligned}$$

Using the fact that the kets are normalized, the result is

$$\langle \underline{x}^2\rangle = \frac{\hbar\omega}{k}(n+{}^1\!/_2). \tag{6.98}$$

It is necessary to calculate the average value of the square of the momentum [equation (6.97b)]

$$\langle n|\,\underline{P}^2\,|n\rangle = -\left(\frac{\hbar m\omega}{2}\right)\langle n|\,\underline{a}^2 - \underline{a}\,\underline{a}^+ - \underline{a}^+\underline{a} + \underline{a}^{+2}\,|n\rangle$$

$$= \left(\frac{\hbar m\omega}{2}\right)\langle n|\,\underline{a}\,\underline{a}^+ + \underline{a}^+\underline{a}\,|n\rangle\,,$$

and, therefore,

$$\langle \underline{P}^2\rangle = \hbar\, m\, \omega\, \left(n + 1/2\right). \tag{6.99}$$

Using equations (6.98) and (6.99) gives

$$(\Delta x)^2\,(\Delta p)^2 = \langle \underline{x}^2\rangle\langle \underline{P}^2\rangle = \frac{\hbar^2\omega^2 m}{k}\left(n + 1/2\right)^2,$$

and since $k = m\omega^2$,

$$(\Delta x)^2\,(\Delta p)^2 = \hbar^2\left(n + 1/2\right)^2. \tag{6.100}$$

Taking the square root of both sides gives the uncertainty relation for the states of the quantum harmonic oscillator:

$$\Delta x\;\Delta p = \hbar\;\left(n + 1/2\right). \tag{6.101}$$

This relationship can also be derived in the Schrödinger representation using a generating function approach like that used to obtain the normalization constant for the Schrödinger representation eigenfunctions. However, the Dirac raising and lowering operator representation greatly reduces the mathematical complexity.

The ground state of the harmonic oscillator, $n = 0$, gives

$$\Delta x\;\Delta p = \frac{\hbar}{2}. \tag{6.102}$$

Thus the ground state has the minimum allowable uncertainty. In a classical system, $\Delta x\,\Delta p$ can equal zero. The particle can be located at $x = 0$, with the particle stationary; that is, $p = 0$. The lowest energy state of the quantum oscillator is the state that is as close to this as possible without violating the uncertainty principle. It is the minimum uncertainty state. Because the system must obey the Uncertainty Principle, the lowest energy state must have a nonzero energy eigenvalue. The energy of the ground state is the sum of the average value of the kinetic energy plus the average value of the potential energy:

$$E = \left\langle\frac{\underline{P}^2}{2m}\right\rangle + \left\langle\frac{k\underline{x}^2}{2}\right\rangle$$

Using equations (6.98) and (6.99) for $n = 0$ gives

$$E = \frac{k}{2}\left(\frac{\hbar\omega}{2k}\right) + \frac{1}{2m}\left(\frac{\hbar m\omega}{2}\right)$$

$$= \frac{\hbar\omega}{4} + \frac{\hbar\omega}{4} = \frac{1}{2}\hbar\omega,$$

the energy of the ground state of the harmonic oscillator. Given the uncertainty relation, the ground state assumes the minimum uncertainty configuration, and this configuration sets the energy of the $n = 0$ state.

C. TIME-DEPENDENT HARMONIC OSCILLATOR WAVE PACKET

In Chapter 3, Section D, the time dependence of a free particle wave packet was discussed. By superimposing free particle momentum eigenstates, a more or less localized wave packet was formed. The packet's motion arose from changing regions of constructive and destructive interference caused by the time evolution of the time-dependent phase factors associated with each eigenfunction in the superposition. Superpositions of bound states can also produce wave packets. Such wave packets will evolve in time in a manner that is influenced by the form of the potential surface.

The harmonic oscillator is the simplest model of a molecular vibration. A superposition of harmonic oscillator states can be used to illustrate an important phenomenon, namely, the time evolution of a vibrational wave packet on a molecular potential surface. If an ultrashort pulse of light is used to excite a molecule from one electronic state to another, the broad bandwidth of the light pulse can span many vibrational levels of the molecular excited-state potential surface. Because the light is exciting a transition between potential surfaces with different minima and shapes, the normal harmonic oscillator selection rule for vibrational transitions, $n = m \pm 1$ [equation (6.96)], does not apply. Therefore, the state that is prepared by the absorption of light can be a superposition of many vibrational eigenstates of the excited electronic state.

To illustrate the nature of a vibrational wave packet prepared by short pulse optical excitation of an electronic excited state, for simplicity, the potential is modeled as harmonic. For a harmonic oscillator, the eigenkets in the Dirac representation are the kets, $|n\rangle$. Because the harmonic oscillator Hamiltonian is time-independent, the time-dependent kets are

$$|n(t)\rangle = |n\rangle\, e^{-iE_n t/\hbar}, \tag{6.103}$$

where the time dependence is contained in the phase factor. The E_n are the harmonic oscillator energy eigenvalues given by equation (6.84). A time-dependent wave packet, $|t\rangle$, is formed from a superposition of the $|n(t)\rangle$:

$$|t\rangle = \sum_n \alpha_n\, |n\rangle\, e^{-i\omega_n t}, \tag{6.104}$$

where $\omega_n = E_n/\hbar$. The α_n are time-independent coefficients. $\alpha_n^*\alpha_n$ is the probability of finding the system in the state $|n\rangle$, given that it is in the superposition state $|t\rangle$. Generally, n is restricted to a limited consecutive set of kets.

The time-dependent position of the superposition can be examined by evaluating the expectation value of the position operator, $\langle t|\,\underline{x}\,|t\rangle$. The position operator [equation (6.92)] is

$$\underline{x} = \left(\frac{\hbar\omega}{2k}\right)^{1/2} \left(\underline{a} + \underline{a}^{+}\right),$$

where ω is the oscillator frequency, k is the Hooke's law spring constant, and $\underline{a}$ and $\underline{a}^+$ are the harmonic oscillator lowering and raising operators, respectively. Then,

$$\langle t | \underline{x} | t \rangle = \sum_m \alpha_m^* e^{i\omega_m t} \sum_n \alpha_n e^{-i\omega_n t} \langle m | \underline{x} | n \rangle \tag{6.105}$$

$$= \sum_{m,n} \alpha_m^* \alpha_n e^{-i(\omega_n - \omega_m)t} \sqrt{\frac{\hbar\omega}{2k}} \langle m | \underline{a} + \underline{a}^+ | n \rangle . \tag{6.106}$$

The bracket, $\langle m | \underline{a} + \underline{a}^+ | n \rangle$, is nonzero only if $m = n \pm 1$. Therefore,

$$\langle t | \underline{x} | t \rangle = \sqrt{\frac{\hbar\omega}{2k}} \left[\sum_n \left\{ \left(\alpha_{n-1}^* \alpha_n e^{-i(\omega_n - \omega_{n-1})t} \sqrt{n} \right) + \left(\alpha_{n+1}^* \alpha_n e^{-i(\omega_n - \omega_{n+1})t} \sqrt{n+1} \right) \right\} \right] . \tag{6.107}$$

Since $\omega_n - \omega_{n-1} = \omega$ and $\omega_n - \omega_{n+1} = -\omega$,

$$\langle t | \underline{x} | t \rangle = \sqrt{\frac{\hbar\omega}{2k}} \left[\sum_n \left\{ \left(\alpha_{n-1}^* \alpha_n e^{-i\omega t} \sqrt{n} \right) + \left(\alpha_{n+1}^* \alpha_n e^{i\omega t} \sqrt{n+1} \right) \right\} \right] . \tag{6.108}$$

Equation (6.108) shows that the expectation value of the position is time-dependent, and the time dependence depends in part on the coefficients in the superposition state, $|t\rangle$.

To simplify the expression on the right-hand side of equation (6.108) so that it is possible to see more clearly the nature of the time dependence, take the superposition to be over states with n large, that is, $n > 1$. Furthermore, take $\alpha_i \equiv \alpha$, the amplitudes of the states in the superposition are equal, otherwise $\alpha_i = 0$. For these conditions equation (6.108) is reduced to

$$\langle t | \underline{x} | t \rangle = \sqrt{\frac{\hbar\omega}{2k}} \alpha^2 \sum_n \sqrt{n} \left(e^{-i\omega t} + e^{i\omega t} \right) , \tag{6.109}$$

and therefore

$$\langle t | \underline{x} | t \rangle = 2\alpha^2 \sqrt{\frac{\hbar\omega}{2k}} \sum_n \sqrt{n} \cos(\omega t) , \tag{6.110}$$

where the sum is over the limited range of states in the superposition. $\langle t | \underline{x} | t \rangle$ is the average value of the position of the harmonic oscillator wave packet. It can be considered, in some sense, the location of the particle described by the superposition, $|t\rangle$. Equation (6.110) shows that the position oscillates sinusoidally at the oscillator frequency, ω, just as a classical harmonic oscillator does. A single harmonic oscillator eigenstate has a time-independent probability distribution and a time-independent expectation value of the position. However, a harmonic oscillator wave packet is time-dependent. The amplitude of the oscillation depends on the states and the amplitude of the states that make up the wave packet as well as the constants ω and k.

As a physical example, consider exciting the I_2 molecule from its ground electronic state to the excited B state at ~565 nm. If a very short pulse of light is used, the broad bandwidth of the pulse will excite a number of vibrational levels of the B state potential surface. These excited vibrational states will form a wave packet, and the bond length of the I_2 molecule will oscillate in a manner described by equation (6.108). It is possible to calculate the extent of the time-dependent change in the bond length—that is, the difference in the length when the bond is extended and contracted. A 20-fs pulse has a bandwidth of $\sim 700\ \text{cm}^{-1}$. The vibrational energy level spacing of states that will be excited within this bandwidth centered at 565 nm is $\sim 69\ \text{cm}^{-1}$. For simplicity, consider the pulse spectrum to be rectangular. Then the states $n = 15$ to $n = 24$ will be excited. Even for a rectangular spectrum that only excites these states, the probability of exciting a vibrational level of the B state depends on the vibrational quantum number. Therefore, the coefficients in equation (6.108) will not be equal. However, the error introduced by assuming the coefficients are identical is small, making it possible to use equation (6.110) to calculate the distance of the back-and-forth travel of the oscillator. The distance traveled is twice the coefficient of the sin term in equation (6.110), that is,

$$4\alpha^2\sqrt{\frac{\hbar\omega}{2k}}\sum_n \sqrt{n},$$

since sin oscillates between $+1$ and -1. For the ten equal amplitude states in the superposition, α^2 is 0.1. From equation (6.4), $k = \mu\omega^2$, where μ is the reduced mass of I_2; $\mu = 1.05 \times 10^{-22}$ g. $\omega = 1.3 \times 10^{13}$ Hz. Using these values, the oscillator travel is 1.06 Å. Therefore, the I_2 wave packet motion is a distance comparable to a molecular bond length. This type of coherent vibrational wave packet motion is responsible for many effects observed in ultrafast optical experiments on molecular systems.

CHAPTER 7

The Hydrogen Atom

The quantum mechanical theory of the hydrogen atom plays a central role in chemistry. It is the simplest of all atoms, and it is the only atom for which the energy eigenvalue problem can be solved exactly. The hydrogen atom can be solved exactly because it is a two-body problem. It is composed of the nucleus, a positively charged proton, and a negatively charged electron. These two particles interact through a Coulombic potential term in the Hamiltonian. The solution of the hydrogen atom problem in the Schrödinger representation yields the eigenfunctions or wavefunctions. These are frequently referred to as orbitals.

The hydrogen atom wavefunctions are the starting point for the description of all atoms and molecules. As will be shown below, the three-dimensional hydrogen atom Schrödinger equation in spherical polar coordinates (r, θ, φ) is separable into three one-dimensional equations, an equation in φ, an equation in θ, and an equation in r. The combination of the solutions to the φ and θ equations, called the spherical harmonics, is the solution of the orbital angular momentum problem. The spherical harmonics determine the shapes of the wavefunctions. These shapes are the familiar s, p, d, and f orbitals that are ubiquitous in the description of atoms and molecules. The solution to the r equation determines the size of the wavefunctions. When the Schrödinger equation is separated into the three one-dimensional equations, because the problem is centrally symmetric, only the equation in r contains the Coulomb potential term. It is the form of the Coulomb potential that makes one atom different from another. The angular parts of the wavefunctions are, to a first approximation, independent of the form of the potential. Therefore, the shapes of the orbitals are approximately the same for all atoms. The fact that the hydrogen atom energy eigenvalue problem can be solved exactly means that the approximate shapes of the orbitals of all atoms are known. The hydrogen atom wavefunctions are important because molecules can be described in terms of a superposition of atomic orbitals with shapes that are known approximately. In addition, the spherical harmonics, the general solution to the orbital angular momentum problem, can be used in the description of the angular momentum states of any atom and in related problems such as molecular rotations.

A. Separation of the Schrödinger Equation

The first step in the solution of the hydrogen atom problem is to write the energy eigenvalue problem as the time-independent Schrödinger equation and then to separate this multidimensional equation into the three one-dimensional equations involving the internal coordinates of the atom. The solution to the hydrogen atom Schrödinger equation actually applies to any atom with only one electron—for example, the helium +1 ion. Therefore, the problem will be set up for a nucleus of any charge and a single electron.

Consider two point particles, a nucleus of charge $+Ze$ with mass m_1 and an electron of charge $-e$ with mass m_2. The nucleus has coordinates x_1, y_1, and z_1, and the electron has coordinates x_2, y_2, and z_2. The potential is the Coulomb attraction between the positively charged nucleus and the negatively charged electron. In Cartesian coordinates it has the form

$$V = -\frac{Ze^2}{4\pi\varepsilon_o r} = -\frac{Ze^2}{4\pi\varepsilon_o\left[(x_2-x_1)^2+(y_2-y_1)^2+(z_2-z_1)^2\right]^{\frac{1}{2}}}, \tag{7.1}$$

where r is the distance and ε_0 is the permittivity of vacuum. The Schrödinger equation, written in Cartesian coordinates, is

$$\frac{1}{m_1}\left(\frac{\partial^2\Psi_T}{\partial x_1^2}+\frac{\partial^2\Psi_T}{\partial y_1^2}+\frac{\partial^2\Psi_T}{\partial z_1^2}\right)+\frac{1}{m_e}\left(\frac{\partial^2\Psi_T}{\partial x_2^2}+\frac{\partial^2\Psi_T}{\partial y_2^2}+\frac{\partial^2\Psi_T}{\partial z_2^2}\right)$$
$$+\frac{2}{\hbar^2}(E_T-V)\,\Psi_T=0. \tag{7.2}$$

The first term is the kinetic energy of the nucleus with mass m_1. The second term is the kinetic energy of the electron with mass m_e. Ψ_T is the total eigenfunction, a function of six coordinates. E_T is the total energy of the atom, including its internal energy and the kinetic energy associated with the translation of the entire atom. V is the potential given in equation (7.1).

The translational motion of the entire atom can be separated from the relative motion of the nucleus and the electron. To accomplish the separation, new coordinates are introduced. x, y, and z are the center-of-mass coordinates, and r, θ, and φ, are the polar coordinates of the second particle relative to the first:

$$x = \frac{m_1x_1+m_2x_2}{m_1+m_2} \tag{7.3a}$$

$$y = \frac{m_1y_1+m_2y_2}{m_1+m_2} \tag{7.3b}$$

$$z = \frac{m_1z_1+m_2z_2}{m_1+m_2} \tag{7.3c}$$

$$r\sin\theta\cos\varphi = x_2-x_1 \tag{7.4a}$$

$$r\sin\theta\sin\varphi = y_2-y_1 \tag{7.4b}$$

$$r\cos\theta = z_2-z_1. \tag{7.4c}$$

Substituting these into Schrödinger equation gives

$$\frac{1}{m_1+m_2}\left(\frac{\partial^2\Psi_T}{\partial x^2}+\frac{\partial^2\Psi_T}{\partial y^2}+\frac{\partial^2\Psi_T}{\partial z^2}\right)+$$

$$\frac{1}{\mu}\left\{\frac{1}{r^2}\frac{\partial}{\partial r}\left(r^2\frac{\partial\Psi_T}{\partial r}\right)+\frac{1}{r^2\sin^2\theta}\frac{\partial^2\Psi_T}{\partial\varphi^2}+\frac{1}{r^2\sin\theta}\frac{\partial}{\partial\theta}\left(\sin\theta\frac{\partial\Psi_T}{\partial\theta}\right)\right\}+$$

$$\frac{2}{\hbar^2}\left[E_T-V(r,\theta,\varphi)\right]\Psi_T=0, \tag{7.5}$$

where μ is the reduced mass,

$$\mu=\frac{m_1 m_2}{m_1+m_2}. \tag{7.6}$$

The quantity in first set of parentheses is the Laplacian operator [∇^2, see equation (5.5)] in Cartesian coordinates. This term involves the translational kinetic energy of the total atom. The quantity on the second line in brackets is the Laplacian in polar coordinates r, θ, and φ. To obtain the Laplacian in spherical polar coordinates, it is necessary to convert the differential operators from Cartesian coordinates to polar coordinates. This term involves the relative motion of the two particles.

In equation (7.5) there are two types of terms: those with operators depending on x, y, and z and those with operators depending on r, θ, and φ. The sum of these two types of terms equals zero, independent of the choice of the variables. This implies that the equation is separable. The equation can be separated using the solution

$$\Psi_T(x, y, z, r, \theta, \varphi)=F(x, y, z)\,\Psi(r, \theta, \varphi), \tag{7.7}$$

which is substituted into equation (7.5), and the result is divided by Ψ_T. This procedure yields two independent equations,

$$\frac{\partial^2 F}{\partial x^2}+\frac{\partial^2 F}{\partial y^2}+\frac{\partial^2 F}{\partial z^2}+\frac{2(m_1+m_2)}{\hbar^2}E_{Tr}F=0 \tag{7.8}$$

and

$$\frac{1}{r^2}\frac{\partial}{\partial r}\left(r^2\frac{\partial\Psi}{\partial r}\right)+\frac{1}{r^2\sin^2\theta}\frac{\partial^2\Psi}{\partial\varphi^2}+\frac{1}{r^2\sin\theta}\frac{\partial}{\partial\theta}\left(\sin\theta\frac{\partial\Psi}{\partial\theta}\right)$$
$$+\frac{2\mu}{\hbar^2}\left[E-V(r,\theta,\varphi)\right]\Psi=0, \tag{7.9}$$

with

$$E_T=E_{Tr}+E. \tag{7.10}$$

The first equation is the free-particle Schrödinger equation for a particle with mass m_1+m_2. E_{Tr} is the translational energy of the atom. The free-particle problem was discussed in detail in Chapter 3. Equation (7.9) describes the relative motion of the nucleus and the electron. The energy, E, is the energy associated with the internal degrees of freedom of the atom. The solutions of equation (7.9) are of interest here. The total energy, E_T, is the sum of atoms translation energy, E_{Tr}, and internal energy, E.

Equation (7.9) is written in terms of the center of mass. Therefore, the Schrödinger equation is for a particle with mass μ moving in a potential $V(r, \theta, \varphi)$. In the absence of an external field, the potential is the Coulomb potential, which is independent of the angles and depends only on the distance, r. Therefore,

$$V = V(r). \tag{7.11}$$

Equation (7.9) depends on the three internal coordinates. It is a three-dimensional differential equation. To separate it into three one-dimensional equations, the three-dimensional wavefunction, Ψ, is written as the product of three one-dimensional functions,

$$\Psi(r, \theta, \varphi) = R(r)\, \Theta(\theta)\, \Phi(\varphi). \tag{7.12}$$

If this is a valid representation of the wavefunction, then using it in equation (7.9) should result in a separation of the equation into the desired three one-dimensional equations. Substituting equation (7.12) into equation (7.9) and dividing by $R\,\Theta\,\Phi$ gives

$$\frac{1}{Rr^2}\frac{d}{dr}\left(r^2\frac{dR}{dr}\right) + \frac{1}{\Phi\, r^2 \sin^2\theta}\frac{d^2\Phi}{d\varphi^2} + \frac{1}{\Theta\, r^2 \sin\theta}\frac{d}{d\theta}\left(\sin\theta\frac{d\Theta}{d\theta}\right) + \frac{8\pi^2\mu}{\hbar^2}[E - V(r)] = 0. \tag{7.13}$$

Multiplying the equation by $r^2 \sin^2\theta$ yields

$$\frac{\sin^2\theta}{R}\frac{d}{dr}\left(r^2\frac{dR}{dr}\right) + \frac{1}{\Phi}\frac{d^2\Phi}{d\varphi^2} + \frac{\sin\theta}{\Theta}\frac{d}{d\theta}\left(\sin\theta\frac{d\Theta}{d\theta}\right) + \frac{8\pi^2\mu r^2\sin^2\theta}{\hbar^2}[E - V(r)] = 0. \tag{7.14}$$

The second term depends only on φ. The equation is equal to zero regardless of the choice of φ, even for fixed r and θ. Therefore, the term in φ must be equal to a constant; otherwise changing the value of φ without some compensating change in r and θ would destroy the equality. Taking the constant to be $-m^2$, a differential equation for Φ is obtained:

$$\frac{1}{\Phi}\frac{d^2\Phi}{d\varphi^2} = -m^2 \tag{7.15}$$

In equation (7.14), the term in φ is replaced by $-m^2$. Dividing this equation by $\sin^2\theta$ gives

$$\frac{1}{R}\frac{d}{dr}\left(r^2\frac{dR}{dr}\right) - \frac{m^2}{\sin^2\theta} + \frac{1}{\Theta\sin\theta}\frac{d}{d\theta}\left(\sin\theta\frac{d\Theta}{d\theta}\right) + \frac{2\mu r^2}{\hbar^2}(E - V(r)) = 0. \tag{7.16}$$

The second and third terms depend only on θ, and the other terms depend only on r.

Therefore, the θ terms are equal to a constant. Setting these terms equal to the constant, $-\beta$, multiplying by Θ, and transposing the $-\beta\Theta$ gives

$$\frac{1}{\sin\theta}\frac{d}{d\theta}\left(\sin\theta\frac{d\Theta}{d\theta}\right)-\frac{m^2\Theta}{\sin^2\theta}+\beta\,\Theta=0. \tag{7.17}$$

This is a differential equation for Θ. Then replacing the θ terms by $-\beta$ in equation (7.16) and multiplying by R/r^2 gives

$$\frac{1}{r^2}\frac{d}{dr}\left(r^2\frac{dR}{dr}\right)-\frac{\beta}{r^2}R+\frac{2\mu}{\hbar^2}\{E-V(r)\}R=0, \tag{7.18}$$

a differential equation for R.

B. SOLUTIONS OF THE THREE ONE-DIMENSIONAL EQUATIONS

The full Schrödinger equation for the internal degrees of freedom of the atom has been separated into three one-dimensional equations, one for each function, Φ, Θ, R. First the Φ equation is solved, and it is found that the solutions to the differential equation are acceptable eigenfunctions only for certain values of the constant m. With these values of m, the Θ equation is solved, and the solutions are found to be acceptable eigenfunctions only for certain values of the constant β. These values of β are used in the R equation, and solutions to the R equation are only acceptable eigenfunctions for certain values of the energy, E.

It is important to note that the actual form of the potential only enters the R equation. The solutions to the other equations depend on the fact that the potential, V, is only a function of r. Φ and Θ are general solutions for any spherically symmetric potential. For example, they are the solutions to the problem of a rotating molecule.

1. Solution of the Φ Equation

The equation for Φ is given in equation (7.15),

$$\frac{d^2\Phi}{d\varphi^2}=-m^2\Phi.$$

Since the second derivative of the function is a negative constant times the function, the solutions are sin and cos. The solutions can be written in exponential form as

$$\Phi_m(\varphi)=\frac{1}{\sqrt{2\pi}}e^{im\varphi}. \tag{7.19}$$

This function solves the differential equation, but it is not an acceptable wavefunction. The point $\varphi=0$ and $\varphi=2\pi$ is the same point in space. One of the four Born conditions for a function to be an acceptable solution to the Schrödinger is that it must be single-valued. For an arbitrary value of m, $e^{im\varphi}$ is not single valued. However,

$$e^{i2\pi}=\cos 2\pi+i\sin 2\pi=1$$

and

$$e^{i2\pi n} = 1$$

if n is a positive or negative integer or zero. If $\varphi = 0$, then $e^{im\varphi} = 1$. If $\varphi = 2\pi$, then $e^{im\varphi} = 1$ only if m is a positive or negative integer or zero. Therefore, for the function to be single-valued, m is restricted to these values. The result is that the solution to the Φ equation is

$$\Phi_m(\varphi) = \frac{1}{\sqrt{2\pi}} e^{im\varphi}, \tag{7.20a}$$

with

$$m = 0,\ \pm 1,\ \pm 2,\ \pm 3,\ \ldots . \tag{7.20b}$$

m is called the magnetic quantum number.

The complex solutions having the same absolute values of m, $|m|$, can be added and subtracted to give real solutions because a superposition of eigenkets having the same eigenvalues are also eigenkets with the same eigenvalues. Then the solutions to equation (7.15) are also

$$\Phi_0(\varphi) = \frac{1}{\sqrt{2\pi}} \qquad m = 0 \tag{7.21a}$$

$$\begin{aligned} \Phi_{|m|}(\varphi) &= \frac{1}{\sqrt{\pi}} \cos |m|\,\varphi \\ \Phi_{|m|}(\varphi) &= \frac{1}{\sqrt{\pi}} \sin |m|\,\varphi \end{aligned} \qquad |m| = 1,\ 2,\ 3,\ \ldots . \tag{7.21b}$$

2. Solution of the Θ Equation

The equation for Θ is given in equation (7.17),

$$\frac{1}{\sin\theta}\frac{d}{d\theta}\left(\sin\theta\frac{d\Theta}{d\theta}\right) - \frac{m^2\Theta}{\sin^2\theta} + \beta\,\Theta = 0.$$

$z = \cos\,\theta$ is used as a substitution. z varies between $+1$ and -1.

$$P(z) = \Theta(\theta)$$

and

$$\sin^2\,\theta = 1 - z^2.$$

Then,

$$\begin{aligned} \frac{d\Theta}{d\theta} &= \frac{dP}{dz}\frac{dz}{d\theta} \\ &= -\frac{dP}{dz}\,\sin\,\theta, \end{aligned}$$

and

$$dz = -\sin\theta\, d\theta$$
$$d\theta = -\frac{1}{\sin\theta} dz.$$

Making these substitutions in equation (7.17) gives the differential equation for $P(z)$:

$$\frac{d}{dz}\left\{(1-z^2)\frac{dP(z)}{dz}\right\} + \left\{\beta - \frac{m^2}{1-z^2}\right\} P(z) = 0. \tag{7.22}$$

The original differential equation is now expressed in equation (7.22) in terms of the variable, z. This will ultimately lead to a form with a known solution.

However, equation (7.22) has a singularity at $z = \pm 1$ because the term

$$\frac{m^2}{1-z^2}$$

blows up at these points. These singularities are called regular points. There is a standard procedure for removing such singularities. It is discussed in some detail in math texts on differential equations. Making the substitutions

$$x = 1 - z, \qquad y = 1 + z$$

and following the standard procedure removes the singularities and yields

$$P(z) = x^{|m|/2}\, y^{|m|/2}\, G(z)$$

and

$$P(z) = (1 - z^2)^{|m|/2}\, G(z). \tag{7.23}$$

Then substituting equation (7.23) into the differential equation, equation (7.22), yields, after some algebra,

$$(1 - z^2)G'' - 2(|m| + 1)zG' + \{\beta - |m|\,(|m| + 1)\}G = 0, \tag{7.24}$$

with

$$G' = \frac{dG}{dz} \quad \text{and} \quad G'' = \frac{d^2G}{dz^2}.$$

The differential equation for $G(z)$ is not singular. Solving equation (7.24) for $G(z)$ gives $P(z)$ using equation 7.23, and $P(z) = \Theta(\theta)$.

The differential equation, equation (7.24), can be solved using the polynomial method in the same manner as in the solution to Schrödinger equation treatment of the harmonic oscillator in Chapter 6. G is expanded as a power series in z,

$$G(z) = a_0 + a_1 z + a_2 z^2 + a_3 z^3 + \ \ldots, \tag{7.25}$$

and G' and G'' are obtained by term-by-term differentiation. The series for G, G', and G'' are substituted into equation (7.24), and the resulting equation is arranged such that each power of z is multiplied by a coefficient. The sum of all of the terms is equal to zero. As in the harmonic oscillator solution, the sum of all of the different powers of z can only vanish if the coefficients of each of the terms vanish separately. With

$$D = \{\beta - |m|\,(|m| + 1)\}\,, \tag{7.26}$$

the coefficients of the first few powers of z are

$$\begin{aligned}
&\{z^0\} \qquad 2\,a_2 + Da_0 = 0\\
&\{z^1\} \qquad 6\,a_3 + (D - 2\,(|m| + 1))a_1 = 0\\
&\{z^2\} \qquad 12\,a_4 + (D - 4\,(|m| + 1) - 2)a_2 = 0\\
&\{z^3\} \qquad 20\,a_5 + (D - 6\,(|m| + 1) - 6)a_3 = 0.
\end{aligned}$$

The term in brackets prior to each equation indicates the power of z that is associated with the coefficient. Each coefficient is equal to zero. Examination of the coefficients yields the recursion formula

$$a_{\nu+2} = \frac{(\nu + |m|)(\nu + |m| + 1) - \beta}{(\nu + 1)(\nu + 2)} a_\nu. \tag{7.27}$$

The recursion formula gives an even series and an odd series. Selecting a value for a_0 and substituting it into equation (7.27) gives a_2. Substituting the value for a_2 gives a_4, and so forth. Beginning with a value for a_1 gives a_3, and so on. To obtain the even series, a_1 is set equal to zero. To obtain the odd series, a_0 is set equal to zero. Any values for a_0 and a_1 can be used to obtain the even and odd series. The a_0 and a_1 values are set by requiring that the wavefunction, $\Theta(\theta)$, be normalized.

Like the series solution found using the polynomial method to solve the Schrödinger equation for the harmonic oscillator, the series defined by the recursion relation, equation (7.27), is a solution to the differential equation, but it does not provide a good wavefunction. The wavefunction must be finite everywhere—that is, from $-1 \leq z \leq 1$. The infinite series diverges for $z = \pm 1$, and therefore it is necessary to terminate the series after a finite number of terms. To break off the even series—or separately, the odd series—at the ν' term, β is chosen such that

$$\begin{aligned}
&\beta = (\nu' + |m|)(\nu' + |m| + 1)\\
&\nu' = 0,\ 1,\ 2,\ \ldots.
\end{aligned} \tag{7.28}$$

The series is even or odd as ν' is even or odd.

Setting

$$\ell = \nu' + |m|\,, \tag{7.29}$$

gives

$$\beta = \ell(\ell + 1). \tag{7.30}$$

The smallest value that ℓ can have is 0 because the smallest value of ν' is 0 and the smallest value of $|m|$ is zero. This is the only combination of ν' and $|m|$ that gives $\ell = 0$. $\ell = 0$ corresponds to an s orbital. The next possible value for ℓ is 1. ν' can be 1 with $|m| = 0$, or ν' can be 0 with the $|m| = 1$, that is, $m = \pm 1$. Therefore, there are three different ways to obtain $\ell = 1$. $\ell = 1$ corresponds to the p orbitals, and there are three of them. In a similar manner, there are five distinct ways to obtain $\ell = 2$ with $m = 0, \pm 1, \pm 2$. These m values correspond to the five different d orbitals. In general, there are $2\ell + 1$ m values for a given ℓ.

The solution to the Θ equation is

$$\Theta(\theta) = (1 - z^2)^{|m|/2}\, G(z). \tag{7.31}$$

$z = \cos\theta$, and $G(z)$ is defined by recursion relation, equation (7.27), with $\beta = \ell(\ell + 1)$. The $\Theta(\theta)$ are the associated Legendre functions, which are discussed further below.

3. Solution of the *R* Equation

The differential equation for $R(r)$ is given in equation (7.18). Substituting the result obtained from the solution of the Θ equation, $\beta = \ell(\ell + 1)$, yields

$$\frac{1}{r^2}\frac{d}{dr}\left(r^2\frac{dR}{dr}\right) + \left[-\frac{\ell(\ell+1)}{r^2} + \frac{2\mu}{\hbar^2}(E - V(r))\right] R = 0 \tag{7.32}$$

with

$$V(r) = -\frac{Ze^2}{4\pi\varepsilon_o r}.$$

To reduce the number of constants that must be carried along in the calculation, the following substitutions are made:

$$\alpha^2 = -\frac{2\mu E}{\hbar^2} \tag{7.33}$$

$$\lambda = \frac{\mu Z e^2}{4\pi\varepsilon_o\hbar^2\alpha} \tag{7.34}$$

With these substitutions, the new independent variable is

$$\rho = 2\alpha r. \tag{7.35}$$

ρ is the distance in units of 2α. With these substitutions, using the appropriate differential operators and

$$S(\rho) = R(r), \tag{7.36}$$

the differential equation becomes

$$\frac{1}{\rho^2}\frac{d}{d\rho}\left(\rho^2\frac{dS}{d\rho}\right) + \left(-\frac{1}{4} - \frac{\ell(\ell+1)}{\rho^2} + \frac{\lambda}{\rho}\right) S = 0, \qquad 0 \le \rho \le \infty. \tag{7.37}$$

A procedure analogous to that employed to solve the Schrödinger equation for the harmonic oscillator is employed to solve equation (7.37). First the asymptotic solution for very large ρ is found; and then using the polynomial method, a solution for all ρ is obtained. To find the asymptotic solution, the first term in equation (7.37) is examined. Taking the derivative using the chain rule gives

$$\frac{1}{\rho^2}\left(\frac{d}{d\rho}\left(\rho^2\frac{dS}{d\rho}\right)\right) = \frac{1}{\rho^2}\left(\rho^2\frac{d^2S}{d\rho^2} + 2\rho\frac{dS}{d\rho}\right)$$
$$= \frac{d^2S}{d\rho^2} + \frac{2}{\rho}\frac{dS}{d\rho}. \tag{7.38}$$

As ρ becomes very large, $\rho \to \infty$, the second term on the right-hand side of equation (7.38) goes to zero. Then, for very large ρ, the first term of equation (7.37) is just

$$\frac{d^2S}{d\rho^2}.$$

The second term in brackets in equation (7.37) reduces to $-{}^1\!/_4$ for very large ρ. Therefore, for very large ρ, equation (7.37) becomes

$$\frac{d^2S}{d\rho^2} = \frac{1}{4}S. \tag{7.39}$$

There are two solutions to this equation,

$$S = e^{+\rho/2} \tag{7.40a}$$

and

$$S = e^{-\rho/2}. \tag{7.40b}$$

Only the second equation is acceptable, since the first equation becomes infinite when $\rho \to \infty$, violating the Born condition that the wavefunction must be finite everywhere. The solution for very large ρ is multiplied by a function $F(\rho)$ to give a solution to the differential equation, equation (7.37), that is good for all ρ:

$$S(\rho) = e^{-\rho/2}\,F(\rho) \tag{7.41}$$

is substituted into equation (7.37), and the resulting equation is divided by $e^{-\rho/2}$ to yield

$$F'' + \left(\frac{2}{\rho} - 1\right)F' + \left(\frac{\lambda}{\rho} - \frac{\ell(\ell+1)}{\rho^2} - \frac{1}{\rho}\right)F = 0, \qquad 0 \le \rho \le \infty. \tag{7.42}$$

F' and F'' are the first and second derivatives of F with respect to ρ, respectively. Equation (7.42) has singularities at the origin, $\rho = 0$. This is a regular point, analogous to the one that occurred in the solution of the Θ equation, equation (7.22). Following standard procedures, it is removed by the substitution,

$$F(\rho) = \rho^\ell\,L(\rho), \tag{7.43}$$

where $L(\rho)$ is a power series in ρ. This substitution removes the singularities and leads to the equation.

$$\rho L'' + (2(\ell + 1) - \rho)L' + (\lambda - \ell - 1)L = 0. \tag{7.44}$$

Solving equation (7.44) will yield L. Substitution of L into equation (7.43) yields F, which in turn is substituted into equation (7.41) to give $S(\rho) = R(r)$.

Equation (7.44) is solved by again using the polynomial method. L is expanded in a power series,

$$L(\rho) = \sum_{\nu} a_\nu \rho^\nu = a_0 + a_1\rho + a_2\rho^2 + \cdots, \tag{7.45}$$

and L' and L'' are found by term by term differentiation. L, L', and L'' are substituted into equation (7.44). The equation is rearranged, and the coefficients of each power of ρ are collected. The sum of all of the terms equals zero. As in the solution to the Schrödinger equation for the harmonic oscillator, in order for the sum to equal zero for all values of ρ, the coefficients of each power of ρ must individually equal zero. The coefficients of the first few powers of ρ are

$$\begin{aligned}
&\{\rho^0\} \qquad (\lambda - \ell - 1)a_0 + 2(\ell + 1)a_1 = 0\\
&\{\rho^1\} \qquad (\lambda - \ell - 1 - 1)a_1 + [4(\ell + 1) + 2]\,a_2 = 0\\
&\{\rho^2\} \qquad (\lambda - \ell - 1 - 2)a_2 + [6(\ell + 1) + 6]\,a_3 = 0.
\end{aligned}$$

Examining the form of these coefficients yields the recursion relation

$$a_{\nu+1} = \frac{-(\lambda - \ell - 1 - \nu)a_\nu}{[2(\nu + 1)(\ell + 1) + \nu(\nu + 1)]}. \tag{7.46}$$

Thus $L(\rho)$ is given by the power series in equation (7.45) with the coefficients given by the recursion formula, equation (7.46). The value of a_0 is arbitrary. It is ultimately determined by the normalization condition for the wavefunction. Substituting a value for a_0 into the recursion yields the value for a_1. Then substituting a_1 gives a value for a_2, and so forth. Notice that equation (7.46) does not give an even and odd series, in contrast to the recursion formulas obtained previously [e.g., equation (7.27)].

As in the treatment of the harmonic oscillator in Chapter 6, Section A, when the power series solution for L is combined with the other pieces to give the complete radial wavefunction, $S(\rho) = R(r)$, it becomes infinite as $\rho \to \infty$. Therefore, it is not an acceptable solution to the Schrödinger equation. The power series $F(\rho)$ behaves like e^ρ when the index in the power series, ν, becomes very large. When this is multiplied by the solution for large ρ, equation (7.40b), the function behaves as $e^{\rho/2}$ for large ρ, and it diverges. To make the solution to the differential equation an acceptable wavefunction, the series must have a finite number of terms. For a finite number of terms, the large ρ solution, $e^{-\rho/2}$, goes to zero as $\rho \to \infty$ faster than any finite power of ρ goes to infinity. Therefore, the function $S(\rho)$, goes to zero as $\rho \to \infty$, and it is an acceptable wavefunction.

To break off the series defined by the recursion relation, equation (7.46), after the n' term, the condition

$$\lambda - \ell - 1 - n' = 0 \tag{7.47}$$

must be met. Since ℓ is an integer and n' is an integer, λ must be an integer for the sum to vanish. n' is called the radial quantum number. Setting

$$\lambda = n$$

with

$$n = n' + \ell + 1, \tag{7.48}$$

n is the total quantum number. It is the quantum number that labels the "shells" of the hydrogen atom. For the hydrogen atom wavefunctions (see below) $3s$, $3p$, and $3d$, $n = 3$.

When the series for $L(\rho)$, defined by the recursion relation, equation (7.46), is broken off after a finite number of terms, the solution to the differential equation for $R(r)$ is an acceptable wavefunction with

$$R(r) = e^{-\rho/2}\rho^{\ell} L(\rho) \tag{7.49}$$

4. The Hydrogen Atom Energy Levels

In solving the $R(r)$ equation, it was found that $\lambda = n$. λ is a collection of constants defined in equation (7.34). λ depends on α, and α^2 contains the energy, E. α^2 is given in equation (7.33). Squaring λ and inserting the definition of α^2 gives

$$n^2 = \lambda^2 = -\frac{\mu Z^2 e^4}{32\pi^2 \varepsilon_o^2 \hbar^2 E},$$

and solving for E yields

$$E_n = -\frac{\mu Z^2 e^4}{8\varepsilon_o^2 h^2 n^2}. \tag{7.50}$$

The subscript on E labels the energy with the quantum number, n. For the hydrogen atom, $Z = 1$. Equation (7.50) gives the nonrelativistic energy levels for any one electron atom. For He^+, $Z = 2$. The energy levels depend inversely on the square of the quantum number, n and on the square of the nuclear charge, Z.

Since $n = n' + \ell + 1$, the smallest value of n is 1. To have $n = 1$, n' and ℓ are both zero. Since ℓ is zero, there is no orbital angular momentum (see Chapter 15). $n = 1$ corresponds to a single state, the $1s$ state. If $n = 2$, there are two possibilities, $n' = 1$ and $\ell = 0$, which is the $2s$ state, and $n' = 0$ and $\ell = 1$, which is the $2p$ state with the total angular momentum quantum number equal to 1. If $n = 3$, there are three possibilities, $n' = 2$ and $\ell = 0$, $n' = 1$ and $\ell = 1$, or $n' = 0$ and $\ell = 2$, corresponding to the $3s$, $3p$, and $3d$ states, respectively. Since there are $2\ell + 1$ m values for a given ℓ, there will be 1 $3s$ state, 3 different $3p$ states, and 5 different $3d$ states.

Defining the Bohr radius as

$$a_0 = \frac{\varepsilon_o h^2}{\pi \mu e^2}, \tag{7.51}$$

E_n becomes

$$E_n = -\frac{Z^2 e^2}{8\pi \varepsilon_o a_0 n^2}. \tag{7.52a}$$

The Bohr radius (a_0) is the characteristic length scale of the hydrogen atom, 5.29×10^{-11} m. The Bohr radius is often expressed with the reduced mass, μ, replaced by the mass of the electron, m_e, so that it is in terms of fundamental constants (see "Physical Constants and Conversion Factors for Energy Units" at end of book). This replacement is as if the mass of the proton is infinite. Replacing μ by m_e changes the value of a_0 in the third descimal place. The ground state of the H atom (i.e., $n = 1$) has energy -13.6 eV. This is the energy that is required to ionize the electron. The energy can also be written in terms of the Rydberg constant, R_{H} for the H atom, which is an experimental number, $R_{\mathrm{H}} = 109{,}677\ \mathrm{cm}^{-1}$. In terms of the Rydberg constant,

$$E_n = -\frac{Z^2}{n^2} R_H hc. \tag{7.52b}$$

R_{H} is closely related to a fundamental constant, the Rydberg constant for infinite mass, R_∞:

$$R_\infty = \frac{m_e e^4}{8\varepsilon_o^2 h^3 c}.$$

$R_\infty = 109{,}737\ \mathrm{cm}^{-1}$. It is the value the Rydberg constant would have if the mass of the H atom nucleus was infinite; that is, μ is replaced by m_e.

C. THE HYDROGEN ATOM WAVEFUNCTIONS

In Sections B.1, B.2, and B.3, the equations for $\Phi_m(\varphi)$, $\Theta_{\ell m}(\theta)$, and $R_{n\ell}(r)$ were solved. The total wavefunction for the internal states of the hydrogen atom and hydrogen like atoms is

$$\Psi_{n\ell m}(\varphi, \theta, r) = \Phi_m(\varphi)\, \Theta_{\ell m}(\theta)\, R_{n\ell}(r) \tag{7.53}$$

$$n = 1, 2, 3, \ldots$$

$$\ell = n-1, n-2, \ldots, 0$$

$$m = \ell, \ell - 1 \cdots - \ell.$$

The solutions to the $\Phi_m(\varphi)$ equation are

$$\Phi_m(\varphi) = \frac{1}{\sqrt{2\pi}} e^{im\varphi} \tag{7.54a}$$

or, if written as real functions instead of complex functions,

$$\Phi_0(\varphi) = \frac{1}{\sqrt{2\pi}} \qquad m = 0$$

$$\Phi_{|m|}(\varphi) = \begin{cases} \frac{1}{\sqrt{\pi}} \cos |m| \varphi \\ \frac{1}{\sqrt{\pi}} \sin |m| \varphi \end{cases} \qquad |m| = 1, 2, 3, \ldots . \tag{7.54b}$$

The solutions to the $\Theta_{\ell m}(\theta)$ equation are

$$\Theta(\theta) = (1 - z^2)^{|m|/2} G(z), \tag{7.55}$$

where $z = \cos\,\theta$, and $G(z)$ is defined by series and recursion relation in equations (7.25) and (7.27), respectively.

The solutions to the $R_{n\ell}(r)$ equation are

$$R_{n\ell}(r) = e^{-\rho/2} \rho^{\ell} L(\rho), \tag{7.56}$$

with $\rho = 2\alpha r$, α is defined in equation (7.33), and $L(\rho)$ is defined by series and recursion relation given in equations (7.45) and (7.46), respectively.

1. Generating Functions for the $\Theta_{\ell m}(\theta)$ Functions

The $\Phi_m(\varphi)$ functions are explicit normalized functions obtained from the solution of the hydrogen atom φ equation. $\Theta_{\ell m}(\theta)$'s are given in terms of a series expansion and are, so far, not normalized. In Chapter 6, the wavefunctions for the harmonic oscillator were expressed in terms of a generating function as wells as in terms of the series expansion obtained from the solution of the Schrödinger equation. The $\Theta_{\ell m}(\theta)$ also can be conveniently obtained from a generating function procedure.

The normalized expression for $\Theta_{\ell m}(\theta)$ is

$$\Theta(\theta) = \sqrt{\frac{(2\ell + 1)}{2} \frac{(\ell - |m|)!}{(\ell + |m|)!}} P_{\ell}^{|m|}(\cos\,\theta). \tag{7.57}$$

The $P_{\ell}^{|m|}$ are a well-known set of functions called the associated Legendre polynomials. (Note that $0! = 1$.)

The associated Legendre polynomials, $P_{\ell}^{|m|}(z)$, can be obtained from the Legendre polynomials, $P_{\ell}(z)$.

$$P_0(z) = 1, \tag{7.58a}$$

and

$$P_{\ell}(z) = \frac{1}{2^{\ell}\ell!} \frac{d^{\ell}(z^2 - 1)^{\ell}}{dz^{\ell}}. \tag{7.58b}$$

In terms of the Legendre polynomials the associated Legendre polynomials are

$$P_{\ell}^{|m|}(z) = (1 - z^2)^{|m|/2} \frac{d^{|m|} P_{\ell}(z)}{dz^{|m|}}. \tag{7.58c}$$

To illustrate the use of the generating functions, the first few Θ's will be obtained.

The s orbital has the Θ function, Θ_{00}; that is, $\ell = 0$ and $m = 0$. Then equations (7.58) give

$$P_0(z) = 1$$
$$P_0^0(z) = 1.$$

Using equation (7.57) to include the normalization constant the result is

$$\Theta_{00}(\theta) = \frac{\sqrt{2}}{2}. \tag{7.59}$$

Θ_{00} is a constant; there is no θ dependence. Since $m = 0$, there is no φ dependence either. The s orbital is spherically symmetric.

The p orbitals have the Θ functions, Θ_{10} ($\ell = 1, m = 0$) and $\Theta_{1\pm1}$ ($\ell = 1, m = \pm1$). To obtain Θ_{10}, using equation (7.58) gives

$$P_1(z) = \frac{1}{2}\frac{d(z^2-1)}{dz} = \frac{1}{2}2z = z$$

and

$$P_1^0(z) = z.$$

Then using equation (7.57) to include the normalization constant, the result is

$$\Theta_{10}(\theta) = \frac{\sqrt{6}}{2}\cos\theta. \tag{7.60}$$

To obtain $\Theta_{1\pm1}$, using equation (7.58) gives

$$P_1(z) = z$$

and

$$P_1^{\pm1}(z) = (1-z^2)^{1/2}\frac{dz}{dz} = (1-z^2)^{1/2}.$$

Then using equation (7.57) to include the normalization constant gives

$$\Theta_{1,\pm1}(\theta) = \frac{\sqrt{3}}{2}\sin\theta. \tag{7.61}$$

The three different p orbitals are obtained by multiplying Θ_{10} by $\Phi_m(\varphi)$ with $m = 0$ and multiplying $\Theta_{1\pm1}$ by the $\Phi_m(\varphi)$ with $m = \pm1$.

The d orbitals have the Θ functions $\Theta_{20}(\ell = 2, m = 0)$, $\Theta_{2\pm1}(\ell = 2, m = \pm1)$, and $\Theta_{2\pm2}(\ell = 2, m = \pm2)$. To obtain Θ_{20}, using equation (7.58) gives

$$P_2(z) = \frac{1}{2^2 2}\frac{d^2(z^2-1)^2}{dz^2} = \frac{1}{2}(3z^2-1)$$

and

$$P_\ell^0(z) = \frac{3}{2}z^2 - \frac{1}{2}.$$

Then using equation (7.57) to include the normalization constant, the result is

$$\Theta_{20} = \frac{\sqrt{10}}{4}(3\cos^2\theta - 1). \tag{7.62}$$

In the same manner, the Θ component of the other four d orbitals ($\Theta_{2,\pm1}$ and $\Theta_{2,\pm2}$) and the Θ components of the seven different f orbitals ($\Theta_{3,\pm3}$, $\Theta_{3,\pm2}$, $\Theta_{3,\pm1}$, and $\Theta_{3,0}$) can be obtained.

2. Generating Functions for the $R_{n\ell}(r)$ Functions

The $R_{n\ell}(r)$ can also be obtained from a generating function procedure. The normalized expression for $R_{n\ell}(r)$ is

$$R_{n\ell}(r) = -\sqrt{\left(\frac{2Z}{na_0}\right)^3 \frac{(n-\ell-1)!}{2n[(n+\ell)!]^3}}\, e^{-\rho/2}\rho^\ell L_{n+\ell}^{2\ell+1}(\rho) \tag{7.63}$$

with

$$\rho = 2\alpha r = \frac{2Z}{a_0 n} r,$$

and a_0 is the Bohr radius given in equation (7.51). The $L_{n+\ell}^{2\ell+1}(\rho)$ are the associated Laguerre polynomials. They can be obtained from the Laguerre polynomials, $L_{n+\ell}$, which are given by

$$L_{n+\ell}(\rho) = e^\rho \frac{d^{n+\ell}(\rho^{n+\ell}e^{-\rho})}{d\rho^{n+\ell}}. \tag{7.64a}$$

Then the associated Laguerre polynomials are

$$L_{n+\ell}^{2\ell+1}(\rho) = \frac{d^{2\ell+1}L_{n+\ell}(\rho)}{d\rho^{2\ell+1}}. \tag{7.64b}$$

To illustrate the use of the generating functions for $R_{n\ell}(r)$, the radial part of the $1s$ and $2s$ wavefunctions will be found. The $R_{n\ell}(r)$ for the $1s$ wavefunction is $R_{10}(r)$, that is, $n = 1$ and $\ell = 0$. Then from equation (7.64a), the Laguerre polynomial is

$$L_1(\rho) = e^\rho \frac{d(\rho e^{-p})}{d\rho} = 1 - \rho,$$

and the associated Laguerre polynomial is

$$L_1^1(\rho) = \frac{d(1-\rho)}{d\rho} = -1.$$

Using equation (7.63) to include the other parts of $R_{n\ell}(r)$ and the normalization constant gives

$$R_{10}(r) = -\sqrt{\left(\frac{2Z}{a_0}\right)^3 \frac{1}{2}}\, e^{-\rho/2}(-1) = 2\left(\frac{Z}{a_0}\right)^{3/2} e^{-\rho/2}, \tag{7.65}$$

the radial part of the 1s wavefunction. The $R_{n\ell}(r)$ for the 2s wavefunction is $R_{20}(r)$, that is, $n = 2$ and $\ell = 0$. Then from equation (7.64a), the Laguerre polynomial is

$$L_2(\rho) = e^{\rho} \frac{d^2(\rho^2 e^{-\rho})}{d\rho^2} = \rho^2 - 4\rho + 2,$$

and the associated Laguerre polynomial is

$$L_2^1(\rho) = \frac{d(\rho^2 - 4\rho + 2)}{d\rho} = 2\rho - 4.$$

Using equation (7.63) to include the other parts of $R_{n\ell}(r)$ and the normalization constant gives

$$R_{20}(r) = \frac{\sqrt{2}}{4}\left(\frac{Z}{a_0}\right)^{3/2} e^{-\rho/2}(2 - \rho), \tag{7.66}$$

the radial part of the 2s wavefunction. Note that when $\rho = 2$, the wavefunction vanishes; there is a node in the radial wavefunction at $\rho = 2$.

The associated Laguerre polynomials are also given by

$$L_{n+\ell}^{2\ell+1}(\rho) = \sum_{k=0}^{n-\ell-1} (-1)^{k+1} \frac{[(n+\ell)!]^2}{(n-\ell-1-k)!(2\ell+1+k)!k!}\rho^k. \tag{7.67}$$

Equation (7.67) can be used with equation (7.63) to obtain the $R_{n\ell}(r)$.

3. The 1s and 2s Total Wavefunctions

The total wavefunction for a hydrogen-like atom is the product of the three one-dimensional functions, that is, $\Psi_{n\ell m}(\varphi, \theta, r) = \Phi_m(\varphi)\, \Theta_{\ell m}(\theta)\, R_{n\ell}(r)$. The 1$s$ wavefunction is

$$\Psi_{1s}(\varphi, \theta, r) = \Psi_{100} = \Phi_0 \Theta_{00} R_{10} = \left(\frac{1}{\sqrt{2\pi}}\right)\left(\frac{\sqrt{2}}{2}\right)\left(2\left(\frac{Z}{a_0}\right)^{3/2} e^{-\rho/2}\right),$$

where the first term in brackets is Φ_0, the second term in brackets is Θ_{00}, and the third term in brackets is R_{10}. For the hydrogen atom, $Z = 1$ and Ψ_{1s} is

$$\Psi_{1s} = \frac{1}{\sqrt{\pi a_0^3}} e^{-r/a_0}. \tag{7.68}$$

The 2s wavefunction is

$$\Psi_{2s}(\varphi, \theta, r) = \Psi_{200} = \Phi_0 \Theta_{00} R_{20} = \frac{1}{4\sqrt{2\pi a_0^3}}(2 - r/a_0)e^{-r/2a_0}. \tag{7.69}$$

The s wavefunctions are spherically symmetric. The $1s$ wavefunction falls off exponentially from the nucleus. The $2s$ wavefunction has an exponential decay from the nucleus multiplied by a polynomial which has a node at $2a_0$. The $3s$ function falls off exponentially multiplied by a quadratic polynomial which has two nodes. Like the particle in the box and the harmonic oscillator, each succeeding energy level has one additional radial node. A particle in an infinite box or the harmonic oscillator only has bound state, so the number of nodes continues to increase as the energy increases. This is also true of the bound states of the hydrogen atom. However, at sufficiently high energy above the ground state (the $1s$ state), $\mu e^4/8\varepsilon_o^2 h^2$, the electron is ionized. For higher energies, the electron is a free particle and will have a continuum of energy levels; the wavefunction will not have nodes. Unbound states were discussed qualitatively for the particle in a finite box in Chapter 5, Section E.

For any state of the hydrogen atom, the probability of finding the electron a distance r away from the nucleus in a thin spherical shell is the radial distribution function given by

$$D_{n\ell}(r) = [R_{n\ell}(r)]^2 r^2\, dr. \tag{7.70}$$

The $R_{n\ell}(r)$ are given by equation (7.63). Figure 7.1 displays the radial distribution functions for the $1s$ and $2s$ wavefunctions. The areas under the curves are normalized. The volume element, $4\pi r^2 dr$, reduces the probability of finding the electron at the nucleus. The $1s$ wavefunction appears as a ball around the nucleus. The $2s$ wavefunction appears as a ball around the nucleus surrounded by a concentric shell at greater distance. The node at $r = 2a_0$ is evident and separates the ball from the concentric shell. The $2s$ orbital has a larger fraction of the probability of finding the electron further from the nucleus than the $1s$ orbital. The $3s$ wavefunction has a ball and two concentric shells. The two shells are separated by the two nodes in the $3s$ wavefunction. The outermost shell has the largest

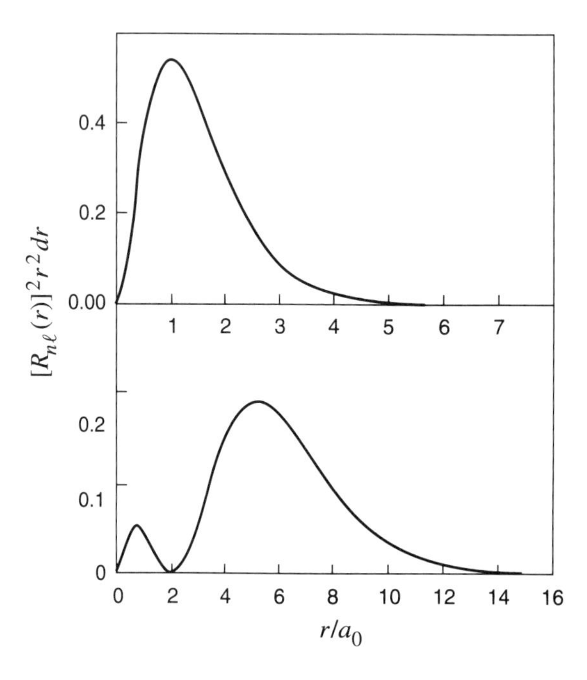

Figure 7.1 The radial distribution functions for the $1s$ and $2s$ states of the hydrogen atom. The areas under the curves are normalized to one.

fraction of probability. The p, d, and f states have similar radial distribution functions. The $2p$ state has no radial nodes, and the $3p$ state has one radial node. The $3d$ state has no radial nodes, and the $4d$ state has one radial node, and so forth. The p, d, and f states do not have the Φ and Θ parts of the wavefunctions equal to constants. Therefore they are not spherically symmetric. In addition to the radial nodes, the p, d, and f states have angular nodes. Detailed illustrations of the hydrogen atom wavefunctions can be found in many undergraduate physical chemistry textbooks.

CHAPTER 8

Time-Dependent Two-State Problem

In previous chapters, several problems were treated that could be solved exactly. The problems that were solved are the momentum and energy of a free particle, the energy of the particle in an infinite box, the harmonic oscillator, and the hydrogen atom. In this chapter, a time-dependent problem that can be solved exactly will be analyzed in considerable detail. The dynamics of coupled states are a prototype for a number of important physical problems. The problem is also useful for illustrating a number of methods that have been discussed previously in the context of a concrete example.

The problem is the quantum mechanical analog of a pair of classical pendulums couple by a weak spring. Figure 8.1 shows two pendulums, A and B, which are connected by a weak spring. In the absence of the spring, the two pendulums will be able to swing independently. However, with the spring, the motions of A and B are coupled. There are two normal modes for this problem; that is, there are two states of motion that are "time-independent" in the sense that once the motion begins, the nature of the motion continues indefinitely. (For this problem it is assumed that there is no friction, so that once the system is set in motion, it will not damp.)

The two normal modes are illustrated in Figures 8.2 and 8.3. Figure 8.2 shows the two pendulums swinging back and forth together. In Figure 8.3 the two pendulums are shown moving in opposite directions. They will move away from each other and then toward each other. In both cases, the displacements from vertical will be identical for the two pendulums at a given time except for the sign of the displacements. In Figure 8.2, the pendulums move with the same sign; their displacements are both negative and then both positive. In Figure 8.3, they move with opposite sign. A is negative when B is positive, and vice versa.

Figure 8.4 illustrates a time-dependent state. If pendulum A is displaced and released, it will initially swing almost as if it is not coupled to B, provided that the spring is very weak. After some time, the amplitude of the displacement of A will decrease and B will

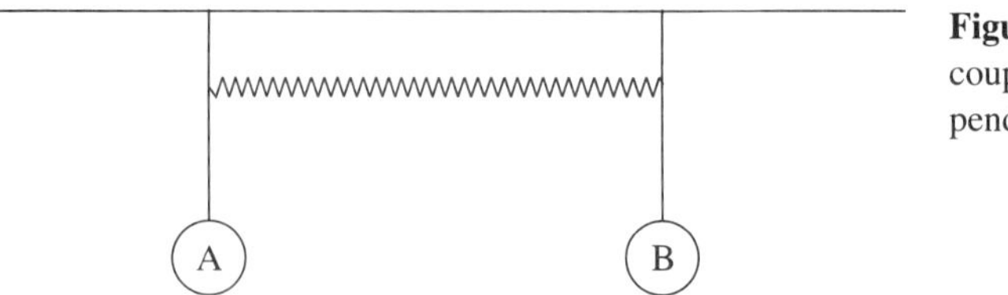

Figure 8.1 Pendulum A is coupled by a weak spring to pendulum B.

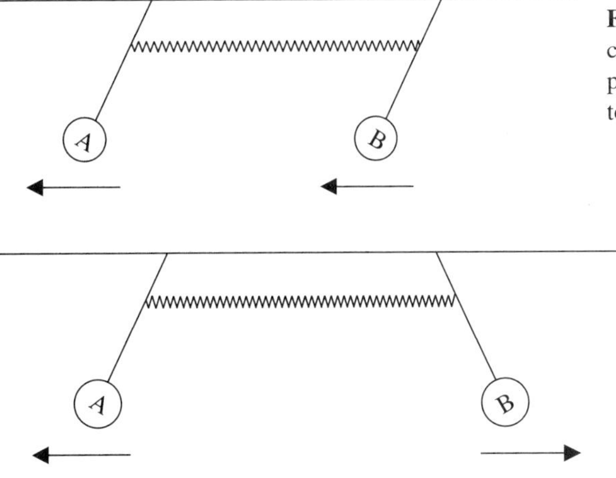

Figure 8.2 A normal mode of the coupled pendulums. The two pendulums swing back and forth together.

Figure 8.3 A normal mode of the coupled pendulums. The two pendulums swing back and forth moving in opposite directions.

swing with a small amplitude. At a longer time, the amplitude of A will be very small, and the amplitude of the displacement of B will be significant. After a period of time, τ, which is, in part, determined by the spring constant, A will be stationary and B will swing with the full amplitude that A had at time $t = 0$. This is illustrated in Figure 8.5. As B continues to swing, its amplitude will decrease and the amplitude of A will increase. At $t = 2\tau$, B will be stationary and A will again be swinging with the full amplitude it had at $t = 0$. For the frictionless system considered here, transfer of the full oscillation back and forth will continue indefinitely.

The classical problem of two pendulums coupled by a weak spring has a well-defined quantum mechanical analog. The analog is two states of a system with equal energy that are coupled by a weak interaction. In the absence of the interaction, the states, $|A\rangle$ and $|B\rangle$, are eigenstates. Therefore, the system is time independent except for the time-dependent phase factors, $e^{i\omega t}$. Weakly coupling the two states is the equivalent of adding the weak spring to the pendulum system. If the system is in the state $|A\rangle$ at time $t = 0$, a time evolution

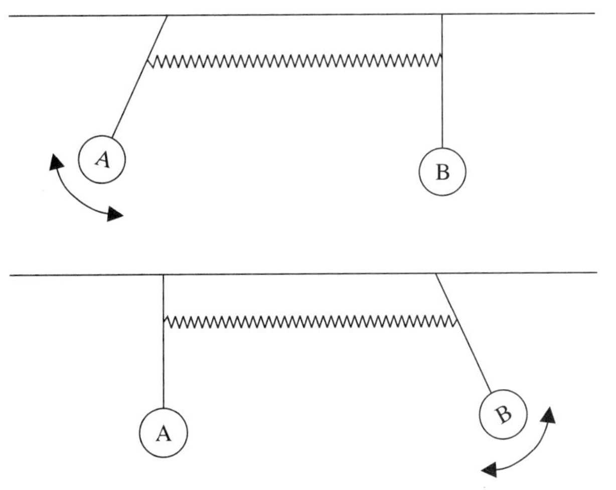

Figure 8.4 A time-dependent state of two pendulums coupled by a weak spring. Initially, pendulum A oscillates back and forth and B is stationary.

Figure 8.5 After some time, A is stationary, and B moves with the full amplitude that A had initially.

will occur. The probability of finding the system in $|A\rangle$ will decrease with time, and the probability of finding the system in $|B\rangle$ will increase. At some time, τ, the system will have unit probability of being found in $|B\rangle$ and zero probability of being found in $|A\rangle$.

The solution to the problem briefly outlined above is an initial description of a number of important physical phenomena that are time-dependent. One of these is electron transfer between two metal atoms, M, joined by a bridging ligand, L. The oxidation state of the two metals is initially the same. An extra electron is place on one of the metals. As indicated in Figure 8.6, the oxidation states are now different with $m = n + 1$. The two metal atoms do not interact directly, but they are coupled since each is covalently bonded to the ligand. The electrons on one metal are couple to the ligand electrons, which are in turn coupled to the other metal. This is referred to as a through bond interaction. So the interactions of the metals with L, couple the states of the metal atoms. The extra electron on the metal on the left in the figure will move to the metal on the right. If the electron is placed on M^{+n} at $t = 0$, the probability of finding it there will decrease with time, and the probability of finding it on the other metal atom will increase. At some time τ, the electron will have transferred from one metal atom to the other. At time 2τ, it will again have unit probability of being found on the initial metal atom. Electron transfer occurs in many processes in chemistry and biology. It can begin with the reduction of one metal or it can be photoinduced through the absorption of a photon.

A second example is vibrational energy flow. A polyatomic molecule has many vibrational modes; each is approximately a quantum harmonic oscillator. If an infrared laser pulse or thermal excitation excites a mode of the molecule, anharmonic interactions (see Chapter 9, Section B) can couple the initially excited mode to other modes of the molecule. In the case depicted in Figure 8.7, the initially excited state, $V_1 = 1$, is degenerate with a state of a different mode, $V_2 = 2$. Weak anharmonic interactions act as the weak spring in the coupled pendulum problem. The probability of finding the initial mode excited will decrease with time, and the probability of finding $V_2 = 2$ excited will increase with time. Under the right conditions, at a time τ the probability of finding $V_2 = 2$ excited will be unity, and the probability of finding $V_1 = 1$ excited will be zero. The vibrational excitation will have transferred from one mode to another. At time 2τ the initial mode will again have unit probability of being excited. Vibrational energy transfer is involved in molecular processes that occur by thermal activation, including chemical reactions.

A third process is electronic excitation transfer. If two identical molecules are near each other, an electronic excitation of one of them can be transferred to the other. A photon is absorbed placing molecule 1 in an electronic excited state (see Figure 8.8). An intermolecular interaction couples the excitation of the first molecule to the second molecule. For a singlet excited state, the coupling is via a transition dipole–transition dipole interaction. This through-space interaction falls off as $1/r^3$, where r is the separation between the molecules. The transition dipole–transition dipole interaction acts as the weak spring. If

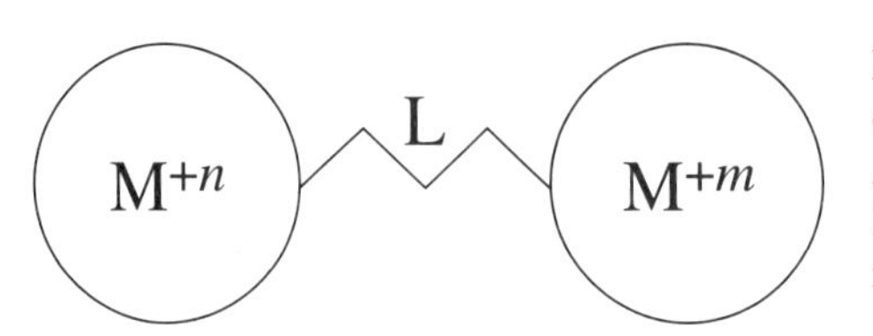

Figure 8.6 Two metal atoms in different oxidation states joined by a bridging ligand, L. At $t = 0$, $m = n + 1$, so the metal atom on the left has one more electron than the metal on the right.

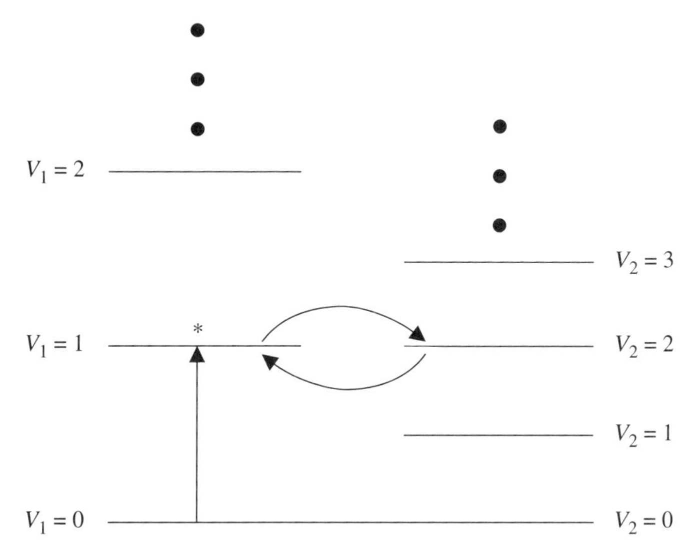

Figure 8.7 A high-frequency mode, V_1, is promoted to its first excited level, $V_1 = 1$ at $t = 0$. A second mode, V_2, has its second excited level, $V_2 = 2$, equienergetic with $V_1 = 1$. The vibrational excitation can oscillate between the two states.

molecule 1 is excited at $t = 0$, as time increases, the probability of finding the excitation on molecule 1 will decrease, and the probability of finding the excitation on molecule 2 will increase. At a time τ later, the probability of finding the excitation on molecule 1 will have decreased to zero and the probability of finding it on molecule 2 will be unity. At time 2τ the excitation will again have unit probability of being found on molecule 1. The excitation can also be a vibrational excitation excited by an infrared pulse of light on molecule 1. The vibrational excitation of molecule 1 can be transferred to molecule 2 and back again.

Each of the examples given above is idealized. The description of the transfer probability does not include the influence of thermal fluctuations of the solvent in which the molecules are dissolved. This influence will be discussed briefly at the end of this chapter.

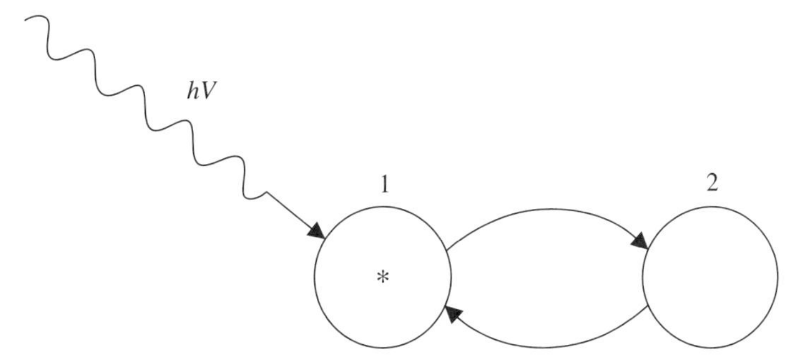

Figure 8.8 A photon is absorbed by molecule 1. The electronic excited state transfers to molecule 2, and then back to molecule 1.

A. ELECTRONIC EXCITATION TRANSFER

To make the discussion concrete, the problem of the quantum mechanical analog of the classical coupled pendulum problem will be discussed in the context of electronic excitation transport. Consider the ground and first electronic excited states of two molecules, A and B (Figure 8.9). The kets $|A\rangle$ and $|B\rangle$ are defined as

$$|A\rangle \equiv \text{molecule A excited, B unexcited}$$

$$|B\rangle \equiv \text{molecule B excited, A unexcited.}$$

The kets are taken to be normalized and orthogonal. If initially there is no intermolecular interaction (no spring) because the molecules are far apart, then

$$\underline{H}\,|A\rangle = E_A\,|A\rangle \tag{8.1a}$$

$$\underline{H}\,|B\rangle = E_B\,|B\rangle\,, \tag{8.1b}$$

with

$$|A\rangle = e^{-iE_A t/\hbar}\,|\alpha\rangle \tag{8.2a}$$

$$|B\rangle = e^{-iE_B t/\hbar}\,|\beta\rangle \tag{8.2b}$$

The kets $|A\rangle$ and $|B\rangle$ are eigenkets of the time-independent Hamiltonian operator, $\underline{H}$. The kets $|\alpha\rangle$ and $|\beta\rangle$ are the spatial parts of the wavefunctions. Each has associated with it a time-dependent phase factor.

If the molecules are placed close to each other, the transition dipole–transition dipole interaction will couple the states of the system. The Hamiltonian operator will have additional terms that represent the intermolecular interaction between the molecules. $\underline{H}$ is still time-independent. $|A\rangle$ is no longer an eigenket of $\underline{H}$. Then

$$\underline{H}\,|A\rangle = \underline{H}e^{-iE_A t/\hbar}\,|\alpha\rangle = e^{-iE_A t/\hbar}\,\underline{H}\,|\alpha\rangle\,, \tag{8.3}$$

since $\underline{H}$ is time-independent. However, the states of the isolated molecules are coupled, so

$$\underline{H}|\alpha\rangle = E_A|\alpha\rangle + \gamma\,|\beta\rangle\,. \tag{8.4}$$

γ is the interaction energy. It represents the strength of the interaction between the molecules and plays the role of the weak spring in the classical pendulum problem. In the absence of an intermolecular interaction, $\gamma = 0$, and α is an eigenket of $\underline{H}$. The form of equation (8.4) will be discussed in detail in Chapter 13, Section E, which

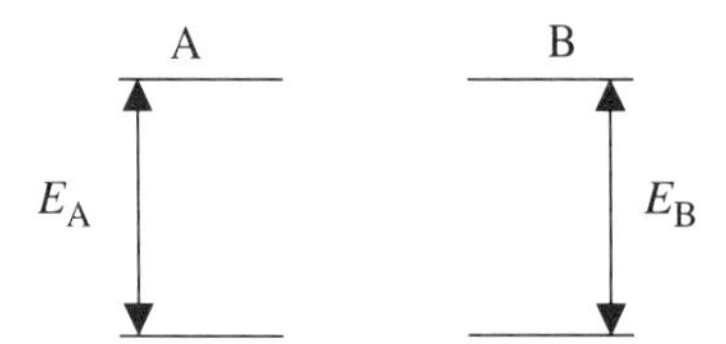

Figure 8.9 Ground and first excited states of two molecules, A and B.

deals with the matrix representation of quantum mechanics. Using equations (8.3) and (8.4) gives

$$\underline{H}\,|A\rangle = (E_A\,|\alpha\rangle + \gamma\,|\beta\rangle)e^{-iE_At/\hbar} \tag{8.5a}$$

$$\underline{H}\,|B\rangle = (E_B\,|\beta\rangle + \gamma\,|\alpha\rangle)e^{-iE_Bt/\hbar}. \tag{8.5b}$$

The time-dependent phase factor plays an important role in this problem.

To this point, E_A and E_B need not be equal. For identical molecules, they will be equal. The results for $E_A \neq E_B$ are given in Section D. Considering identical molecules,

$$E_A = E_B = E_0. \tag{8.6}$$

Furthermore, since the zero of energy is arbitrary, it can be chosen such that

$$E_0 = 0. \tag{8.7}$$

For this choice of E_0, which does not change the results of the subsequent calculations but does simplify the algebra, equations (8.5) reduce to

$$\underline{H}\,|\alpha\rangle = \gamma\,|\beta\rangle \tag{8.8a}$$

$$\underline{H}\,|\beta\rangle = \gamma\,|\alpha\rangle\,. \tag{8.8b}$$

There are two kets describing two states of the system. If the system is in state $|\alpha\rangle$, molecule A is excited and molecule B is in its ground state. If the system is in state $|\beta\rangle$, molecule B is excited and molecule A is in its ground state. Other situations can be described by superposition. The most general state of the system is

$$|t\rangle = C_1\,|A\rangle + C_2\,|B\rangle\,. \tag{8.9}$$

The coefficients C_1 and C_2 can be time-dependent. The kets $|A\rangle$ and $|B\rangle$ have time-dependent parts of them given by $e^{iEt/\hbar}$, but in this case $E = E_A = E_B = E_0 = 0$. Therefore, $|A\rangle = |\alpha\rangle$ and $|B\rangle = |\beta\rangle$, which are time-independent, and any time dependence must be contained in the coefficients C_1 and C_2.

To investigate the time dependence, the time-dependent Schrödinger equation can be employed:

$$i\hbar\frac{\partial}{\partial t}\,|t\rangle = \underline{H}\,|t\rangle\,.$$

Substituting $|t\rangle$ [equation (8.9)] into the time-dependent Schrödinger equation gives

$$i\hbar\,(\dot{C}_1\,|\alpha\rangle + \dot{C}_2\,|\beta\rangle) = C_1\gamma\,|\beta\rangle + C_2\gamma\,|\alpha\rangle\,, \tag{8.10}$$

where

$$\dot{C}_i = \frac{\partial C_i}{\partial t}.$$

Left-multiplying equation (8.10) by $\langle\alpha|$ and using normalization and the orthogonality of the kets gives

$$i\hbar\dot{C}_1 = \gamma C_2. \tag{8.11a}$$

Then left-multiplying equation (8.10) by $\langle\beta|$ gives

$$i\hbar\dot{C}_2 = \gamma C_1. \tag{8.11b}$$

Equation (8.11) are called equations of motion of the coefficients. Since $\langle\alpha|$ and $\langle\beta|$ are time-independent, the time dependence will be determined by the time dependence of C_1 and C_2, which are obtained by solving the pair of coupled differential equations, equations (8.11).

The equations of motion for the coefficients, C_i, can be readily solved. Taking the derivative of equation (8.11a) with respect to t and dividing by $i\hbar$ gives

$$\ddot{C}_1 = -\frac{i\gamma}{\hbar}\dot{C}_2. \tag{8.12}$$

Then solving equation (8.11b) for $\dot{C}_2$ and substituting the result into equation (8.12) yields

$$\ddot{C}_1 = -\frac{\gamma^2}{\hbar^2}C_1. \tag{8.13}$$

Equation (8.13) is now one equation in one unknown for C_1. It states that the derivative of a function is equal to the function multiplied by a negative constant. The solutions are sin and cos or a linear combination of the two. The most general solution is

$$C_1 = Q\sin\ \gamma t/\hbar + R\cos\ \gamma t/\hbar; \tag{8.14a}$$

substituting this result in equation (8.11a), C_2 is

$$C_2 = i[Q\cos\ \gamma t/\hbar - R\sin\ \gamma t/\hbar], \tag{8.14b}$$

where Q and R are constants. To preserve normalization, $|t\rangle$ must be normalized.

$$\langle t\,|\,t\rangle = 1 = \left(C_1^*\langle A| + C_2^*\langle B|)(C_1|A\rangle + C_2|B\rangle\right)$$

Using orthogonality and normalization of the kets $|A\rangle$ and $|B\rangle$,

$$\langle t\,|\,t\rangle = C_1^*C_1 + C_2^*C_2.$$

Therefore, $|t\rangle$ is normalized if

$$C_1^*C_1 + C_2^*C_2 = 1. \tag{8.15}$$

Equation (8.15) yields the condition

$$R^2 + Q^2 = 1. \tag{8.16}$$

To proceed further, it is necessary to specify an initial condition. At the outset, it was stated that molecule A is excited initially, so the coefficient of $|B\rangle$ in equation (8.9) must be zero. The probability of find molecule B excited is zero at $t = 0$; that is, $C_2^*(0)C_2(0) = 0$. Then, from equation (8.14b),

$$Q = 0. \tag{8.17a}$$

Taking R to be real, from equation (8.16)

$$R = 1. \tag{8.18b}$$

Therefore, for the initial condition such that molecule A is excited and molecule B is not excited at time $t = 0$, the time-dependent coefficients are

$$C_1 = \cos(\gamma t/\hbar) \tag{8.19a}$$

$$C_2 = -i \sin(\gamma t/\hbar). \tag{8.19b}$$

C_1 and C_2 are the probability amplitudes for finding molecule A excited and molecule B excited, respectively. The time-dependent state of the system is

$$|t\rangle = \cos(\gamma t/\hbar)\,|A\rangle - i\,\sin(\gamma t/\hbar)\,|B\rangle\,. \tag{8.20}$$

B. PROJECTION OPERATORS

Consider

$$|A\rangle\langle A|\,. \tag{8.21}$$

This is the projection operator for the state $|A\rangle$. Operating the projection operator on the time-dependent state $|t\rangle$,

$$|A\rangle\langle A \mid t\rangle = C_1\,|A\rangle\,, \tag{8.22}$$

gives the piece of $|t\rangle$ which is $|A\rangle$. In general, if

$$|S\rangle = \sum_i C_i\,|i\rangle\,, \tag{8.23}$$

applying the projection operator $|k\rangle\langle k|$ to $|S\rangle$ gives

$$|k\rangle\langle k \mid S\rangle = C_k\,|k\rangle\,. \tag{8.24}$$

For normalized kets $|S\rangle$ and $|i\rangle$ in equation (8.23), the coefficient C_k is the probability amplitude for finding the system in the state $|k\rangle$, given that it is in the superposition state $|S\rangle$.

Now consider

$$\langle S \mid k\rangle\langle k \mid S\rangle\,.$$

This is a closed bracket, so it is a number.

$$\langle S \mid k\rangle\langle k \mid S\rangle = C_k^* C_k = |C_k|^2\,, \tag{8.25}$$

where $C_k^* C_k$ is the absolute value squared of the probability amplitude of the particular ket $|k\rangle$, given that the system is in the superposition state $|S\rangle$. The absolute value squared of the probability amplitude is the probability. Therefore, $\langle S \mid k\rangle\langle k \mid S\rangle$ is the probability of finding the system in the state $|k\rangle$, given that the system is in the superposition state $|S\rangle$.

Projection operators can be used to examine the time dependence of the state $|t\rangle$. Given that the system is in the state $|t\rangle$, the time-dependent probabilities of finding the system in the states $|A\rangle$ and $|B\rangle$—that is, P_A and P_B—are

$$P_A = \langle t \mid A\rangle\langle A \mid t\rangle = C_1^* C_1 = \cos^2 \gamma t/\hbar \tag{8.26a}$$

$$P_B = \langle t \mid B\rangle\langle B \mid t\rangle = C_2^* C_2 = \sin^2 \gamma t/\hbar. \tag{8.26b}$$

The probability of finding molecule A excited is 1 at $t = 0$, and the probability of finding molecule B excited is 0. These probabilities oscillate. The total probability of finding an excitation of either molecule is always 1 since $\cos^2 + \sin^2 = 1$. When

$$\gamma t/\hbar = \pi/2$$

P_A is 0 and P_B is 1. Thus the time for the excitation to transfer from molecule A to molecule B is

$$t = h/4\gamma. \tag{8.27}$$

At twice this time, molecule A is excited ($P_A = 1$) and molecule be is B is in its ground state ($P_B = 0$). The excitation oscillates sinusoidally between the two molecules. The period of the oscillation is determined by the strength of the interaction, γ. At intermediate times, there is some probability of find the excitation on A and some probability of finding it on B. These probabilities are given by equations (8.26). The transfer of the electronic excitation back and forth between the two molecules is analogous to the transfer of the classical pendulum motion back and forth between the two pendulums.

C. STATIONARY STATES

It is informative to examine two specific normalized and orthogonal superpositions of the states $|A\rangle$ and $|B\rangle$,

$$|+\rangle = \frac{1}{\sqrt{2}}(|A\rangle + |B\rangle) \tag{8.28a}$$

$$|-\rangle = \frac{1}{\sqrt{2}}(|A\rangle - |B\rangle). \tag{8.28b}$$

Operating with the Hamiltonian $\underline{H}$ on the state $|+\rangle$ using the relations in equations (8.5) and (8.8) gives

$$\underline{H}\,|+\rangle = \frac{1}{\sqrt{2}}(\underline{H}\,|A\rangle + \underline{H}\,|B\rangle)$$
$$= \frac{1}{\sqrt{2}}(\gamma\,|B\rangle + \gamma\,|A\rangle)$$
$$\underline{H}\,|+\rangle = \gamma\,|+\rangle\,. \tag{8.29a}$$

Similarly,

$$\underline{H}\,|-\rangle = \frac{1}{\sqrt{2}}(\underline{H}\,|A\rangle - \underline{H}\,|B\rangle)$$
$$= \frac{1}{\sqrt{2}}(\gamma\,|B\rangle - \gamma\,|A\rangle)$$
$$\underline{H}\,|-\rangle = -\gamma\,|-\rangle\,. \tag{8.29b}$$

Therefore, the kets $|+\rangle$ and $|-\rangle$ are eigenkets of the Hamiltonian operator with eigenvalues γ and $-\gamma$, respectively. (The manner in which the kets $|+\rangle$ and $|-\rangle$ were selected so that they are eigenkets will be discussed in detail in Chapter 13, Section E.)

The kets $|+\rangle$ and $|-\rangle$ are eigenkets of the Hamiltonian operator, which is the energy operator. There are two eigenvalues, $E_0 + \gamma$ and $E_0 - \gamma$. Therefore, this system has two possible excited state energies. They are equally spaced about E_0. Their separation is 2γ (see Figure 8.10). In the absence of the intermolecular interaction, the excited states of the two molecules were degenerate in energy. The intermolecular interaction breaks the degeneracy, giving rise to two distinct energy levels. This splitting is referred to as the dimer splitting. It can be observed spectroscopically for a molecular dimer—that is, two closely spaced molecules or an associated pair. For example, the "special pair" of chlorophyll molecules in the photosynthetic reaction center has a large dimer splitting that plays an important role in the initial steps in photosynthesis.

The two states $|+\rangle$ and $|-\rangle$ given in equations (8.28) are delocalized. There is equal probability of finding the electronic excitation on molecules A and B. This can be seen by noting that the squares of the coefficients of $|A\rangle$ and $|B\rangle$ are the same in each superposition; for example,

$$\langle + \mid A\rangle\langle A \mid +\rangle = \frac{1}{2}.$$

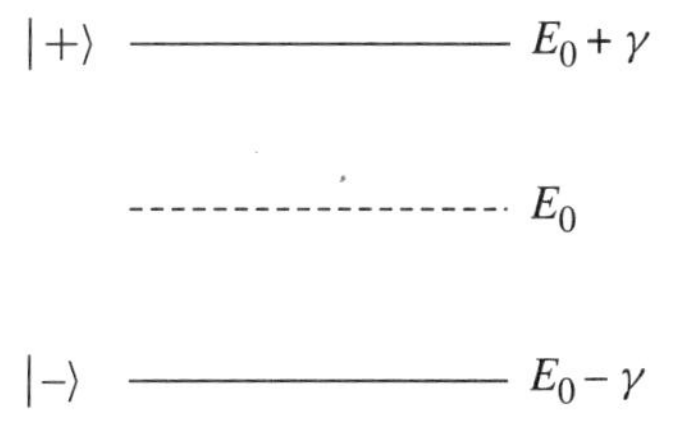

$|+\rangle$ ———————— $E_0 + \gamma$

- - - - - - - - - - - - - - - E_0

$|-\rangle$ ———————— $E_0 - \gamma$

Figure 8.10 The excited-state energy levels of two interacting molecules. The interaction strength is γ. The splitting of the levels is 2γ. These are the excited state energy levels of a dimer.

The delocalized eigenstates associated with excitations of the system are the time-independent stationary states of the system.

If the system is prepared initially in the state $|A\rangle$, it is not in an eigenstate.

$$\begin{aligned} \underline{H}\,|t\rangle &= \underline{H}\,[C_1\,|A\rangle + C_2\,|B\rangle] \\ &= \gamma[C_2\,|A\rangle + C_1\,|B\rangle] \\ &\neq K\,|t\rangle\,. \end{aligned}$$

Operating $\underline{H}$ on $|t\rangle$ does not yield $|t\rangle$ multiplied by a constant K since $C_1 \neq C_2$ as shown in equations (8.19). When the system is not prepared in an eigenstate, the system is time-dependent, and the probability of finding the excitation on one of the molecules oscillates. This is the equivalent of the wave packet problem that was discussed in Chapter 3, Section D. A single momentum eigenstate is delocalized over all space and is time-independent aside from a phase factor. However, a superposition of eigenstates forms a wave packet that is time-dependent. In the two-state problem, the entire space involves the two molecules, so the time-dependent state is a wave packet that moves between the two molecules. The dimer eigenstates are like the eigenstates of the free particle and the quantum equivalent of the two normal modes of the coupled pendulums.

Given that the system is in the state $|t\rangle$, projection operators can be used to determine the probability of finding the system in the eigenstates $|+\rangle$ and $|-\rangle$.

$$\langle t \mid +\rangle\langle + \mid t\rangle = [(C_1^*\,\langle A| + C_2^*\,\langle B|)\frac{1}{\sqrt{2}}(|A\rangle + |B\rangle)] \cdot [C.C.], \tag{8.30}$$

where [C.C.] means the complex conjugate of the first term in brackets. Using orthogonality and normalization yields

$$\begin{aligned} &= \frac{1}{\sqrt{2}}(C_1^* + C_2^*)\frac{1}{\sqrt{2}}(C_1 + C_2) \\ &= \frac{1}{2}[C_1^*C_1 + C_2^*C_2 + C_1^*C_2 + C_2^*C_1]. \end{aligned}$$

Substituting the expressions for C_1 and C_2 from equations (8.19) and their complex conjugates gives

$$\begin{aligned} &= \frac{1}{2}[\cos^2(\gamma t/\hbar) + \sin^2(\gamma t/\hbar) - i\,\cos(\gamma t/\hbar)\sin(\gamma t/\hbar) \\ &\quad + i\,\sin(\gamma t/\hbar)\cos(\gamma t/\hbar)]. \end{aligned} \tag{8.31}$$

The imaginary terms cancel, and $\cos^2 + \sin^2 = 1$. Therefore,

$$\langle t \mid +\rangle\langle + \mid t\rangle = \frac{1}{2}. \tag{8.32a}$$

In the same manner it can be determined that

$$\langle t \mid -\rangle\langle - \mid t\rangle = \frac{1}{2}. \tag{8.32b}$$

Therefore, the probability of finding the system in either of the states $|+\rangle$ or $|-\rangle$ is equal and time-independent. Since a measurement always yields an eigenvalue, an energy measurement will give $+\gamma$ half of the time and $-\gamma$ half of the time. $|t\rangle$ is not an eigenstate and exhibits a time-dependent oscillation, which is causal in nature. However, an observation, the energy measurement, forces the system into an eigenstate with the excitation delocalized over both molecules.

Since $|t\rangle$ is an equal mixture of $|+\rangle$ and $|-\rangle$ with eigenvalues $+\gamma$ and $-\gamma$, the combined results of many measurements will yield the average value (i.e., the expectation value), which is 0. [The expectation value is E_0. The zero of energy was chosen such that $E_0 = 0$ in equation (8.7).] The expectation value for the state $|t\rangle$ can be calculated directly:

$$\langle t|\, \underline{H}\, |t\rangle = (C_1^* \langle A| + C_2^* \langle B|)\underline{H}(C_1 |A\rangle + C_2 |B\rangle). \tag{8.33}$$

Substituting equations (8.2) gives the result

$$= C_1^* C_1 \langle \alpha|\, \underline{H}\, |\alpha\rangle + C_2^* C_1 \langle \beta|\, \underline{H}\, |\alpha\rangle + C_1^* C_2 \langle \alpha|\, \underline{H}\, |\beta\rangle + C_2^* C_2 \langle \beta|\, \underline{H}\, |\beta\rangle, \tag{8.34}$$

and applying equations (8.8),

$$= \gamma C_2^* C_1 + \gamma C_1^* C_2. \tag{8.35}$$

Then using equations (8.19) for C_1 and C_2 gives

$$= \gamma(i \sin(\gamma t/\hbar) \cos(\gamma t/\hbar) - i \sin(\gamma t/\hbar) \cos(\gamma t/\hbar)). \tag{8.36}$$

Therefore,

$$\langle t|\, \underline{H}\, |t\rangle = 0. \tag{8.37}$$

The expectation value is time-independent and equal to $E_0 = 0$.

D. THE NONDEGENERATE CASE AND THE ROLE OF THERMAL FLUCTUATIONS

In equation (8.6), E_A was taken to be equal to E_B; the energy levels of the states involved are identical. This simplified the problem and led to the result that there is a 100% oscillation of the probability between the two states, $|A\rangle$ and $|B\rangle$; that is, there is a 100% transfer of the excitation from molecule A to molecule B and back again. If the states do not have identical energy, the same procedure is followed except that $E_A \neq E_B$ (see Figure 8.11). No shift in the zero of energy will cause the time-dependent phase factors to vanish. Following procedures that are basically identical to the treatment of the degenerate case, the solutions for the nondegenerate case with molecule A initially excited are

$$C_1 = \frac{\sqrt{\Delta E^2 + 4\gamma^2} - \Delta E}{2\sqrt{\Delta E^2 + 4\gamma^2}} e^{i\left(\left(\sqrt{\Delta E^2+4\gamma^2}+\Delta E\right)/2\hbar\right)t} + \frac{\sqrt{\Delta E^2 + 4\gamma^2} + \Delta E}{2\sqrt{\Delta E^2 + 4\gamma^2}} e^{-i\left(\left(\sqrt{\Delta E^2+4\gamma^2}-\Delta E\right)/2\hbar\right)t} \tag{8.38a}$$

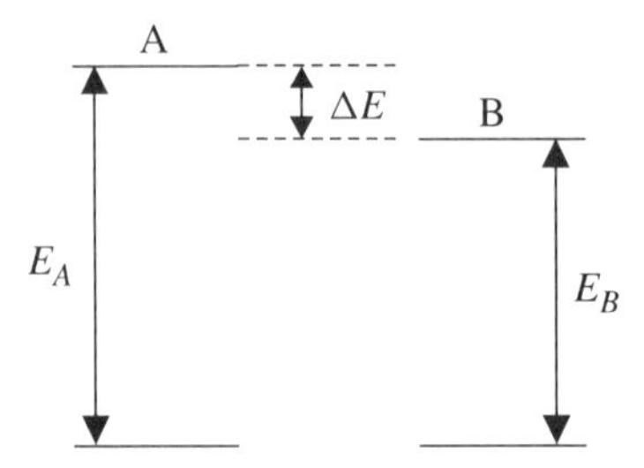

Figure 8.11 The excited-state energy levels of two molecules, A and B. ΔE is the difference in the energies of the excited states.

$$C_2 = -\frac{\gamma}{\sqrt{\Delta E^2 + 4\gamma^2}} e^{i\left(\left(\sqrt{\Delta E^2+4\gamma^2}-\Delta E\right)/2\hbar\right)t} + \frac{\gamma}{\sqrt{\Delta E^2 + 4\gamma^2}} e^{-i\left(\left(\sqrt{\Delta E^2+4\gamma^2}+\Delta E\right)/2\hbar\right)t}. \tag{8.38b}$$

The time dependence of the coefficients in the superposition state $|t\rangle$ now depend on both the strength of the intermolecular interaction, γ, and on the energy difference, ΔE. If γ is much larger than ΔE, $\gamma \gg \Delta E$, the solutions reduce to

$$C_1 = \frac{1}{2}e^{i(\gamma t/\hbar)} + \frac{1}{2}e^{-i(\gamma t/\hbar)} = \cos(\gamma t/\hbar) \tag{8.39a}$$

$$C_2 = -\frac{1}{2}e^{i(\gamma t/\hbar)} + \frac{1}{2}e^{-i(\gamma t/\hbar)} = -i\sin(\gamma t/\hbar), \tag{8.39b}$$

which are identical to equations (8.19). However, when $\gamma \ll \Delta E$, the solutions become

$$C_1 \cong 1$$
$$C_2 \cong 0.$$

Thus when the difference in energy between the states is very large compared to the strength of the interaction that couples the states, there is essentially no transfer of probability. The initial excitation of molecule A remains on molecule A.

The probabilities of finding the excitation on A and B are obtained from the projection operators; they are

$$P_A = C_1^*C_1 = \frac{\Delta E^2 + 2\gamma^2}{\Delta E^2 + 4\gamma^2} + \frac{2\gamma^2}{\Delta E^2 + 4\gamma^2}\cos\frac{\sqrt{\Delta E^2 + 4\gamma^2}}{\hbar}t \tag{8.40a}$$

$$P_B = C_2^*C_2 = \frac{2\gamma^2}{\Delta E^2 + 4\gamma^2}\left(1 - \cos\frac{\sqrt{\Delta E^2 + 4\gamma^2}}{\hbar}t\right). \tag{8.40b}$$

From these expressions it is clear that when ΔE and γ are comparable, the transfer of excitation is not complete. The first term on the right-hand side of equation (8.40a) is a constant. P_A does not fall below a value determined by this constant. As ΔE becomes large compared to γ, this constant approaches 1, and little transfer of excitation occurs. Another important feature of equation (8.40) is the transfer time. The frequency of oscillation depends both on γ and ΔE. As ΔE becomes larger, the frequency of the probability oscillation increases, but the amount of probability that is transferred decreases. For $\Delta E > \gamma$, the frequency of oscillation is determined by ΔE, but a negligible amount of probability is transferred.

The results presented for both the degenerate and nondegenerate cases are for the ideal situation in which both γ and ΔE have fixed values. In the three examples discussed at the beginning of the chapter, the molecules are generally in solvents. The ideal case can be approached at very low temperatures ($\sim$1 K) in solids or in gas-phase molecular beam experiments. However, in room temperature liquids or solids, the type of treatment given above does not take into account thermal fluctuations of the solvent environment that surrounds the molecules of interest.

Molecules in solvents experience constantly changing intermolecular interactions. These time-dependent solute–solvent interactions cause time-dependent fluctuations of the eigenstates and eigenvalues of the molecules of interest. Therefore, molecules that are degenerate in the absence of a solvent have randomly fluctuating energy differences. This gives rise to a $\Delta E(t)$. The range of the ΔE's that are sampled depends on the strength of the solute–solvent interactions and the temperature. At room temperature, the span of $\Delta E(t)$ can be very large. Furthermore, the solute–solute interaction in the excitation transport problem (metal–metal interaction in the electron transfer problem and the vibration–vibration interaction in the vibrational energy transfer problem) is time-dependent, that is, $\gamma(t)$.

When ΔE and γ are constants (ΔE may be 0), the excitation exhibits a well defined oscillation of probability between the two molecules. This is referred to as coherent transfer. However, when there are strong interactions with a fluctuating solvent, $\Delta E(t)$ and $\gamma(t)$ cause the probability functions, equations (8.40), to no longer behave coherently. If the majority of the time, $\Delta E(t) \gg \gamma(t)$, the probability of transfer is close to zero, even though the molecules are identical. Occasionally, $\Delta E(t) \leq \gamma(t)$. For the brief period of time for which this condition is met, probability evolves approximately as in equations (8.40). This is then followed by a period of little probability transfer until the energy gap, $\Delta E(t)$, is again small. The net result is that small amounts of probability "hop" from molecule A to molecule B, rather than the probability flowing in an oscillatory manner. The excitation can still transfer, but it makes a sudden hop rather than a continuous oscillation. The thermal fluctuations destroy the coherence of the process. This type of transfer is called incoherent. It is the type of transfer that commonly occurs at room temperature.

E. AN INFINITE SYSTEM—EXCITONS

In the system discussed above, two molecules are coupled by an intermolecular interaction. If the molecules are identical (i.e., $\delta E = 0$), there is complete transfer of an excitation (an electron or a vibration) from one molecule to another. Many systems have more than two molecules and may have so many molecules that the systems are basically infinite in extent. Crystals are a particularly important class of infinite systems because the order (symmetry) of the crystal lattice makes it possible to treat the problem of electronic excitation transport, vibrational transport, or electron transport in a straightforward manner that takes advantage of the lattice symmetry properties. In the two-molecule problem, an excitation of the system can transfer between the two molecules, and the eigenstates are comprised of two delocalized states [equation (8.28)]. The motion and the eigenstates span the entire system. In an infinite lattice of molecules or atoms, an excitation can move throughout the entire crystal, and there are an infinite number of delocalized eigenstates. For an excitation or an electron in an infinite lattice, the motion and the eigenstates span the entire system.

The nature of the eigenstates and the motion of an excitation of a perfect, infinite lattice will be illustrated with the simplest problem, a one-dimensional lattice with nearest-neighbor interactions only. Even large crystals are not actually infinite. They have surfaces that makes them imperfect in the sense that a molecule right at the surface does not have the same environment and intermolecular interactions as a molecule in the interior of the crystal. While surfaces of crystals are important in many problems in chemistry and physics, for a macroscopic crystal, the vast majority of molecules are far from a surface. A molecule in the interior of a crystal has an environment that is identical to another molecule in the lattice which can be reached by moving over an integral number of lattice sites. To avoid the surface problem in treating the bulk properties of interest here, a cyclic boundary condition is used. In a one-dimensional crystal composed of n molecules, labeled 0 to $n-1$, molecule 0 and molecule $n-1$ are taken to be adjacent. Thus, there is no surface.

1. Eigenstates for an Excitation in a One-Dimensional Perfect Lattice

The normalized ket representing the ground state of the ith molecule in the lattice is $|\varphi_i\rangle$. The excited state of this molecule is represented by the normalized ket $|\varphi_i^e\rangle$. Then the ground state of the one-dimensional lattice with n molecules and a cyclic boundary condition can be written as the product function of the ground-state kets of the individual molecules. (Here, the molecular functions are taken to be orthogonal. Actually, they are not strictly orthogonal. Taking the functions to be orthogonal does not change the nature of the results. See Chapter 17, Section B.) The ground state ket is

$$|\Phi^g\rangle = |\varphi_0\rangle\,|\varphi_1\rangle\,|\varphi_2\rangle \cdots |\varphi_{n-1}\rangle\,. \tag{8.41}$$

The ket representing the excited state of the lattice with jth molecule in the excited state and all other molecules in their ground states is

$$|\Phi_j^e\rangle = |\varphi_0\rangle\,|\varphi_1\rangle\,|\varphi_2\rangle \cdots |\varphi_j^e\rangle \cdots |\varphi_{n-1}\rangle\,. \tag{8.42}$$

Anyone of the n molecules can be the excited molecule. Therefore, there are n degenerate functions of the form equation (8.42); the only difference being the label of the molecule that is excited. If the zero of energy is taken to be the state with all of the molecules in their ground states, and if the energy of a single molecule in its excited state is E^e, then, in the absence of intermolecular interactions, the kets $|\Phi_j^e\rangle$ are a set of n-fold degenerate eigenstates. In the two-state problem with the molecules A and B identical, without the intermolecular interaction, γ, there are two degenerate states of the system, $|A\rangle$ and $|B\rangle$. With γ, the eigenstates are two orthogonal superpositions of $|A\rangle$ and $|B\rangle$ [equation (8.28)], and the degeneracy is removed [equation (8.29) and Figure 8.10]. In the crystal lattice problem, there are a very large number of degenerate states in the absence of intermolecular interactions. For simplicity, only a nearest-neighbor interaction will be considered; that is, when molecule i is in its excited state, the state is $|\Phi_i^e\rangle$, and this state interacts only with the states $|\Phi_{i+1}^e\rangle$ and $|\Phi_{i-1}^e\rangle$. Again, the strength of the intermolecular interaction is γ. Even a very small crystal will contain 10^{20} molecules. If the techniques of degenerate perturbation theory (Chapter 9, Section C) or the matrix formulation of quantum mechanics (Chapter 13) are used, it would be necessary to deal with a determinant that is $10^{20} \times 10^{20}$ to obtain the eigenvalues and eigenvectors. Clearly, this is an intractable problem using these methods, even if very fast computers are employed.

There is another approach that takes advantage of the symmetry of the crystal lattice that makes the solution possible. The nature of a lattice is that it is periodic. If a lattice is translated by a lattice spacing—α, or 2α, or 3α, and so on—the system looks identical. Starting at any point in a lattice, for a displacement by any integral number of lattice spacings, the potential is unchanged. The potential in one unit cell of the lattice is identical to the potential following any integral number of unit cell displacements. A lattice is said to have a periodic potential. Since the potential is unchanged by a lattice translation, the Hamiltonian is also unchanged. Using group theory arguments based on this symmetry property of lattices, Bloch derived what is known as the Bloch theorem of solid-state physics. If the Hamiltonian is unchanged by an any integral number of lattice translations, the eigenvectors of the Hamiltonian must also be unchanged by lattice translations. For a one-dimensional lattice, the size of the lattice is $L = \alpha n$, where n is the number of unit cells in the lattice (labeled 0 to $n - 1$). Bloch proved that for a single lattice translation the state at position $(x + \alpha)$ is related to the state at α by

$$|\psi_p(x + \alpha)\rangle = e^{2\pi i p\alpha/L} |\psi_p(x)\rangle \tag{8.43}$$

$$= e^{ik\alpha} |\psi_p(x)\rangle, \tag{8.44}$$

where p is an integer that runs from 0 to $n - 1$, and $k = 2\pi p/L$. This result is true for any number of lattice translations, so α is replaced by $j\alpha$ on both sides of the equations, where j is an integer that runs from 0 to $n - 1$. Since any number of lattice translations produces an equivalent function, the final result is a superposition of the kets with each of the n possible translations:

$$|\psi(k)\rangle = \frac{1}{\sqrt{n}} \sum_{j=0}^{n-1} e^{ik\alpha j} |\Phi_j^e\rangle \tag{8.45}$$

In equation (8.45), the ket from equation (8.42) has been used explicitly. Equation (8.45) is a superposition of the kets $|\Phi_j^e\rangle$ in which there is an equal amplitude for each value of j in the superposition. $e^{ik_\alpha j}$ is a position dependent phase factor. There are n different orthonormal $|\psi(k)\rangle$ arising from the n different values of the integer p, which give n different values of k. The $|\psi(k)\rangle$ differ by the number of nodes contained in the function. The number of nodes ranges from zero, the $k = 0$ state, to n nodes, one between each molecule. k is the lattice wave vector.

To illustrate the nature of equation (8.45), consider the two-molecule problem discussed at the beginning of the chapter. For this problem, $n = 2$, so p can take on values 0 and 1, and j can take on values 0 and 1. $L = 2\alpha$. k can have two values. For $p = 0$, $k = 0$, and for $p = 1$, $k = 2\pi/2\alpha$. Substituting in equation (8.45), the $k = 0$ ket is

$$|\psi(k=0)\rangle = \frac{1}{\sqrt{2}} \sum_{j=0}^{1} |\Phi_j^e\rangle \tag{8.46a}$$

$$= \frac{1}{\sqrt{2}} \left(|\Phi_0^e\rangle + |\Phi_1^e\rangle \right). \tag{8.46b}$$

Equation (8.46) is the sum of two kets. $|\Phi_0^e\rangle$ has molecule 0 excited and molecule 1 in the ground state. $|\Phi_1^e\rangle$ has molecule 1 excited and molecule 0 in the ground state. These kets are the same as kets $|A\rangle$ and $|B\rangle$, respectively, in the two-state problem. Equation (8.46) is the same as the two-state problem eigenket $|+\rangle$ given in equation (8.28a). The ket with $k = \pi/\alpha$ is

$$|\psi(k=\pi/\alpha)\rangle = \frac{1}{\sqrt{2}} \sum_{j=0}^{1} e^{i\pi j} |\Phi_j^e\rangle \tag{8.47a}$$

$$= \frac{1}{\sqrt{2}} \left(|\Phi_0^e\rangle - |\Phi_1^e\rangle \right), \tag{8.47b}$$

since for $j = 1$, the exponential is $\exp(i\pi) = -1$. Equation (8.47) is the same as the two-state problem eigenket $|-\rangle$ given in equation (8.28b).

The Bloch theorem states that the eigenkets of a Hamiltonian with a periodic potential must have the form given in equation (8.45). The two-state problem is the smallest such problem, but the result applies to any number of lattice sites with a cyclic (periodic) boundary condition. Equation (8.45) is for one dimension, but the Bloch theorem also applies in three dimensions. The exponential contains three lattice constants, the sum is over three lattice indices, and k (usually called the wave vector) is a three-dimensional vector.

For the one-dimensional lattice problem with nearest-neighbor interactions only, the Hamiltonian operator can be written as

$$\underline{H} = \underline{H}_M + \underline{H}_{j,j\pm 1}, \tag{8.48}$$

where $\underline{H}_M$ is the Hamiltonian for the molecules in the absence of intermolecular interactions, and $\underline{H}_{j,j\pm 1}$ is the nearest-neighbor intermolecular interaction contribution to the Hamiltonian. $\underline{H}_{j,j\pm 1}$ couples the state with molecule j excited and all other molecules in their ground states to the state in which molecule $j+1$ is excited with all other molecules in their ground states, and to the state in which molecule $j-1$ is excited with all other molecules in their ground states. $\underline{H}_M$ is the sum of the Hamiltonians for each molecule:

$$\underline{H}_M = \underline{H}_{M_1} + \underline{H}_{M_2} + \cdots + \underline{H}_{M_j} + \cdots + \underline{H}_{M_{n-1}}. \tag{8.49}$$

$\underline{H}_M$ operating on the ket $|\Phi_j^e\rangle$, in which the jth molecules are excited, gives

$$\underline{H}_M |\Phi_j^e\rangle = E^e |\Phi_j^e\rangle, \tag{8.50}$$

because the ground-state energy was taken to be zero. Then,

$$\underline{H}_M \left|\psi(k)\right\rangle = \underline{H}_M \frac{1}{\sqrt{n}} \sum_{j=0}^{n-1} e^{ik\alpha j} \left|\Phi_j^e\right\rangle \tag{8.51a}$$

$$= \frac{1}{\sqrt{n}} \sum_{j=0}^{n-1} e^{ik\alpha j} \underline{H}_M \left|\Phi_j^e\right\rangle \tag{8.51b}$$

$$= E^e \left|\psi(k)\right\rangle . \tag{8.51c}$$

In the absence of intermolecular interactions, the energy is equal to the energy of the single excited state in the system and is independent of the wave vector k. The system is still n-fold degenerate.

As in the two-state problem, the inclusion of intermolecular interactions breaks the excited state degeneracy. $\underline{H}_{j,j\pm1}$ operating on the ket, $|\Phi_j^e\rangle$, in which the jth molecule is excited, gives

$$\underline{H}_{j,j\pm1} \left|\Phi_j^e\right\rangle = \gamma \left|\Phi_{j+1}^e\right\rangle + \gamma \left|\Phi_{j-1}^e\right\rangle , \tag{8.52}$$

where γ is the strength of the nearest-neighbor coupling. Then

$$\underline{H}_{j,j\pm1} \left|\psi(k)\right\rangle = \underline{H}_{j,j\pm1} \frac{1}{\sqrt{n}} \sum_{j=0}^{n-1} e^{ik\alpha j} \left|\Phi_j^e\right\rangle \tag{8.53a}$$

$$= \frac{1}{\sqrt{n}} \sum_{j=0}^{n-1} e^{ik\alpha j} \underline{H}_{j,j\pm1} \left|\Phi_j^e\right\rangle \tag{8.53b}$$

$$= \frac{1}{\sqrt{n}} \sum_{j=0}^{n-1} \left[e^{ik\alpha j} \gamma \left|\Phi_{j+1}^e\right\rangle + e^{ik\alpha j} \gamma \left|\Phi_{j-1}^e\right\rangle \right] . \tag{8.53c}$$

Each of the terms in the square brackets can be multiplied by

$$e^{ik\alpha} e^{-ik\alpha} = 1,$$

which gives

$$\underline{H}_{j,j\pm1} \left|\psi(k)\right\rangle = \frac{1}{\sqrt{n}} \sum_{j=0}^{n-1} \left[e^{ik\alpha j} e^{ik\alpha} e^{-ik\alpha} \gamma \left|\Phi_{j+1}^e\right\rangle + e^{ik\alpha j} e^{-ik\alpha} e^{ik\alpha} \gamma \left|\Phi_{j-1}^e\right\rangle \right] \tag{8.54a}$$

$$= \frac{1}{\sqrt{n}} \sum_{j=0}^{n-1} \left[e^{-ik\alpha} \gamma e^{ik\alpha(j+1)} \left|\Phi_{j+1}^e\right\rangle + e^{ik\alpha} \gamma e^{ik\alpha(j-1)} \left|\Phi_{j-1}^e\right\rangle \right] . \tag{8.54b}$$

In equation (8.54b), one of the kets has the index $j+1$, and it is multiplied by the exponential with the index $j+1$. The other ket has the index $j-1$, and it is multiplied by the exponential with the index $j-1$. Because of the cyclic boundary condition, the sum over all j is still a sum over all possible locations for the excited state. In the first term in the square brackets, when $j = n - 1$, the index is one greater than the maximum, which is the location 0. In the second term in the square bracket, when $j = 0$, the index is one less than the minimum, which is the location $n - 1$. Thus, including the normalization constant, the sum over j of each ket with its associated j-dependent exponential combine to give a factor of $|\psi(k)\rangle$. Rewriting this result gives

$$\underline{H}_{j,j\pm1}\,|\psi(k)\rangle = e^{-ik\alpha}\gamma\,|\psi(k)\rangle + e^{ik\alpha}\gamma\,|\psi(k)\rangle \tag{8.55a}$$

$$= \gamma(e^{ik\alpha} + e^{-ik\alpha})\,|\psi(k)\rangle \tag{8.55b}$$

$$= 2\gamma\cos(k\alpha)\,|\psi(k)\rangle\,. \tag{8.55c}$$

Therefore, the $|\psi(k)\rangle$ are eigenkets of the total Hamiltonian [equation (8.48)].

Combining equations (8.55c) and (8.51c) gives the wave-vector-dependent energies of the eigenstates as

$$E(k) = E^e + 2\gamma\cos(k\alpha). \tag{8.56}$$

k runs over a full cycle, which is taken to be from $-\pi/\alpha$ to π/α. The energy ranges from $E^e - 2\gamma$ to $E^e + 2\gamma$. Therefore, there is a band of states with bandwidth 4γ. For n molecules in the lattice, there are n states in the band, each with a quantum number k. The states k and $-k$ are degenerate. These excitations are called excitons, and the band of energies is called an exciton band. Figure 8.12 is a plot of the exciton energy versus k. In Figure 8.12, the values of k run from $-\pi/\alpha$ to π/α. k has units of inverse length, and it plays an important role in the description of phenomena associated with crystals, such as x-ray diffraction. In many contexts, k is referred to as a reciprocal lattice vector, and it defines a point on a reciprocal lattice. For a three-dimensional lattice, k is a three-dimensional vector in reciprocal space. The range of k plotted in the figure is called the first Brillouin zone.

Each k-state is a delocalized excitation of the lattice with equal probability of finding the excitation on any molecule. The delocalized exciton states are analogous to the eigenstates of the two state problem [equation (8.28)], in which the probability of finding the excitation on either of the molecules is equal. The excitons are delocalized probability amplitude waves. The intermolecular interactions change the nature of an excitation from a excited state localized on a single molecule to an excitation that is spread out over the entire lattice. The wave vector k ranges in value from $k = 0$ to $k = \pm\pi/\alpha$. The $k = 0$ state has no nodes and the $k = \pm\pi/\alpha$ states have n nodes with a nodal spacing equal to the lattice spacing. If the time-dependent phase factors are included (Chapter 3, Section D and Chapter 5, Section A), the exciton waves have a phase velocity, and the difference between $+k$ and $-k$ is the direction of the phase velocity. The exciton probability amplitude waves are analogous to the momentum eigenstates of a free particle (Chapter 3); but instead of being delocalized over all space and having a continuous range of values of k, the exciton states are delocalized over the whole lattice and the values of k are discrete. However, even for a small crystal with 10^{20} molecules, the number of states in the band is so large that the band is effectively continuous.

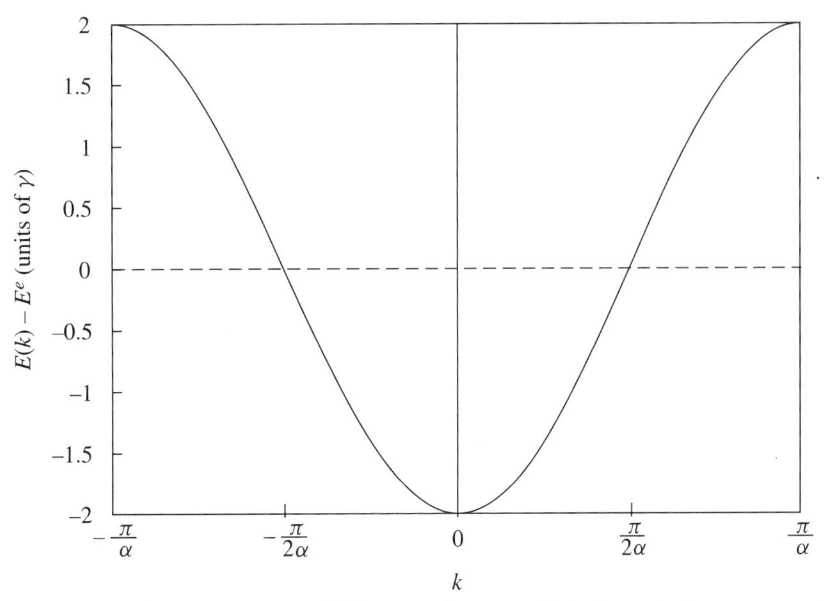

Figure 8.12 The exciton energy versus k. The zero of energy, E^e, is the excited-state energy in the absence intermolecular interactions. γ is taken to be negative, so that $k = 0$ is at the bottom of the band. The bandwidth is 4γ.

2. Exciton Transport

Because of the similarities between the natures of the exciton eigenstates and the eigenstates of true particles such as electrons or photons, excitons are considered to be quasi-particles with a quasi-momentum $\hbar k$. Like the free particles discussed in Chapter 3, exciton k states can be superimposed to form wave packets. The exciton wave packets are more or less localized and can move through the crystal lattice with a group velocity, V_g. For the one-dimensional lattice with nearest-neighbor interactions, the dispersion relation is obtained from $E(k)$ [equation (8.56)]:

$$\omega(k) = \frac{1}{\hbar}E(k) = \frac{1}{\hbar}[E^e + 2\gamma\cos(k\alpha)]. \tag{8.57}$$

Then

$$V_g = -\frac{2\gamma\alpha}{\hbar}\sin(k\alpha). \tag{8.58}$$

Since γ is an energy and α is a distance, V_g has units of velocity. The sign of the group velocity is determined by the sign of k. The excitons can move along the lattice in a positive or negative direction. The maximum velocity occurs when the slope of the dispersion is maximum. From the expression for the group velocity and from Figure 8.12, it can be seen that the maximum group velocity occurs at $k = \pm\pi/2\alpha$, the middle of the band, and that

V_g is zero at the top and bottom of the band. In a three-dimensional lattice, V_g is a vector. In general there will be a different γ and α for each direction.

The results show that an exciton can move through a crystal lattice like a particle. Except for the form of the dispersion relation, the transport is completely analogous to the motion of a free particle. If the excitation of the lattice is a molecular vibration rather than an electronic excited state, then the quasi-particle is called a vibron, and the band is a vibron band. An electron in a insulating lattice, such as a molecular crystal, is called a polaron, and the resulting band is a polaron band. While the nature of the interactions that give rise to γ will differ, and the size of γ will differ, the treatments of these various quasi-particles are identical. Very similar analysis is also used to describe the states of electrons in metals. The bands that that are responsible for electrical conductivity are called conduction bands. In all cases, the lattice symmetry gives rise to the energy bands in crystals.

The exciton motion with a well-defined group velocity is referred to as coherent transport. Such transport depends on the k-states that comprise the wave packet maintaining a well-defined phase relationship. Coherent exciton transport occurs at very low temperature ($\sim$1 K) in molecular crystals. However, the same considerations that were discussed at the end of Section D apply to exciton transport. As the temperature is increased, thermal excitation of the lattice increases. The mechanical motions of the lattice are also described as quasi particles called phonons. Lattice motions cause the site energy, E^e, and the intermolecular interactions to vary in time. The influence of phonons on excitons or electrons is referred to as phonon scattering. Phonon scattering destroys the coherent wave packet character of the quasi-particle (or real particle in the case of electrons). When the rate of scattering of the quasi-particle by phonons becomes fast compared to the time required to move a single lattice site, the quasi-particle is localized. It is described as a superposition of all k states, which localizes it on a single site, but it can still move by hopping from one lattice site to another. This is referred to as incoherent transport. The quasi-particle will execute a random walk on the lattice. This is the nature of exciton or electron transport on a room temperature lattice. When a voltage is applied to a metal wire at room temperature, the electrons acquire a slight bias in their random walk, moving them in the direction of the positive electrode. The electric field accelerates the electrons, but rapid electron–phonon scattering dissipates the electron velocity, transferring the energy to the crystal lattice. The result is resistive heating of the wire. The interatomic interactions give rise to the conduction band and electron transport, but electron–phonon scattering produces incoherent transport and electrical resistance. In some metals at low temperatures, a complex type of electron–phonon interaction mitigates electron–phonon scattering. The result is a type of coherent electron transport referred to as superconductivity.

CHAPTER 9

PERTURBATION THEORY

In the last several chapters, a number of problems were treated that can be solved exactly. However, for almost all problems that require quantum mechanics to describe real systems, it is not possible to solve eigenvalue problems exactly to obtain observables. For this reason, a wide variety of approximation techniques have been developed, and the development of such techniques is an ongoing field of research. The need to solve problems approximately arises in classical mechanics as well as in quantum mechanics. It is not possible to solve a general three body problem in either classical or quantum theory.

The fact that approximation methods are used does not necessarily mean that the results are inaccurate. Furthermore, in many cases, the results of an approximate calculation may be important in explaining qualitatively how nature works even when highly accurate predictions of observables are not possible. One of the approaches for the approximate solution of quantum mechanical problems is called perturbation theory. Perturbation theory involves a hierarchy of methods. Perturbation theory can be carried out to various orders—for example, first order, second order, third order, and so on. By treating problems to sufficiently high order, it is possible, in principle, to obtain accurate results. Methods have been developed for performing infinite-order perturbation theory, although the results are still approximate. Infinite-order treatments can prevent divergences that can occur when a series expansion is truncated at finite order.

In this chapter, the method of perturbation theory will be developed for nondegenerate and degenerate states. The treatment will involve the energy eigenvalue problem, but the identical method can be used for any eigenvalue problem. Results for first- and second-order perturbation theory are given and several examples are presented. In the next chapter, perturbation theory will be used to treat the helium atom, and the results will then be improved using an additional procedure based on the variational method.

A. PERTURBATION THEORY FOR NONDEGENERATE STATES

Consider the eigenvalue problem

$$\underline{H}\,|\varphi_n\rangle = E_n\,|\varphi_n\rangle\,. \tag{9.1}$$

$\underline{H}$ is known, but the eigenvalue problem, which yields the eigenvalues, E_n, and the kets, $|\varphi_n\rangle$, cannot be solved mathematically. To apply perturbation theory, it is necessary for the Hamilton to be expressible in the form

$$\underline{H} = \underline{H}^0 + \lambda\underline{H}' + \lambda^2\underline{H}'' + \cdots . \tag{9.2}$$

$\underline{H}^0$ is called the zeroth-order Hamiltonian. λ is an expansion parameter. In the limit that λ goes to zero, the problem reduces to

$$\underline{H}^0 \left|\varphi_n^0\right\rangle = E_n^0 \left|\varphi_n^0\right\rangle . \tag{9.3}$$

The essential feature of equation (9.3) is that it must be exactly solvable. Therefore, the zeroth-order eigenvalues, E_n^0, and the zeroth-order eigenkets, $\left|\varphi_n^0\right\rangle$, are known exactly. Initially, nondegenerate perturbation theory will be developed. Thus the states, $\left|\varphi_n^0\right\rangle$, are nondegenerate; no two of them have the same eigenvalues, E_n^0. Equation (9.3) is referred to as the unperturbed eigenvalue problem.

$$\lambda\underline{H}' + \lambda^2\underline{H}'' + \cdots \tag{9.4}$$

are the perturbation pieces of the Hamiltonian, $\underline{H}$. $\underline{H}'$ is the first-order part of the perturbation because it is first order in λ. $\underline{H}''$ is the second-order part of the perturbation because it is second order in λ; that is, its coefficient in the expansion of $\underline{H}$ in powers of λ is λ^2, and so forth, for higher-order perturbations.

An example of a problem that can be treated with perturbation theory is the influence of an E field on the hydrogen atom (the Stark effect for the hydrogen atom). The E field is the expansion parameter. When $E \to 0$, the problem reduces to the normal H atom. The solutions to the H atom eigenvalue problem are known. The H atom Hamiltonian, eigenvalues, and wavefunctions are the necessary zeroth-order problem that can be solved exactly. When an E field is applied, the additional terms in the Hamiltonian are the perturbations.

The eigenkets, which are obtained from the solution to the zeroth-order problem, $\left|\varphi_n^0\right\rangle$—that is, $\left|\varphi_0^0\right\rangle,\ \left|\varphi_1^0\right\rangle,\ \left|\varphi_2^0\right\rangle \cdots$ [equation (9.3)]—are a complete orthonormal set of functions,

$$\left\langle\varphi_n^0 \mid \varphi_m^0\right\rangle = \delta_{mn}, \tag{9.5}$$

with eigenvalues $E_0^0,\ E_1^0,\ E_2^0,\ \ldots$. The nature of a perturbation is that it is small. That is, the wavefunctions and eigenvalues will not be far removed from the zeroth-order wavefunctions and eigenvalues. The application of a small perturbation will not cause a large change in a system. The wavefunctions and the eigenvalues, which are the solutions to equation (9.1), are expanded as

$$\left|\varphi_n\right\rangle = \left|\varphi_n^0\right\rangle + \lambda\left|\varphi_n'\right\rangle + \lambda^2\left|\varphi_n''\right\rangle + \cdots \tag{9.6}$$

$$E_n = E_n^0 + \lambda E_n' + \lambda^2 E_n'' + \cdots . \tag{9.7}$$

If the perturbation is small, the series may converge rapidly. Equations (9.2), (9.6), and (9.7) are series expansion for $\underline{H}$, $\left|\varphi_n\right\rangle$, and E_n. These are substituted into the eigenvalue equation, equation (9.1), and after collecting like powers of λ the result is

$$\begin{aligned}&\left(\underline{H}^0\left|\varphi_n^0\right\rangle - E_n^0\left|\varphi_n^0\right\rangle\right) + \left(\underline{H}^0\left|\varphi_n'\right\rangle + \underline{H}'\left|\varphi_n^0\right\rangle - E_n^0\left|\varphi_n'\right\rangle - E_n'\left|\varphi_n^0\right\rangle\right)\lambda \\ &+ \left(\underline{H}^0\left|\varphi_n''\right\rangle + \underline{H}'\left|\varphi_n'\right\rangle + \underline{H}\left|\varphi_n^0\right\rangle - E_n^0\left|\varphi_n''\right\rangle - E_n'\left|\varphi_n'\right\rangle - E_n''\left|\varphi_n^0\right\rangle\right)\lambda^2 \\ &+ \cdots = 0.\end{aligned} \tag{9.8}$$

The first term in brackets is zeroth order in λ. The second term in brackets is first order in λ. The third term in brackets is second order in λ, and so forth.

Equation (9.8) is a series in powers of λ. For this series to be equal to zero for any value of λ, the coefficients of the individual powers of λ must equal zero. This condition produces a set of equations, one for each coefficient. The zeroth-order equation is

$$\underline{H}^0 \left|\varphi_n^0\right\rangle - E_n^0 \left|\varphi_n^0\right\rangle = 0. \tag{9.9}$$

The first-order equation is

$$\underline{H}^0 \left|\varphi_n'\right\rangle + \underline{H}' \left|\varphi_n^0\right\rangle - E_n^0 \left|\varphi_n'\right\rangle - E_n' \left|\varphi_n^0\right\rangle = 0. \tag{9.10}$$

The second-order equation is

$$\underline{H}^0 \left|\varphi_n''\right\rangle + \underline{H}' \left|\varphi_n'\right\rangle + \underline{H} \left|\varphi_n^0\right\rangle - E_n^0 \left|\varphi_n''\right\rangle - E_n' \left|\varphi_n'\right\rangle - E_n'' \left|\varphi_n^0\right\rangle = 0. \tag{9.11}$$

Equations such as these can be written for any order in λ. The solutions to the zeroth-order equation, equation (9.9), are known. Knowledge of the solutions to the zeroth-order equation is a necessary condition for the development and application of perturbation theory.

1. The First-Order Solutions

a. *Correction to the Energies*

Equation (9.10) can be rewritten as

$$\underline{H}^0 \left|\varphi_n'\right\rangle - E_n^0 \left|\varphi_n'\right\rangle = (E_n' - \underline{H}') \left|\varphi_n^0\right\rangle. \tag{9.12}$$

In this equation, $\underline{H}^0$, the $\left|\varphi_n^0\right\rangle$, and the E_n^0 are known. It is necessary to obtain expressions for E_n', the first-order correction to the energy, and $\left|\varphi_n'\right\rangle$, the first-order correction to the eigenket. $\left|\varphi_n'\right\rangle$ can be expanded in the complete orthonormal set of zeroth-order wavefunctions,

$$\left|\varphi_n'\right\rangle = \sum_i c_i \left|\varphi_i^0\right\rangle. \tag{9.13}$$

This expansion is substituted into equation (9.12). First, substituting into the first term on the left-hand side of equation (9.12) gives

$$\underline{H}^0 \left|\varphi_n'\right\rangle = \sum_i c_i \underline{H}^0 \left|\varphi_i^0\right\rangle = \sum_i c_i E_i^0 \left|\varphi_i^0\right\rangle. \tag{9.14}$$

The right-hand side arises because the $\left|\varphi_i^0\right\rangle$ are eigenkets of $\underline{H}^0$; therefore, operating with $\underline{H}^0$ yields the eigenvalues, E_i^0. Substituting the series expansion, equation (9.13), and the result, equation (9.14), into equation (9.12) yields

$$\sum_i c_i \left(E_i^0 - E_n^0\right) \left|\varphi_i^0\right\rangle = \left(E_n' - \underline{H}'\right) \left|\varphi_n^0\right\rangle. \tag{9.15}$$

Left-multiplying by $\left\langle\varphi_n^0\right|$ gives

$$\left\langle \varphi_n^0 \right| \sum_i c_i \left(E_i^0 - E_n^0\right) \left|\varphi_i^0\right\rangle = \left\langle \varphi_n^0 \right| \left(E_n' - \underline{H}'\right) \left|\varphi_n^0\right\rangle. \tag{9.16}$$

Since the center of the bracket on the left-hand side contains only numbers, they can be brought out of the bracket to give

$$\sum_i c_i \left(E_i^0 - E_n^0\right) \left\langle \varphi_n^0 \mid \varphi_i^0 \right\rangle = \left\langle \varphi_n^0 \right| \left(E_n' - \underline{H}'\right) \left|\varphi_n^0\right\rangle. \tag{9.17}$$

Because the zeroth-order states are not degenerate, the left-hand side of equation (9.17) is zero. If $i \neq n$, then

$$\left\langle \varphi_n^0 \mid \varphi_i^0 \right\rangle = 0$$

by orthogonality. If $i = n$, the bracket is nonzero, but

$$E_n^0 - E_n^0 = 0.$$

For any i, one term or the other on the left-hand side of equation (9.17) is zero. Thus equation (9.17) becomes

$$\left\langle \varphi_n^0 \right| \left(E_n' - \underline{H}'\right) \left|\varphi_n^0\right\rangle = 0. \tag{9.18}$$

Therefore, the first-order correction to the energy is

$$E_n' = \left\langle \varphi_n^0 \right| \underline{H}' \left|\varphi_n^0\right\rangle, \tag{9.19}$$

and the energy to first order is

$$E_n = E_n^0 + \lambda E_n'.$$

Usually, the expansion parameter is included as part of $\underline{H}'$; then λ becomes part of E_n', and the energy, to first order, is written as

$$E_n = E_n^0 + E_n'. \tag{9.20}$$

The bracket in equation (9.19) is frequently abbreviated as

$$H_{nn}' = \left\langle \varphi_n^0 \right| \underline{H}' \left|\varphi_n^0\right\rangle, \tag{9.21}$$

and λ is included in $\underline{H}'$.

Equation (9.19) shows that the first-order correction to the energy for the zeroth-order state, $\left|\varphi_i^0\right\rangle$, is the expectation value of the perturbation. Thus, the first-order correction to the energy is the first-order perturbation piece of the Hamiltonian averaged over the zeroth-order state of the system. Since the $\left|\varphi_i^0\right\rangle$are known and $\underline{H}'$ is known, the first-order correction to the energy can be calculated.

b. Correction to the Eigenkets

The first-order correction to the eigenkets can also be found in terms of the zeroth-order kets by determining the coefficients, c_i, in equation (9.13). Knowing the c_i gives the first-order correction to the eigenkets as a series in the known zeroth-order kets. Employing equation (9.15) again, but this time left-multiplying by $\langle \varphi_j^0 |$, gives

$$\langle \varphi_j^0 | \sum_i c_i (E_i^0 - E_n^0) | \varphi_i^0 \rangle = \langle \varphi_j^0 | (E_n' - \underline{H}') | \varphi_n^0 \rangle . \tag{9.22}$$

In summing over i on the left-hand side of the equation, the orthogonality of the bra and the ket will yield zero for all terms except $i = j$. For the $i = j$ term, the equation becomes

$$c_j (E_j^0 - E_n^0) = \langle \varphi_j^0 | (E_n' - \underline{H}') | \varphi_n^0 \rangle , \tag{9.23}$$

since the center of the bracket contains only numbers and normalization yields one for the scalar product of the bra and ket on the left-hand side. Since E_n' is a number, its operation on the ket does not change it, and orthogonality of the bra and ket gives zero for this piece of the right-hand side. The result is

$$c_j (E_j^0 - E_n^0) = - \langle \varphi_j^0 | \underline{H}' | \varphi_n^0 \rangle . \tag{9.24}$$

Then the desired coefficients needed to define the $|\varphi_n'\rangle$ in terms of the zeroth-order kets $|\varphi_i^0\rangle$ using equation (9.13) are

$$c_j = \frac{\langle \varphi_j^0 | \underline{H}' | \varphi_n^0 \rangle}{(E_n^0 - E_j^0)} , \qquad j \neq n. \tag{9.25}$$

This is frequently written in abbreviated form as

$$c_j = \frac{H_{jn}'}{(E_n^0 - E_j^0)} , \tag{9.26}$$

where H_{jn}' represents the bracket with the operator $\underline{H}'$ and the bra $\langle \varphi_j^0 |$ and the ket $|\varphi_n^0\rangle$.

The eigenket, corrected to first order, requires the sum given in equation (9.13). Therefore, to first order, the nth eigenket is

$$|\varphi_n\rangle = |\varphi_n^0\rangle + \sum_j{}' \frac{H_{jn}'}{(E_n^0 - E_j^0)} |\varphi_j^0\rangle , \tag{9.27}$$

where the primed summation means $j \neq n$. λ has been absorbed into $\underline{H}_{jn}'$. The $|\varphi_n\rangle$ are normalized to first order. In principle, the $|\varphi_n\rangle$ contain all of the zeroth-order kets because the sum is over all $j \neq n$ and the first term on the right-hand side is $|\varphi_n^0\rangle$. However, $\underline{H}_{jn}'$ may be zero for many values of j. This situation occurs in the example given below of a quartic perturbation of a harmonic oscillator. The energy difference $(E_n^0 - E_j^0)$, which appears in the denominator of equation (9.27) is frequently referred to as the energy denominator. For states j far removed from the state n of interest, the energy denominator may become so large that higher terms in the sum are negligible. Thus it may be possible, under the appropriate conditions, to truncate the sum even though the H_{jn}' are nonzero.

2. Second-Order Correction to the Energies and the Eigenkets

The second-order corrections to the energies and the eigenkets are obtained in a manner that is analogous to the method used to develop the first-order corrections. Equation (9.11), which is the coefficient of λ^2 in the expansion equation (9.8), is employed. Both $|\varphi_n'\rangle$ and $|\varphi_n''\rangle$ are expanded in the complete set of zeroth-order eigenkets, as in equation (9.13) and substituted into equation (9.8). As in the first-order problem, this expression is manipulated to provide solutions for the unknowns.

After considerable algebra, the second-order correction to the energy is

$$E_n'' = \lambda^2 \sum_i{}' \frac{H_{ni}' H_{in}'}{(E_n^0 - E_i^0)} + \lambda^2 H_{nn}''. \tag{9.28}$$

The prime on the sum indicates that the summation excludes $i = n$. The first term is the second-order correction arising from the first-order piece of the Hamiltonian. There are two brackets containing $\underline{H}'$, each of which contribute a λ, making the term over all second order. The second term occurs if there is an additional part of the Hamiltonian which is explicitly second order in the perturbation parameter λ. For example, there is a first-order and second-order Stark effect. The expansion parameter for the influence of an external electric field on a system is E, the strength of the electric field. When E goes to zero, the system is unperturbed. The first-order Stark effect is linear in E. However, it is possible to calculate the influence of the first-order Stark effect to second or higher orders of perturbation theory. At second order, the first-order Stark effect will give rise to the first term of equation (9.28). The second-order Stark effect is explicitly second order in the E field; that is, the energy depends on E^2. This does not have a first-order contribution, but at second order it gives rise to the second term in equation (9.28). Usually the expansion parameter for each order is included as part of the Hamiltonian for that order. Then, the second-order correction to the energy is given as

$$E_n'' = \sum_i{}' \frac{H_{ni}' H_{in}'}{(E_n^0 - E_i^0)} + H_{nn}'' \tag{9.29}$$

The second-order correction to the eigenket is

$$|\varphi_n''\rangle = \sum_k{}' \left[\sum_m{}' \frac{H_{km}' H_{mn}'}{(E_n^0 - E_k^0)(E_n^0 - E_m^0)} - \frac{H_{nn}' H_{kn}'}{(E_n^0 - E_k^0)^2} \right] |\varphi_k^0\rangle$$

$$+ \sum_k{}' \frac{H_{kn}''}{(E_n^0 - E_k^0)} |\varphi_k^0\rangle. \tag{9.30}$$

λ^2 has been absorbed into the first- and second-order pieces of the Hamiltonian.

Combining the zeroth-, first-, and second-order terms, the energies and the eigenkets to second order are

$$E = E^0 + H_{nn}' + \sum_i{}' \frac{H_{ni}' H_{in}'}{(E_n^0 - E_i^0)} + H_{nn}'' \tag{9.31}$$

$$
\begin{aligned}
|\varphi_n\rangle = {} & |\varphi_n^0\rangle + \sum_j{}' \frac{H'_{jn}}{(E_n^0 - E_j^0)} |\varphi_j^0\rangle \\
& + \sum_k{}' \left[\sum_m{}' \frac{H'_{km} H'_{mn}}{(E_n^0 - E_k^0)(E_n^0 - E_m^0)} - \frac{H'_{nn} H'_{kn}}{(E_n^0 - E_k^0)^2} \right] |\varphi_k^0\rangle \\
& + \sum_k{}' \frac{H''_{kn}}{(E_n^0 - E_k^0)} |\varphi_k^0\rangle .
\end{aligned} \tag{9.32}
$$

B. EXAMPLES—PERTURBED HARMONIC OSCILLATOR AND THE STARK EFFECT FOR THE RIGID PLANE ROTOR

1. Perturbed Harmonic Oscillator

In Chapter 6, the energy eigenvalue problem for the quantum harmonic oscillator was solved exactly. The eigenvalues and eigenkets were found. The vibrational modes of molecules are also quantum oscillators. However, they are not strictly harmonic. Figure 9.1 is a diagram of the potential energy of a diatomic molecule. The dashed line is the energy of the atoms separated at infinity; that is, there is no chemical bond. x is the separation of the atoms. The value of x at the minimum is the equilibrium bond length. Near the bottom of the well, the potential is approximately harmonic. If molecular vibrations were harmonic oscillators, there would be equally spaced vibrational energy levels, the eigenvalues of the harmonic oscillator. However, in real molecules, the levels are not equally spaced.

One approach for going beyond the harmonic oscillator as a model for molecular vibrations is to expand the potential in powers of x so that there will be contributions to

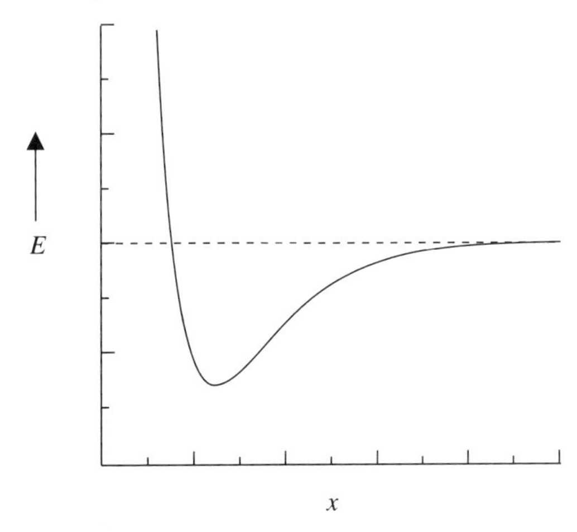

Figure 9.1 Potential energy for a vibrational mode of a diatomic molecule. Near the bottom, the potential is approximately harmonic.

the potential energy greater than the x^2 harmonic term. Perturbation theory can be used to determine the effect of adding anharmonicity to the potential. To illustrate the nature of an anharmonic potential on molecular vibrations and the application of perturbation theory, the energy to first order for the Hamiltonian

$$\underline{H} = \frac{p^2}{2m} + \frac{1}{2}k\underline{x}^2 + c\underline{x}^3 + q\underline{x}^4 \tag{9.33}$$

will be found. The first term is the kinetic energy, and the second term is the harmonic oscillator potential arising from a Hooke's law form of the force (see Chapter 6). The third and fourth terms are cubic and quartic contributions to the potential, respectively. c is the cubic force constant, and q is the quartic force constant. c and q play the role of the expansion parameter λ. If c and q go to zero, the harmonic oscillator problem is recovered.

The zeroth-order Hamiltonian is the harmonic oscillator Hamiltonian

$$\underline{H}^0 = \frac{p^2}{2m} + \frac{1}{2}k\underline{x}^2,$$

which can be written using raising and lowering operators [equation (6.88)] as

$$\underline{H}^0 = \frac{1}{2}\hbar\,\omega(\underline{a}\,\underline{a}^+ + \underline{a}^+\underline{a}). \tag{9.34}$$

The zeroth-order kets are the eigenkets $|n\rangle$ found in Chapter 6. The zeroth-order energy is

$$E^0 = \left(n + \frac{1}{2}\right)\hbar\omega_0. \tag{9.35}$$

The subscript zero on ω_0 indicates that this is the frequency of the oscillator to zeroth order. The first-order perturbation piece of the Hamiltonian is

$$\underline{H}' = c\underline{x}^3 + q\underline{x}^4. \tag{9.36}$$

The first-order correction to the energy is

$$\begin{aligned} H'_{nn} &= \langle n|\,\underline{H}'\,|n\rangle \\ &= \langle n|\,c\underline{x}^3 + q\underline{x}^4\,|n\rangle \\ &= c\,\langle n|\,\underline{x}^3\,|n\rangle + q\,\langle n|\,\underline{x}^4\,|n\rangle\,. \end{aligned} \tag{9.37}$$

The operator $\underline{x}$ in terms of raising and lower operators is given in equation (6.92), that is,

$$\underline{x} = \left(\frac{\hbar\omega_0}{2k}\right)^{1/2}(\underline{a} + \underline{a}^+). \tag{9.38}$$

Consider first the cubic term of the perturbation. The operator $\underline{x}^3$ is

$$\underline{x}^3 \propto (\underline{a} + \underline{a}^+)^3.$$

$\underline{x}^3$ will have terms involving

$$\underline{a}^3,\ \underline{a}^2\underline{a}^+,\ \underline{a}\,\underline{a}^+\underline{a},\ \ldots,\underline{a}^{+3}.$$

There are no terms with the same number of raising and lowering operators, a^+ and a. When $\underline{x}^3$ operates on the $|n\rangle$ in equation (9.37), the results are $|n+3\rangle$, $|n+1\rangle$, $|n-1\rangle$, and $|n-3\rangle$. The closed bracket is $\langle n|m\rangle$ with $m \neq n$. Since the zeroth-order eigenkets are orthogonal, this bracket vanishes. Therefore,

$$\langle n|\,\underline{x}^3\,|n\rangle = 0. \tag{9.39}$$

The first-order correction to the energy arising from the cubic perturbation is zero. However, equation (9.27) shows that the cubic perturbation does contribute a first-order correction to the eigenket because the correction involves a sum over all zeroth-order kets. Although the sum is over the infinite set of harmonic oscillator eigenkets, the first-order cubic correction is nonzero only for the four kets given immediately above.

The first-order quartic correction to the energy is

$$\langle n|\,\underline{x}^4\,|n\rangle = \frac{\hbar^2\omega_0^2}{4k^2}\,\langle n|\,(\underline{a}+\underline{a}^+)^4\,|n\rangle\,. \tag{9.40}$$

$(\underline{a}+\underline{a}^+)^4$ has terms that range from ones with four raising operators to ones with four lowering operators. Some of the terms have the same number of raising and lowering operators. Therefore, $\langle n|\,\underline{x}^4\,|n\rangle \neq 0$. The nonzero terms can be evaluated using the expressions for the result of operating a raising or lowering operator on a ket $|n\rangle$ given in Chapter 6, Section B. The nonzero terms are

$$\begin{aligned}
&\langle n|\,\underline{a}\,\underline{a}\,\underline{a}^+\underline{a}^+\,|n\rangle = (n+1)(n+2)\\
&\langle n|\,\underline{a}^+\underline{a}^+\underline{a}\,\underline{a}\,|n\rangle = n(n-1)\\
&\langle n|\,\underline{a}\,\underline{a}^+\underline{a}\,\underline{a}^+\,|n\rangle = (n+1)^2\\
&\langle n|\,\underline{a}^+\underline{a}\,\underline{a}^+\underline{a}\,|n\rangle = n^2\\
&\langle n|\,\underline{a}\,\underline{a}^+\underline{a}^+\underline{a}\,|n\rangle = n(n+1)\\
&\langle n|\,\underline{a}^+\underline{a}\,\underline{a}\,\underline{a}^+\,|n\rangle = (n+1)n.
\end{aligned}$$

The bracket on the right-hand side of equation (9.40) is the sum of these terms:

$$\langle n|\,(\underline{a}+\underline{a}^+)^4\,|n\rangle = 6\left(n^2+n+\frac{1}{2}\right). \tag{9.41}$$

Since the cubic contribution to the first-order correction to the energy is zero, the first-order correction, equation (9.37), is given by using equation (9.41) in equations (9.40) and (9.37).

$$H'_{nn} = \frac{q\,\hbar^2\omega_0^2}{k^2}\frac{3}{2}\left(n^2+n+\frac{1}{2}\right). \tag{9.42}$$

In the limit that the quartic force constant, q, goes to zero, the perturbation vanishes, as required.

Using $\omega_0 = \sqrt{k/m}$, so that $k^2 = \omega_0^4 m^2$, the perturbed oscillator energy to first order is

$$E = \left(n + \frac{1}{2}\right)\hbar\omega_0 + q\frac{3}{2}\left(n^2 + n + \frac{1}{2}\right)\frac{\hbar^2}{m^2\omega_0^2}. \tag{9.43}$$

Unlike a harmonic oscillator, equation (9.43) demonstrates that the anharmonic oscillator does not have equally spaced energy levels. For molecules, the level spacing decreases as n increases. Therefore, q is negative. Within in the context of this model for the anharmonicity of molecular vibrations, q and ω_0 can be determined by measuring the vibrational energy separations of the 0–1 and 1–2 levels of a vibrational mode spectroscopically since the mass, m (actually the reduced mass of the molecular mode), is known. This description of a molecular anharmonic oscillator can be used to describe the first few vibrational levels of real molecules. However, if n is made sufficiently large in equation (9.43), the second term will become larger than the first term, and the energies of succeeding levels will decrease as n is increased further. In real molecules the energy levels become increasingly closely spaced until the dissociation energy is reached, causing the bond to break. Above the dissociation energy, the energy of the resulting particles is a continuum. The dissociation energy is the dotted line in Figure 9.1. Simple model potentials such as a Morse potential have this correct shape at high energy, in contrast to the potential given in equation (9.33).

2. The Stark Effect for the Rigid Plane Rotor

Molecules rotating in the gas phase have quantized rotational angular momenta that give rise to quantized rotational energy levels. Even a diatomic molecule will rotate with two angular coordinates, φ and θ. The solutions to the rotational angular momentum problem are the same as the solutions to the hydrogen atom $\Phi(\varphi)$ and $\Theta(\theta)$ equations that give rise to the quantum numbers m and ℓ. Real molecules do not have rigid bonds between atoms. The bond length is not fixed, so the rotational energy levels depend on the vibrational state of the molecule. In addition, the centrifugal force of the rotation can stretch the bond. These factors give rise to rotational–vibrational coupling.

A simplified rotational problem is that of the rigid plane rotor. For a diatomic, the bond length is fixed, and the molecule is constrained to rotate in a plane. A molecule rotating on a flat frictionless surface would act as a plane rotor. The rigid plane rotor is a one-dimensional angular momentum problem. The Hamiltonian only has a kinetic energy term because there is no potential. The plane rotor energy eigenvalue problem can be written in the Schrödinger representation as

$$\frac{-\hbar^2}{2I}\frac{d^2\psi(\varphi)}{d\varphi^2} = E\psi(\varphi). \tag{9.44}$$

The Hamiltonian operator here is identical to the one for a free particle moving in one dimension [equation (5.36)], except the mass is replaced by I, the moment of inertia. Rearranging gives

$$\frac{d^2\psi(\varphi)}{d\varphi^2} = \frac{-2I\,E}{\hbar^2}\psi(\varphi). \tag{9.45}$$

This equation is identical in form to the $\Phi(\varphi)$ equation solved in Chapter 7 as part of the analysis of the hydrogen atom. The wavefunctions and the eigenvalues are

$$\psi_m^0(\varphi) = \frac{1}{\sqrt{2\pi}} e^{im\varphi} \tag{9.46}$$

$$E_m^0 = \frac{m^2\hbar^2}{2I}, \tag{9.47}$$

with $m = 0, \pm1, \pm2, \pm3, \ldots$. The superscript 0 on the wavefunction and the energy indicate that these will be used as the zeroth-order solutions in the perturbation theory calculation of the Stark effect.

If a rigid plane rotor has a permanent dipole moment, $\vec{\mu}$, and it is placed in an external electric field $\vec{E}$, there is an additional term in the Hamiltonian, $-\vec{\mu} \cdot \vec{E}$, which accounts for the interaction of the dipole with the field. The Hamiltonian is

$$\underline{H} = \frac{-\hbar^2}{2I}\frac{d^2}{d\varphi^2} - \vec{\mu} \cdot \vec{E}. \tag{9.48}$$

The perturbation piece of the Hamiltonian is

$$\underline{H'} = -\vec{\mu} \cdot \vec{E} = -\mu E \cos\varphi, \tag{9.49}$$

where φ is the angle between the molecular dipole and the direction of the external electric field. The electric field strength, E, serves the role of the expansion parameter. When E goes to zero, the zeroth-order Hamiltonian is recovered, and the zeroth-order solutions are equations (9.46) and (9.47).

The first-order correction to the energy is

$$\begin{aligned} E'_m = H'_{mm} &= \langle \psi_m^0 | \underline{H'} | \psi_m^0 \rangle \\ &= -\mu E \int_0^{2\pi} \psi_m^{0*} \cos(\varphi)\, \psi_m^0 \, d\varphi. \end{aligned} \tag{9.50}$$

Substituting the zeroth-order wavefunctions, equation (9.46), gives

$$H'_{mm} = \frac{-\mu E}{2\pi} \int_0^{2\pi} e^{-im\varphi} e^{im\varphi} \cos(\varphi)\, d\varphi \tag{9.51}$$

$$= \frac{-\mu E}{2\pi} \int_0^{2\pi} \cos(\varphi)\, d\varphi \tag{9.52}$$

$$= 0, \tag{9.53}$$

since the integral of $\cos(\varphi)$ over one full cycle is zero. The first-order correction to the energy vanishes. This result arises from the symmetry of the first-order correction for the Stark effect on a rigid plane rotor. H'_{mm} is the expectation value of the perturbation for the zeroth-order states. For the plane rotor, the dipole will be aligned with and against the field an equal amount, so the expectation value is zero.

The fact that the first-order correction is zero does not imply that higher-order corrections are also zero. The second-order correction due to the first-order perturbation is

$$E''_m = \sum_n{}' \frac{H'_{mn} H'_{nm}}{(E^0_m - E^0_n)}$$
$$= \left(\frac{\mu E}{2\pi}\right)^2 \frac{\sum_n \int_0^{2\pi} e^{-im\varphi} e^{in\varphi} \cos(\varphi)\, d\varphi \int_0^{2\pi} e^{-in\varphi} e^{im\varphi} \cos(\varphi)\, d\varphi}{(E^0_m - E^0_n)}. \tag{9.54}$$

Using

$$\cos(\varphi) = \frac{1}{2}(e^{i\varphi} + e^{-i\varphi}), \tag{9.55}$$

equation (9.54) becomes

$$E''_m = \frac{1}{4}\left(\frac{\mu E}{2\pi}\right)^2 \frac{\sum_n \int_0^{2\pi} e^{i(n-m\pm 1)\varphi} d\varphi \int_0^{2\pi} e^{i(m-n\pm 1)\varphi}\, d\varphi}{(E^0_m - E^0_n)}. \tag{9.56}$$

The integrals are zero unless the arguments of the exponentials are zero. This will only occur for

$$n = m \pm 1. \tag{9.57}$$

Therefore, there are only two terms in the sum over n, and for these two terms the product of the two integrals yields $4\pi^2$. Then

$$E''_m = \frac{1}{4}\left(\frac{\mu^2 E^2}{E^0_m - E^0_{m-1}} + \frac{\mu^2 E^2}{E^0_m - E^0_{m+1}}\right). \tag{9.58}$$

Using equation (9.47) for the zeroth-order energies, the second-order correction to the energy is

$$E''_m = \frac{I\mu^2 E^2}{\hbar^2(4m^2 - 1)}. \tag{9.59}$$

The energy of a rigid plane rotor in an electric field to second order is

$$E_m = \frac{\hbar^2 m^2}{2I} + \frac{I\mu^2 E^2}{\hbar^2(4m^2 - 1)}. \tag{9.60}$$

The energy is proportional to E^2. E^2 serves the role of the expansion parameter for the second-order term in the perturbation expansion. In the limit that the E-field goes to zero, the energy goes to the zeroth-order energy. For $m = 0$, the perturbation correction to the energy is negative. For all other values of m, the correction to the energy is positive. This dependence on m is analogous to the classical mechanics result. Classically, a plane rotor with insufficient energy to rotate in the field tends to align parallel to the field, lowering the energy. However, once the rotor has sufficient energy to rotate, the net alignment is antiparallel because the rotor is speeded up as it becomes parallel with the field, and it is slowed down as it becomes antiparallel to the field. Therefore, it spends more time antiparallel than parallel, which increases the energy.

The Stark effect on the plane rotor depends on its dipole moment, μ. Real molecules, which rotate in three dimensions and have permanent dipole moments, will also display a Stark effect on the rotational states. Molecules with permanent dipole moments also have Stark effect changes in vibrational and electronic energy levels. Molecules without dipole moments can also exhibit a Stark effect that is intrinsically second order in the E field. When molecules (which are intrinsically polarizable) are placed in an E field, the field will induce a dipole moment. The induced dipole interacts with the field to produce an interaction energy that varies as E^2. In the context of perturbation theory, this interaction gives rise to a term H''_{nn} in equation (9.29).

C. PERTURBATION THEORY FOR DEGENERATE STATES

In the development of perturbation theory given in Section A, the zeroth-order eigenstates were taken to be nondegenerate. If

$$\underline{H}\,|\varphi_1\rangle = E\,|\varphi_1\rangle$$

$$\underline{H}\,|\varphi_2\rangle = E\,|\varphi_2\rangle\,,$$

with E the same eigenvalue for both $|\varphi_1\rangle$ and $|\varphi_2\rangle$, then $|\varphi_1\rangle$ and $|\varphi_2\rangle$ are degenerate. A superposition of degenerate eigenkets,

$$|\phi\rangle = c_1\,|\varphi_1\rangle + c_2\,|\varphi_2\rangle\,,$$

also has the same eigenvalue, that is,

$$\underline{H}\,|\phi\rangle = E\,|\phi\rangle\,.$$

Any superposition of degenerate eigenstates is also an eigenstate, with same eigenvalue. This was proven in Chapter 2, Section C. Eigenstates are linearly independent if

$$c_1\,|\varphi_1\rangle + c_2\,|\varphi_2\rangle + \cdots + c_n\,|\varphi_n\rangle = 0$$

only if all of the c's equal 0. If there are n linearly independent eigenstates belonging to a certain eigenvalue, then the eigenvalue is n-fold degenerate. n orthonormal sets of the $|\varphi_i\rangle$ can be formed. These are also linearly independent and are an n-fold degenerate set of eigenkets with the same eigenvalue as the original set. In fact, an infinite number of such n-fold degenerate orthonormal sets of eigenkets can be formed.

When some of the zeroth-order states to be used in a perturbation calculation are degenerate, a complication arises in the development of perturbation theory. When the perturbation goes to zero, the perturbed kets must become the well-defined set of zeroth-order kets. However, for a set of degenerate zeroth-order states, there are an infinite number of sets of zeroth-order states. If the perturbation breaks the degeneracy of the set of zeroth-order degenerate kets, as the perturbation approaches zero, the perturbed kets will approach a particular set of degenerate zeroth kets. It is not immediately clear which is the correct set of degenerate zeroth-order kets to use in the perturbation expansion of the eigenkets of the full Hamiltonian. Therefore, unlike perturbation theory for nondegenerate zeroth-order states, when there are degenerate states, there is an ambiguity in the choice of the proper zeroth-order kets. This ambiguity must be resolved in the development of perturbation theory for degenerate states.

Consider the problem in which the Hamiltonian has only zeroth-order and first-order pieces.

$$(\underline{H}^0 + \lambda \underline{H}') \left| \varphi_j \right\rangle = E_j \left| \varphi_j \right\rangle . \tag{9.61}$$

When $\lambda \rightarrow 0$,

$$\underline{H}^0 \left| \varphi_j^0 \right\rangle = E_j^0 \left| \varphi_j^0 \right\rangle , \tag{9.62}$$

but one of the eigenvalues, E_i^0, is m-fold degenerate. The m zeroth-order eigenkets belonging to the m-fold degenerate set are labeled

$$\left| \varphi_1^0 \right\rangle , \left| \varphi_2^0 \right\rangle \cdots \left| \varphi_m^0 \right\rangle . \tag{9.63}$$

These are orthonormal with

$$E_1^0 = E_2^0 = \cdots = E_m^0 \equiv E_1^0 . \tag{9.64}$$

For the perturbed states $1 - m$, as λ approaches zero, the perturbed state approaches the unperturbed state, that is,

$$\left| \varphi_i \right\rangle \rightarrow \left| \psi_i^0 \right\rangle ,$$

where $\left| \psi_i^0 \right\rangle$ is a zeroth-order eigenket with eigenvalue E_1^0. However, the $\left| \psi_i^0 \right\rangle$ is, in general, some linear combination of the $\left| \varphi_i^0 \right\rangle$; that is,

$$\left| \psi_i^0 \right\rangle = c_1 \left| \varphi_1^0 \right\rangle + c_2 \left| \varphi_2^0 \right\rangle + \cdots + c_m \left| \varphi_m^0 \right\rangle . \tag{9.65}$$

$\left| \psi_i^0 \right\rangle$ is the zeroth-order approximation to $\left| \varphi_i \right\rangle$, but the coefficients, c_j, in equation (9.65) are unknown.

To obtain the corrections to the energy and the corrections to the kets for the degenerate zeroth-order states, E and $\left| \varphi_i \right\rangle$ are expanded:

$$E = E_1^0 + \lambda E' + \cdots , \tag{9.66}$$

but

$$\left| \varphi_i \right\rangle = \sum_{j=1}^{m} c_j \left| \varphi_j^0 \right\rangle + \lambda \left| \varphi_i' \right\rangle + \cdots . \tag{9.67}$$

In equation (9.6) used for the expansion of the ket in nondegenerate perturbation theory, only $\left| \varphi_i' \right\rangle$ was unknown. However, in equation (9.67), $\left| \varphi_i' \right\rangle$ and the c_j's are unknowns.

Substituting equations (9.66) and (9.67) into equation (9.61) and collecting the coefficients of each power of λ gives the zeroth-order equation

$$\underline{H}^0 \sum_{j=1}^{m} c_j \left|\varphi_j^0\right\rangle = E_1^0 \sum_{j=1}^{m} c_j \left|\varphi_j^0\right\rangle \tag{9.68}$$

and the first-order equation

$$(\underline{H}^0 - E_1^0) \left|\varphi_i'\right\rangle = \sum_{j=1}^{m} c_j (E' - \underline{H}') \left|\varphi_j^0\right\rangle . \tag{9.69}$$

In the perturbation theory of nondegenerate states, the zeroth-order equation, equation (9.9), contains only known quantities. In contrast, the c_j's in equation (9.68) are unknowns.

$\left|\varphi_i'\right\rangle$ in equation (9.69) is expanded in the complete set of zeroth-order kets including the degenerate kets, $1 - m$, and all other kets,

$$\left|\varphi_i'\right\rangle = \sum_k A_k \left|\varphi_k^0\right\rangle . \tag{9.70}$$

To evaluate the right-hand side of equation (9.69), it is necessary to determine $\underline{H}'|\varphi_j^0\rangle$:

$$\underline{H}' \left|\varphi_j^0\right\rangle = \sum_k \left|\varphi_k^0\right\rangle\!\left\langle\varphi_k^0\right| \underline{H}' \left|\varphi_j^0\right\rangle , \tag{9.71}$$

where the projection operator, $\left|\varphi_k^0\right\rangle\!\left\langle\varphi_k^0\right|$, gives the piece of $\underline{H}'|\varphi_j^0\rangle$ that is $\left|\varphi_k^0\right\rangle$, and the sum over k gives the expansion of $\underline{H}'|\varphi_j^0\rangle$ in terms of the complete set of kets, $\left|\varphi_k^0\right\rangle$. This can be written as

$$\underline{H}' \left|\varphi_j^0\right\rangle = \sum_k H_{kj}' \left|\varphi_k^0\right\rangle , \tag{9.72}$$

where the H_{kj}''s are the brackets on the right-hand side of equation (9.71). The H_{kj}''s can be evaluated because they involve the known perturbation piece of the Hamiltonian and the known zeroth-order kets. Using equation (9.72), the term on the right-hand side of equation (9.69) involving the sum over $j = 1$ to m of $c_j \underline{H}'|\varphi_j^0\rangle$ becomes

$$\sum_{j=1}^{m} c_j H' \left|\varphi_j^0\right\rangle = \sum_{j=1}^{m} \sum_k c_j H_{kj}' \left|\varphi_k^0\right\rangle . \tag{9.73}$$

Substituting this result and the expansion, equation (9.70), into the first-order equation, equation (9.69), gives

$$\sum_k (E_k^0 - E_1^0) A_k \left|\varphi_k^0\right\rangle = \sum_{j=1}^{m} E' c_j \left|\varphi_j^0\right\rangle - \sum_k \left(\sum_{j=1}^{m} c_j H_{kj}' \right) \left|\varphi_k^0\right\rangle . \tag{9.74}$$

Left-multiplying by $\left\langle\varphi_i^0\right|$ yields

$$\sum_k (E_k^0 - E_1^0) A_k \left\langle\varphi_i^0 \mid \varphi_k^0\right\rangle = \sum_{j=1}^{m} E' c_j \left\langle\varphi_i^0 \mid \varphi_j^0\right\rangle - \sum_k \left(\sum_{j=1}^{m} c_j H_{kj}' \right) \left\langle\varphi_i^0 \mid \varphi_k^0\right\rangle . \tag{9.75}$$

1. Correction to the Energies

There are two cases in the evaluation of equation (9.75): $i \leq m$ and $i > m$. First consider $i \leq m$. On the left-hand side, there is one term in the sum, $k = i$; the other terms vanish by orthogonality. However, since $i \leq m$

$$E_i^0 = E_1^0.$$

Therefore, the left-hand side of equation (9.75) vanishes. The first term on the right-hand side is nonzero when $j = i$, and the bracket is 1 by normalization. The second term is nonzero when $k = i$, and the bracket is 1 by normalization. These results yield

$$\sum_{j=1}^{m} H'_{ij} c_j - E' c_i = 0. \tag{9.76}$$

Equation (9.76) is a system of simultaneous equations for the c_i's, one equation for each index of c_i. The system of equations has the form

$$\begin{aligned} (H'_{11} - E')c_1 + H'_{12}c_2 + \cdots + H'_{1m}c_m &= 0 \\ H'_{22}c_1 + (II'_{22} - E')c_2 + \cdots + H'_{2m}c_m &= 0 \\ &\vdots \\ H'_{m1}c_1 + H'_{m2}c_2 + \cdots + (H'_{mm} - E')c_m &= 0. \end{aligned} \tag{9.77}$$

This system of equations can be solved for the c_i's. One solution is the trivial solution

$$c_1 = c_2 = \cdots = c_m = 0.$$

Aside from the trivial solution, the system of equations only has a solution if the determinant of the coefficient of the unknowns vanishes, that is,

$$\begin{vmatrix} (H'_{11} - E') & H'_{12} & \cdots & H'_{1m} \\ \vdots & (H'_{22} - E') & \cdots & H'_{2m} \\ & \vdots & & \vdots \\ H'_{m1} & H'_{m2} & \cdots & (H'_{mm} - E') \end{vmatrix} = 0. \tag{9.78}$$

The $H'_{jk} = \langle \varphi_j^0 | H' | \varphi_k^0 \rangle$ are known, so the unknowns in the determinant are the E's, the corrections to the energies of the initially degenerate states $1-m$. Expanding the determinant gives an mth degree equation in the E's, which has solutions

$$E'_1, E'_2, \cdots, E'_m.$$

Thus, to first order, the energies of the states that have degenerate zeroth-order states are

$$E_i = E_1^0 + E'_i + \cdots, \tag{9.79}$$

where the E_i' are obtained from the solution of the mth-degree equation and the expansion parameter, λ, has been absorbed into the H_{jk}'. If the perturbation removes all of the degeneracies, there are now m different E_i's that reduce to E_1^0 when the perturbation goes to zero. In some instances, the perturbation will not remove all of the degeneracies, so that some of the E_i's may still have the same total energy.

To find $|\psi_i^0\rangle$, the correct zeroth-order ket obtained in the limit that the perturbation goes to zero, the system of equations [equation (9.77)] is employed. Since solution of the mth-degree equation gives the E_i''s, the only unknowns are the c_i's, which are the coefficients of the initial set of zeroth-order kets in the superposition that defines $|\psi_i^0\rangle$, equation (9.65). A particular E_i is substituted into the system of equations. This gives n equations for the n unknown c_i. There are actually only $n-1$ conditions because the system of equations is homogeneous. The normalization condition,

$$c_1^* c_1, + c_2^* c_2, + \cdots + c_m^* c_m = 1, \tag{9.80}$$

provides the additional condition necessary to obtain the first set of c_i. The next value of E' is substituted into equations (9.77); and using normalization, the second set of c_i is found. This procedure is continued until all of the correct zeroth-order kets are obtained as superpositions of the initial set of zeroth-order kets.

Once the correct zeroth-order functions are obtained, the first-order correction to the kets can be found. Returning to equation (9.75), take $\langle\varphi_i^0|$ to have $i = k > m$. This gives

$$\left(E_k^0 - E_1^0\right) A_k = -\sum_{j=1}^{m} c_j H_{kj}' \tag{9.81}$$

since $k > m$ the bracket in the first term on the right-hand side of equation (9.75) vanishes by orthogonality, the sums are nonzero only for $i = k$, and the nonzero brackets are equal to 1 due to normalization. Then A_k in equation (9.70) is

$$A_k = \frac{\sum\limits_{j=1}^{m} c_j H_{kj}'}{\left(E_1^0 - E_k^0\right)} \quad k > m. \tag{9.82}$$

The approximation to the eigenket for the Hamiltonian in equation (9.61) to first order is

$$|\varphi_i\rangle = |\psi_i^0\rangle + \lambda \sum_{k>m} \frac{\sum\limits_{j=1}^{m} c_j H_{kj}'}{(E_1^0 - E_k^0)} |\varphi_k^0\rangle + \cdots . \tag{9.83}$$

The first term on the right-hand side is the correct zeroth-order ket, and the second term is the first-order perturbation correction to the zeroth-order ket.

3. The Stark Effect of the Rigid Plane Rotor Revisited

In Section B.2, the Stark effect on the rigid plane rotor was treated using nondegenerate perturbation theory. However, the states $\pm m$ are degenerate. All states except $m = 0$ are doubly degenerate. Therefore, the problem is actually a degenerate perturbation problem.

For a pair of degenerate zeroth-order states $|q\rangle$ and $|r\rangle$, it can be shown by expansion of the determinant, equation (9.78), that to first order the states will remain degenerate if

$$\langle q|\,\underline{H}'\,|q\rangle = \langle r|\,\underline{H}'\,|r\rangle \quad \text{and} \quad \langle q|\,\underline{H}'\,|r\rangle = 0.$$

Both of these conditions are met for the rigid plane rotor because all three brackets are equal to zero. Degenerate perturbation theory to second order in $\underline{H}'$ can be derived in a manner analogous to the one used for the first-order problem. For a pair of degenerate states, the determinant is

$$\begin{vmatrix} \sum_n{}' \dfrac{\left|\langle q|\,\underline{H}'\,|n\rangle\right|^2}{E_q - E_n} - E'' & \sum_n{}' \dfrac{\langle q|\,\underline{H}'\,|n\rangle\,\langle n|\,\underline{H}'\,|r\rangle}{E_q - E_n} \\ \sum_n{}' \dfrac{\langle r|\,\underline{H}'\,|n\rangle\,\langle n|\,\underline{H}'\,|q\rangle}{E_q - E_n} & \sum_n{}' \dfrac{\left|\langle r|\,\underline{H}'\,|n\rangle\right|^2}{E_q - E_n} - E'' \end{vmatrix} = 0, \tag{9.84}$$

where E'' is the second-order correction to the energy, and the prime on the summation means that both $n = q$ and $n = r$ are excluded from the summations. From this second-order determinate, the conditions for the removal of the degeneracy can be obtained. If both

$$\sum_n{}' \frac{\left|\langle q|\,\underline{H}'\,|n\rangle\right|^2}{E_q - E_n} = \sum_n{}' \frac{\left|\langle r|\,\underline{H}'\,|n\rangle\right|^2}{E_q - E_n} \tag{9.85}$$

and

$$\sum_n{}' \frac{\langle q|\,\underline{H}'\,|n\rangle\,\langle n|\,\underline{H}'\,|r\rangle}{E_q - E_n} = 0, \tag{9.86}$$

then the states are still degenerate.

The first condition is met for all pairs of degenerate states of the rigid plane rotor. The second condition is met for all pairs of degenerate states except $m = \pm 1$. It was found in equation (9.57) that brackets like those that occur above will vanish unless $m = n \pm 1$. In equation (9.86) the term

$$\frac{\langle 1|\,\underline{H}'\,|0\rangle\,\langle 0|\,\underline{H}'\,|-1\rangle}{E_q - E_n} \neq 0.$$

Therefore, the nondegenerate treatment used to second order for the rigid plane rotor is adequate except for the states $m = \pm 1$. For those two states, the perturbation breaks the degeneracy, and equation (9.60) does not give the actual second-order correction to the energy.

To obtain the second-order correction to the energy for the states $m = \pm 1$, the determinate is expanded, and the resulting quadratic equation is solved for E''. The energies of the two zeroth-order states with $m = \pm 1$ to second order are nondegenerate and have energies

$$E_+ = \frac{\hbar^2}{2\mathrm{I}} + \frac{5\mathrm{I}\mu^2 E^2}{6\hbar^2} \tag{9.87a}$$

$$E_- = \frac{\hbar^2}{2I} - \frac{I\mu^2 E^2}{6\hbar^2}. \tag{9.87b}$$

E_+ and E_- are associated with the correct zeroth-order functions

$$\psi_+(\varphi) = \frac{1}{\sqrt{2}}\left(\psi_1^0(\varphi) + \psi_{-1}^0(\varphi)\right) \tag{9.88a}$$

and

$$\psi_-(\varphi) = \frac{1}{\sqrt{2}}\left(\psi_1^0(\varphi) - \psi_{-1}^0(\varphi)\right), \tag{9.88b}$$

respectively. $\psi_m^0(\varphi)$ is given in equation (9.46). The methods for obtaining the energy corrections and the correct superposition of the zeroth functions is similar to those used in the matrix formulation of quantum mechanics, which will be discussed in Chapter 13.

CHAPTER

10 THE HELIUM ATOM: PERTURBATION TREATMENT AND THE VARIATION PRINCIPLE

In Chapter 7, the hydrogen atom energy eigenvalue problem was solved exactly. Such an exact solution is not possible for any other atom or molecule. While the Schrödinger equation can be formulated for any atom or molecule, in general, it is not possible to solve even a three-body problem exactly. Therefore, the study of the electronic energy eigenstates of atoms and molecules relies on approximation techniques. The lowest energy state of the helium atom, the $1s$ state, is nondegenerate. Therefore, its energy can be calculated approximately using perturbation theory for nondegenerate states as developed in Chapter 9. In this chapter, first-order perturbation theory will be used to calculate the energy of the $1s$ state of He and two-electron ions such as Li^+. Another approach for obtaining approximate energy eigenvalues and wavefunctions involves the use of the variational theorem. This theorem is presented, and then its use is demonstrated by calculating an improved value of the He $1s$ energy.

A. PERTURBATION THEORY TREATMENT OF THE HELIUM ATOM GROUND STATE

In treating the hydrogen atom, the first step was the separation of the center of mass motion from the internal degrees of freedom of the atom. The same procedure can be performed for any atom. Here, the simplifying assumption will be made that the nucleus is at rest. This assumption introduces a negligible error, and it gives the Hamiltonian for a two-electron atom as

$$\underline{H} = -\frac{\hbar^2}{2m_o}\nabla_1^2 - \frac{\hbar^2}{2m_o}\nabla_2^2 - \frac{Ze^2}{4\pi\varepsilon_o r_1} - \frac{Ze^2}{4\pi\varepsilon_o r_2} + \frac{e^2}{4\pi\varepsilon_o r_{12}}. \tag{10.1}$$

The first and second terms on the right-hand side of equation (10.1) are the kinetic energy operators for electrons 1 and 2, respectively. m_0 is the electron mass. The second and third terms are the attraction to the nucleus of electrons 1 and 2, respectively. r_1 and r_2 are the distances from the nucleus of electrons 1 and 2. Z is the nuclear charge, which is 2 for helium. The last term is the electron–electron repulsion with r_{12} the distance between the two electrons.

Making the substitutions

$$a_0 = \frac{\varepsilon_o h^2}{\pi m_o e^2}$$
$$r_1 = a_0 R_1$$
$$r_2 = a_0 R_2$$
$$r_{12} = a_0 R_{12}$$
$$\frac{\partial^2}{\partial x_1^2} = \frac{1}{a_0^2}\frac{\partial^2}{\partial X_1^2}, \text{ etc.}$$

gives

$$\underline{H} = -\frac{\hbar^2}{2m_o}\frac{1}{a_0^2}(\nabla_1^2 + \nabla_2^2) - \frac{Ze^2}{4\pi\varepsilon_o a_0 R_1} - \frac{Ze^2}{4\pi\varepsilon_o a_0 R_2} + \frac{e^2}{4\pi\varepsilon_o a_0 R_{12}}, \tag{10.2}$$

where the Laplacians are written in terms of X_i, Y_i, and Z_i. In units of energy,

$$\frac{e^2}{4\pi\varepsilon_o a_0},$$

$\underline{H}$ takes the simplified form

$$\underline{H} = -\frac{1}{2}\left(\nabla_1^2 + \nabla_2^2\right) - \frac{Z}{R_1} - \frac{Z}{R_2} + \frac{1}{R_{12}}. \tag{10.3}$$

To employ perturbation theory, it is necessary to have a piece of the full Hamiltonian, the zeroth-order Hamiltonian, $\underline{H}^0$, for which there is an exact solution to the eigenvalue problem. Take

$$\underline{H}^0 = -\frac{1}{2}\left(\nabla_1^2 + \nabla_2^2\right) - \frac{Z}{R_1} - \frac{Z}{R_2}. \tag{10.4}$$

$\underline{H}_0$ contains the kinetic energy terms for the two electrons and the attraction of the two electrons to the nucleus. It does not include the electron–electron repulsion term. The electron–electron repulsion term is taken as the perturbation, that is,

$$\underline{H}' = \frac{1}{R_{12}}. \tag{10.5}$$

The zeroth-order Hamiltonian is for a fictitious problem in which two electrons interact with the nucleus but do not interact with each other. There is no explicit parameter that can be taken to zero to obtain the zeroth-order Hamiltonian from the full Hamiltonian—for example, the E field in the Stark effect problem (Chapter 9, Section B.2). However, mathematically an expansion parameter multiplying the electron–electron repulsion term, which takes on values from one to zero, can be included in equation (10.3).

To employ perturbation theory, exact solutions to the zeroth-order eigenvalue problem,

$$\underline{H}^0\psi^0 = E^0\psi^0, \tag{10.6}$$

are required. ψ^0 and E^0 are the eigenfunctions and eigenvalues of $\underline{H}^0$. $\underline{H}^0$ describes the interaction of two electrons with the nucleus without electron–electron interaction. In the absence of electron–electron interaction, the each electron behaves as if the other electron doesn't exist. Therefore, each electron will have a zeroth-order wavefunction that is a $1s$ hydrogen-like wavefunction for an atom with nuclear charge, Z. Taking

$$\psi^0 = \psi^0(1)\psi^0(2), \tag{10.7}$$

and

$$E^0 = E^0(1) + E^0(2), \tag{10.8}$$

where 1 and 2 indicate the coordinates of electron 1 and 2, respectively, the zeroth-order equation (10.6) can be separated. Equations (10.7) and (10.8) are substituted into equation (10.6) using equation (10.4) for $\underline{H}^0$. The result is divided by $\psi^0(1)\,\psi^0(2)$, and the following equations result for electron 1 and electron 2:

$$\frac{1}{2}\nabla_1^2\psi^0(1) + \left(E^0(1) + \frac{Z}{R_1}\right)\psi^0(1) = 0 \tag{10.9a}$$

$$\frac{1}{2}\nabla_2^2\psi^0(2) + \left(E^0(2) + \frac{Z}{R_2}\right)\psi^0(2) = 0. \tag{10.9b}$$

These are the Schrödinger equations for two hydrogen-like atoms with nuclear charge Z. The solutions to these equations are the hydrogen wavefunctions for a one-electron atom with nuclear charge Z. Then, the zeroth-order wavefunction for the perturbation theory treatment of the $1s$ state of the helium atom will involve the $1s$ hydrogen-like solutions to equations (10.9a) and (10.9b):

$$\psi^0(1) = \frac{1}{\sqrt{\pi}}Z^{3/2}e^{-ZR_1} \tag{10.10a}$$

$$\psi^0(2) = \frac{1}{\sqrt{\pi}}Z^{3/2}e^{-ZR_2} \tag{10.10b}$$

The zeroth-order wavefunction is the product of the two 1 electron functions,

$$\psi^0(1,2) = \psi^0(1)\,\psi^0(2) = \frac{Z^3}{\pi}e^{-ZR_1}e^{-ZR_2}, \tag{10.11}$$

and the zeroth-order energy is the sum of the two 1 electron energies

$$E^0 = E^0(1) + E^0(2) = 2Z^2E_{1s}(H). \tag{10.12}$$

$E_{1s}(H)$ is the energy of the $1s$ state of hydrogen.

The first-order correction to the energy arising from the perturbation, the electron–electron repulsion term, is

$$E' = H'_{nn} = H'_{1s,1s}$$

$$= \int\int \psi^{0*} \underline{H}' \psi^0 \, d\tau_1 d\tau_2$$

$$= \frac{e^2}{4\pi\varepsilon_o a_0} \frac{Z^6}{\pi^2} \int\int \frac{e^{-2ZR_1} e^{-2ZR_2}}{R_{12}} d\tau_1 d\tau_2. \quad (10.13)$$

The conventional units for the electron–electron repulsion term [see equation (10.2)] have been restored in equation (10.13). The differential operators $d\tau_1$ and $d\tau_2$ in spherical coordinates are

$$d\tau_1 = \sin\,\theta_1 R_1^2 d\varphi_1 d\theta_1 \, dR_1 \quad (10.14a)$$

$$d\tau_2 = \sin\theta_2 R_2^2 d\varphi_2 d\theta_2 \, dR_2. \quad (10.14b)$$

The evaluation of the integral in equation (10.13) cannot be performed in a simple manner. Evaluation of integrals of this type arise when hydrogen-like wavefunctions are employed in problems that involve electron–electron repulsion. The procedure used to evaluate the integral will be outlined briefly.

With the nucleus at the center, R_1 and R_2 can be considered vectors from the nucleus to the positions of electrons 1 and 2, respectively. The angle between the vectors is γ, and R_{12} can be written as

$$R_{12} = \sqrt{R_1^2 + R_2^2 - 2R_1 R_2 \cos\gamma}. \quad (10.15)$$

Let $R_>$ be the greater of R_1 and R_2, and let $R_<$ be the lesser of R_1 and R_2. Then

$$R_{12} = R_> \sqrt{1 + x^2 - 2x\cos\,\gamma}, \quad (10.16a)$$

where

$$x = \frac{R_<}{R_>}. \quad (10.16b)$$

In terms of the above, the perturbation operator in equation (10.13) can be written as

$$\frac{1}{R_{12}} = \frac{1}{R_>} \frac{1}{\sqrt{1 + x^2 - 2x\cos(\gamma)}}. \quad (10.17)$$

$1/R_{12}$ has been written in terms of $\cos(\gamma)$. The Legendre polynomials (Chapter 7, Section B.2) are a complete set of functions in $\cos(\gamma)$. Therefore, the right-hand side of equation (10.17) can be expanded in terms of the Legendre polynomials, that is,

$$\frac{1}{R_{12}} = \frac{1}{R_>} \sum_n a_n P_n(\cos(\gamma)). \quad (10.18)$$

The coefficients in the expansion can be evaluated to yield

$$a_n = x^n$$

and

$$\frac{1}{R_{12}} = \frac{1}{R_>} \sum_n x^n P_n(\cos(\gamma)). \tag{10.19}$$

Equation (10.19) expresses $1/R_{12}$ in terms of the relative angle between the vectors R_1 and R_2. The absolute direction of the vectors, R_1 and R_2 can be expressed in terms of the polar angles, θ_1, φ_1 and θ_2, φ_2. The Legendre polynomials in equation (10.19) can be expanded in terms of θ's and φ's of the two particles using the spherical harmonics of the two particles (for the solutions to the H atom's θ and φ equations, see Chapter 7),

$$P_n^{|m|}(\cos\,\theta_1)\,e^{im\varphi_1}$$
$$P_n^{|m|}(\cos\,\theta_2)\,e^{im\varphi_2},$$

since the spherical harmonics are a complete set of functions in θ and φ. Using the spherical harmonics, $1/R_{12}$ is

$$\frac{1}{R_{12}} = \sum_\ell \sum_m \frac{(\ell - |m|)!}{(\ell + |m|)!} \frac{R_<^\ell}{R_>^{\ell+1}} P_\ell^{|m|}(\cos\,\theta_1)\, P_\ell^{|m|}(\cos\,\theta_2) e^{im(\varphi_1 - \varphi_2)}. \tag{10.20}$$

This is a general result. It expresses $1/R_{12}$ in terms of $R_<$, $R_>$, and an expansion in the spherical harmonics with indices (quantum numbers), ℓ and m. As discussed in Chapter 7, ℓ can take on values of 0, 1, 2, . . . , and m can take on values from ℓ to $-\ell$ in integer steps.

Equation (10.20), is used to evaluate the perturbation integral, equation (10.13). The zeroth-order wavefunction is a product of $1s$ hydrogen-like functions. Although not shown explicitly because they are equal to unity, the angular parts of the zeroth-order wavefunctions are the spherical harmonics for particle 1 and 2 with $\ell = 0$ and $m = 0$. These are

$$P_0^0(\cos\,\theta_1)e^{im\varphi_1}$$

and

$$P_0^0(\cos\,\theta_2)e^{im\varphi_2}.$$

Since the spherical harmonics are orthogonal functions, all of the terms in equation (10.20) when it is used in equation (10.13) will vanish except the term with $\ell = 0$ and $m = 0$. Recalling that $0! = 1$, the summation in equation (10.13) reduces to $1/R_>$, and the integral needed to obtain the perturbation correction to the energy is

$$E' = \frac{e^2}{4\pi\varepsilon_o a_0} \frac{Z^6}{\pi^2} \int\!\!\int \frac{e^{-2ZR_1} e^{-2ZR_2}}{R_>} d\tau_1\, d\tau_2. \tag{10.21}$$

Performing the integral over angles yields $(4\pi)^2$, and

$$E' = 16Z^6 \frac{e^2}{4\pi\varepsilon_o a_0} \int_0^\infty \int_0^\infty \frac{e^{-2ZR_1} e^{-2ZR_2}}{R_>} R_1^2\, dR_1\, R_2^2\, dR_2. \tag{10.22}$$

This integral can be rewritten as

$$E' = \frac{16Z^6 e^2}{4\pi\varepsilon_o a_0}\int_0^\infty$$

$$e^{-2ZR_1}\left[\frac{1}{R_1}\int_0^{R_1} e^{-2ZR_2}R_2^2\,dR_2 + \int_{R_1}^\infty e^{-2ZR_2}R_2\,dR_2\right]R_1^2\,dR_1. \quad (10.23)$$

The integrals inside the square bracket are over R_2. The first integral inside the bracket has $R_1 > R_2$. Consequently, $1/R_> = 1/R_1$, and it can be taken outside of the integral. Since $R_1 > R_2$, R_2 must be between 0 and R_1, which sets the limits of integration. The second integral inside the bracket has $R_2 > R_1$. Thus, $1/R_> = 1/R_2$, which cancels one of the powers of R_2 in the differential operator, $R_2^2\,dR$. Since $R_2 > R_1$, R_2 must be between R_1 and ∞, which sets the limits of integration. Therefore, the two integrals inside the brackets correspond to the integral over R_2 from 0 to ∞, as required. After the integrals over R_2 are performed and the limits of integration are substituted in, the result is a function of R_1. This function is multiplied by the other terms in R_1, and the remaining integral over R_1 is carried out. The result is

$$E' = \frac{5}{8}Z\frac{e^2}{4\pi\varepsilon_o a_0}. \quad (10.24)$$

Equation (10.24) is the first-order perturbation theory correction to the energy of the helium atom or helium-like ions arising from the treatment of the electron–electron repulsion term in the Hamiltonian as a perturbation. Written in terms of the $1s$ energy of the hydrogen atom, the correction to the energy is

$$E' = \left(-\frac{5}{4}Z\right)E_{1s}(H). \quad (10.25)$$

Therefore, to first order the energy of the ground state of the helium atom ($Z = 2$) or helium-like ions is

$$E = E^0 + E' = \left(2Z^2 - \frac{5}{4}Z\right)E_{1s}(H). \quad (10.26)$$

For helium, $E = -74.8$ eV using $E_{1s}(H) = -13.6$ eV.

Equation (10.26) is the binding energy of the two electrons in the helium atom or helium like ion. Experimentally, the ionization energy of the two electrons is measured. This is the energy required to remove the two electrons from the nucleus, a positive number, which is the negative of the binding energy. It is the sum of the first and second ionization energies. Calculated values using equation (10.26) and experimentally determined values are listed in Table 10.1. For helium, first-order perturbation theory provides a reasonable value for the ground state energy given the simplicity of the analytical method used. All of the experimental and theoretical values in the table differ by 4 to 5 eV. As the nuclear charge is increased, the percentage error decreases. For helium, treating electron–electron repulsion, the interaction of two particles with charges $-e$, as a small perturbation compared to the interaction of two particles, one with charge $-e$ and one with charge $+2e$, is not a particularly good approximation. However, as the nuclear charge increases, electron–electron repulsion becomes an increasingly smaller influence, leading to more accurate results using first-order perturbation theory. For helium, the theoretically calculated energy can be improved by using higher-order perturbation theory or by using other methods, such as the variational approach discussed below.

TABLE 10.1
Experimental and Perturbation Theory Values of the Sum of the First and Second Ionization Energies of Helium and Helium-like Ions, and the Percent Error

Atom	Experimental Value (eV)	Calculated Value (eV)	% Error
He	79.00	74.80	5.3
Li^{+}	198.09	193.80	2.2
Be^{+2}	371.60	367.20	1.2
B^{+3}	599.58	595.00	0.76
C^{+4}	882.05	877.20	0.55

B. THE VARIATIONAL THEOREM

The variational theorem is the basis of an approximation method that does not require an exact solution to a zeroth-order problem to provide a set of zeroth-order eigenvectors and eigenvalues.

The variational theorem:
If ϕ is any function such that

$$\int \phi^* \phi \, d\tau = 1 \tag{10.27}$$

(ϕ is normalized) and if the lowest eigenvalue of the operator $\underline{H}$ is E_0, then

$$\langle \phi | \, \underline{H} \, | \phi \rangle = \int \phi^* \underline{H} \, \phi \, d\tau \geq E_0. \tag{10.28}$$

The theorem states that for any function, the expectation value of an operator is greater than or equal to the true lowest eigenvalue of the operator. The equality sign will only hold if the function is actually the eigenfunction of the operator with the lowest eigenvalue.

Proof: Consider

$$\langle \phi | \, \underline{H} - E_0 \, | \phi \rangle = \langle \phi | \, \underline{H} \, | \phi \rangle - \langle \phi | \, E_0 \, | \phi \rangle \tag{10.29}$$

$$= \langle \phi | \, \underline{H} \, | \phi \rangle - E_0 \tag{10.30}$$

Let the orthonormal eigenkets of $\underline{H}$ be $|\varphi_i\rangle$, that is,

$$\underline{H} \, |\varphi_i\rangle = E_i \, |\varphi_i\rangle \, .$$

Expanding $|\phi\rangle$ in terms of the $|\varphi_i\rangle$ gives

$$|\phi\rangle = \sum_i c_i \, |\varphi_i\rangle \, . \tag{10.31}$$

Substituting equation (10.31) into the left hand side of equation (10.29) yields

$$\langle\phi| \, \underline{H} - E_0 \, |\phi\rangle = \sum_i \bar{c}_i \, \langle\varphi_i| \, (\underline{H} - E_0) \sum_j c_j \, |\varphi_j\rangle \tag{10.32}$$

$$= \sum_j \bar{c}_j c_j \, \langle\varphi_j| \, (\underline{H} - E_0) \, |\varphi_j\rangle \, . \tag{10.33}$$

The $|\varphi_j\rangle$ are eigenkets of $(\underline{H} - E_0)$. Operating to the right on $|\varphi_j\rangle$ returns $|\varphi_j\rangle$. Then the double sum in equation (10.32) collapses into the single sum in equation (10.33) because the $|\varphi_j\rangle$ are orthogonal. Operating $\underline{H}$ on $|\varphi_j\rangle$ returns E_j, and since the $|\varphi_j\rangle$ are normalized the closed bracket is

$$\langle\phi| \, \underline{H} - E_0 \, |\phi\rangle = \sum_j \bar{c}_j c_j (E_j - E_0). \tag{10.34}$$

Recognizing that $\bar{c}_j c_j \geq 0$ (a number times its complex conjugate is positive or zero) and that $E_j \geq E_0$ (an eigenvalue is greater than or equal to the lowest eigenvalue), so that $(E_j - E_0) \geq 0$, gives

$$\sum_i \bar{c}_j c_j (E_j - E_0) \; \geq \; 0. \tag{10.35}$$

Consequently, from equation (10.34) the bracket is

$$\langle\phi| \, \underline{H} - E_0 \, |\phi\rangle \; \geq \; 0, \tag{10.36}$$

and from equations (10.29) and (10.30) the bracket can be written as

$$\langle\phi| \, \underline{H} - E_0 \, |\phi\rangle = \langle\phi| \, \underline{H} \, |\phi\rangle - E_0.$$

Therefore,

$$\langle\phi| \, \underline{H} \, |\phi\rangle = \int \phi^* \underline{H} \, \phi \, d\tau \geq E_0. \tag{10.37}$$

The equality sign only holds if

$$|\phi\rangle = |\varphi_0\rangle \, .$$

The variational theorem furnishes a useful approach to the approximate solution of eigenvalue problems. It provides a method for the determination of approximate eigenvalues and eigenfunctions. The theorem states that for the lowest eigenstate of a system, the approximate eigenvalue that is calculated with any function will be greater than or equal to the true eigenvalue. Thus, a function that results in a lower calculated approximate eigenvalue than another function is a better approximation to the true eigenfunction, and the calculated value is closer to the true eigenvalue.

To find a function ϕ, which is a useful approximation, a normalized trial function

$$\phi(\lambda_1, \lambda_2, \cdots)$$

is chosen. ϕ is a function of one or more parameters, λ. The expectation value

$$J = \int \phi^* \underline{H} \phi \, d\tau \tag{10.38}$$

is calculated. J is a function of the λ's. J is minimized with respect to the λ's. The result is an approximation to the lowest eigenvalue, and ϕ, with the values of the λ's that give the minimum J, is the corresponding approximation to lowest eigenfunction. The variational method can be used for eigenstates other than the lowest. Once the approximate function for the lowest eigenstate is found, another normalized function is selected which is orthogonal to the lowest approximate eigenstate. This second function is minimized and is an approximation to the second lowest eigenstate, and the calculated J is an approximation to the second lowest eigenvalue. This procedure can be repeated for higher and higher states. However, the error is cumulative, and it increases rapidly.

C. VARIATION TREATMENT OF THE HELIUM ATOM GROUND STATE

To perform a variation treatment of the helium atom, any trial function can be selected. Here a simple function that bears a direct relationship to the helium atom energy eigenvalue problem is used as a trial function, that is,

$$\phi = \frac{Z'^3}{\pi} e^{-Z'(R_1+R_2)}. \tag{10.39}$$

Equation (10.39) is the zeroth-order function that was used in the perturbation theory treatment of the helium atom, but the fixed nuclear charge Z is replaced with a variable parameter, Z'. Z' is the parameter that will be varied to minimize the energy in the variational calculation.

The Hamiltonian can be written using the system of units defined to obtain equation (10.3) in perturbation theory treatment of the helium atom:

$$\underline{H} = -\frac{1}{2}\left(\nabla_1^2 + \nabla_2^2\right) - \frac{Z}{R_1} - \frac{Z}{R_2} + \frac{1}{R_{12}}. \tag{10.40}$$

The Hamiltonian can be rewritten by adding and subtracting

$$\frac{Z'}{R_1} + \frac{Z'}{R_2},$$

to yield

$$\underline{H} = \left[-\frac{1}{2}(\nabla_1^2 + \nabla_2^2) - \frac{Z'}{R_1} - \frac{Z'}{R_2}\right] - (Z - Z')\left(\frac{1}{R_1} + \frac{1}{R_2}\right) + \frac{1}{R_{12}}. \tag{10.41}$$

Using this form of $\underline{H}$,

$$J = \int \phi^* \underline{H} \phi \, d\tau. \tag{10.42}$$

can be readily calculated.

The first term of $\underline{H}$ in square brackets when equation (10.41) is used in equation (10.42) gives $2Z'^2 E_{1s}(H)$ because this piece of the calculation is the same as calculating the zeroth-order perturbation energy except that the function ϕ with nuclear charge Z' is used instead of the zeroth-order function, equation (10.11). Then

$$J = 2Z'^2 E_{1s}(H) - \ (Z - Z') \left[\int \int \frac{\phi^2}{R_1} \, d\tau_1 d\tau_2 + \int \int \frac{\phi^2}{R_2} \, d\tau_1 d\tau_2 \right] + \int \int \frac{\phi^2}{R_{12}} \, d\tau_1 d\tau_2. \tag{10.43}$$

The sum of the two integrals in the square brackets has a value that is twice the value of the first integral because the two integrals are identical except for interchange of subscripts. Performing the first integral over angles gives $(4\pi)^2$ and the resulting integrals to obtain the value of the term in square brackets:

$$2 \left[16\pi^2 \frac{Z'^6}{\pi^2} \int_0^{\infty} e^{-2Z'R_1} R_1 \, dR_1 \int_0^{\infty} e^{-2Z'R_2} R_2^2 \, dR_2 \right] = 2Z', \tag{10.44a}$$

which in conventional units is

$$2Z' \frac{e^2}{4\pi\varepsilon_o a_0}. \tag{10.44b}$$

The last term in equation (10.43) was evaluated during the perturbation theory calculation of the helium ground-state energy [equation (10.13)] except here Z is replaced by Z'. The result, as in equation (10.24) (in conventional units), is

$$\frac{5}{8} Z' \frac{e^2}{4\pi\varepsilon_o a_0}. \tag{10.45}$$

Substituting expressions (10.44b) and (10.45) in to equation (10.43) yields

$$\begin{aligned} J &= 2Z'^2 E_{1s}(H) - (Z - Z') 2Z' \frac{e^2}{4\pi\varepsilon_o a_0} + \frac{5}{8} Z' \frac{e^2}{4\pi\varepsilon_o a_0} \\ &= \left[2Z'^2 + 4Z'(Z - Z') - \frac{5}{4} Z' \right] E_{1s}(H) \\ &= \left[-2Z'^2 + 4ZZ' - \frac{5}{4} Z' \right] E_{1s}(H), \end{aligned} \tag{10.46}$$

where

$$E_{1s}(H) = -\frac{1}{2} \frac{e^2}{4\pi\varepsilon_o a_0}.$$

J, given in equation (10.46), is an expression for the lowest-energy eigenvalue of the helium atom and helium-like ions in terms of the variation parameter Z'. To obtain the best value of the energy given the trial function that was employed, J is minimized with respect to Z' by setting its derivative equal to zero:

$$\frac{\partial J}{\partial Z'} = \left(-4Z' + 4Z - \frac{5}{4}\right) E_{1s}(H) = 0. \tag{10.47}$$

Solving for Z' yields

$$Z' = Z - \frac{5}{16}. \tag{10.48}$$

Substituting the expression for Z' into equation (10.46) yields an expression for the approximate energy of the ground state of the helium atom and helium-like ions:

$$E = 2Z'^2 E_{1s}(H). \tag{10.49}$$

$Z' = Z - 5/16$ is the "effective" nuclear charge that minimizes the energy in the approximation to the lowest-energy eigenvalue. Note that equation (10.49) is same as the zeroth-order energy [equation (10.12)] used in the perturbation theory treatment of helium except that the nuclear charge, Z, is replaced by the effective nuclear charge, Z'. Using Z' in the expression for ϕ [equation (10.39)] gives an approximation to the eigenfunction associated with the lowest eigenvalue.

Calculated values using equation (10.49) and experimentally determined values of the sum of the first and second ionization energies, along with percent error, are listed in Table 10.2 for the helium atom and helium like ions. The sum of the ionization energies is the negative of the binding energy of the two electrons calculated with equation (10.49), and therefore the values are positive. Comparing Table 10.2 to Table 10.1, it can be seen that the variation calculations provide more accurate results than the first-order perturbation theory calculations. The errors in Table 10.2 are approximately 2 eV in each calculation, while the errors in Table 10.1 are 4 to 5 eV. By performing variational calculations using trial functions with more variation parameters, it is possible to obtain results to any desired degree of accuracy for the helium atom.

TABLE 10.2
Experimental and Variation Theory Values of the Sum of the First and Second Ionization Energies of Helium and Helium-like Ions, and the Percent Error

Atom	Experimental Value (eV)	Calculated Value (eV)	% Error
He	79.00	77.46	1.9
Li^{+}	198.09	196.46	0.82
Be^{+2}	371.60	369.86	0.47
B^{+3}	599.58	597.66	0.32
C^{+4}	882.05	879.86	0.15

CHAPTER 11

TIME-DEPENDENT PERTURBATION THEORY

The time-dependent Schrödinger equation provides one approach for the solution of time-dependent quantum mechanical problems. If the Hamiltonian is time-independent, the time-dependent Schrödinger equation can be separated into time-dependent and time-independent parts (see Chapter 5, Section A). The time-independent equation is the Schrödinger form of the energy eigenvalue problem. The time-dependent equation gives rise to a time-dependent phase factor. Three time-dependent problems with time-independent Hamiltonians—that is, the motion of a free particle wave packet (Chapter 3, Section D), a harmonic oscillator wave packet (Chapter 6, Section C) and the time-dependent two-state problem (Chapter 8)—were discussed previously.

In many time-dependent problems, the Hamiltonian is explicitly time-dependent. For example, the absorption and emission of radiation by atoms and molecules is a time-dependent problem. In the absence of the radiation field, the atomic or molecular Hamiltonian is time-independent and has a set of energy eigenstates. However, the part of the Hamiltonian that accounts for the radiation field involves oscillating electric and magnetic fields. The problem of absorption and emission of radiation is treated in Chapter 12. Another example is scattering of molecules in the gas phase. The isolated molecules have time-independent Hamiltonians. However, as the molecules approach each other, intermolecular interactions arise which are distance-dependent. Since the distance is changing with time, the intermolecular interaction terms in the Hamiltonian are time-dependent. A simple example of an atom-molecule grazing collision is discussed in Section B of this chapter. In general, problems in which the Hamiltonian is explicitly time-dependent cannot be solved exactly. Time-dependent perturbation theory provides a useful approach to such problems if the time-dependent perturbation is not too severe.

A. DEVELOPMENT OF TIME-DEPENDENT PERTURBATION THEORY

Frequently, the Hamiltonian, $\underline{H}$, in the time-dependent Schrödinger equation

$$\underline{H}\,|\Psi\rangle = i\hbar\,\frac{\partial\,|\Psi\rangle}{\partial t} \tag{11.1}$$

can be divided into two parts,

$$\underline{H} = \underline{H}^0 + \underline{H}', \tag{11.2}$$

where $\underline{H}^0$ is independent of time, and $\underline{H}'$ is time-dependent. The solutions to

$$\underline{H}^0 \left|\Psi^0\right\rangle = i\hbar \frac{\partial \left|\Psi^0\right\rangle}{\partial t} \tag{11.3}$$

are

$$\left|\Psi_n^0(q,t)\right\rangle = \left|\psi_n^0(q)\right\rangle e^{-iE_n t/\hbar}. \tag{11.4}$$

The q are the spatial coordinates. The $\left|\psi_n^0(q)\right\rangle$ are the complete set of time-independent eigenkets that are the solutions to the time-independent energy eigenvalue problem. The $e^{-iE_n t/\hbar}$ are the time-dependent phase factors, which are the time-dependent parts of the eigenkets because the Hamiltonian is time-independent.

To solve equation (11.1) with the Hamiltonian given in equation (11.2), the solution is expanded in terms of the complete set of kets, the $\left|\Psi_n^0(q,t)\right\rangle$:

$$\left|\Psi(q,t)\right\rangle = \sum_n c_n(t) \left|\Psi_n^0(q,t)\right\rangle. \tag{11.5}$$

The right-hand side of equation (11.5) is substituted into equation (11.1) to give

$$\sum_n c_n \underline{H}^0 \left|\Psi_n^0\right\rangle + \sum_n c_n \underline{H}' \left|\Psi_n^0\right\rangle = i\hbar \sum_n \dot{c}_n \left|\Psi_n^0\right\rangle + i\hbar \sum_n c_n \frac{\partial \left|\Psi_n^0\right\rangle}{\partial t}. \tag{11.6}$$

The first term on the left-hand side of equation (11.6) is equal to the last term on the right-hand side because each pair of terms (left-hand side equals right-hand side) in the sums is the Schrödinger equation for the Hamiltonian, $\underline{H}^0$, and the nth eigenket of $\underline{H}^0$ multiplied by a constant c_n. Therefore, equation (11.7) reduces to

$$\sum_n c_n \underline{H}' \left|\Psi_n^0\right\rangle = i\hbar \sum_n \dot{c}_n \left|\Psi_n^0\right\rangle. \tag{11.7}$$

The $\left|\psi_n^0(q)\right\rangle$ are taken to be orthonormal; consequently, left-multiplying by $\left\langle\Psi_m^0\right|$ gives

$$\sum_n c_n \left\langle\Psi_m^0\right| \underline{H}' \left|\Psi_n^0\right\rangle = i\hbar\, \dot{c}_m, \tag{11.8}$$

and, therefore

$$\dot{c}_m = -\frac{i}{\hbar} \sum_n c_n \left\langle\Psi_m^0\right| \underline{H}' \left|\Psi_n^0\right\rangle. \tag{11.9}$$

Equation (11.9) is exact. It is a system of coupled differential equations for the c_m since the $\left|\Psi_n^0\right\rangle$ are taken to be the known solutions, equation (11.4), to equation (11.3). If the system of coupled differential equations can be solved, an exact solution to the time-dependent problem is obtained. In the time-dependent two state problem (Chapter 8), just such a treatment led to a pair of coupled differential equations [equations (8.11)]. Their solutions gave the results discussed in Chapter 8. However, if the system of equations cannot be solved exactly, the approximate method of time-dependent perturbation theory can be used.

Take the system initially to be in a particular eigenstate, $\left|\Psi_\ell^0\right\rangle$, of the time-independent piece of the Hamiltonian, $\underline{H}^0$. If the time-dependent piece of the Hamiltonian, $\underline{H}'$, has only a small influence on the eigenkets of the time-independent Hamiltonian—that is, it is a weak time-dependent perturbation—then

$$c_\ell^* c_\ell \cong 1 \tag{11.10}$$

for all time. This condition basically defines when a perturbation is considered weak. The probability of finding the system in the initial state does not change significantly, and, therefore the probability of finding the system in states other than $\left|\Psi_\ell^0\right\rangle$ is always small. The time-dependent perturbation, $\underline{H}'$, changes the system by transferring probability from the initial state to other states. To apply time-dependent perturbation theory, the probability of finding the system in states other than the initial state is small. Nonetheless, the small transfer of probability can be significant. For example, consider a system of 10^{20} molecules initially in their ground electronic state. Application of a weak radiation field can excite 10^{16} of them to an electronic excited state from which they fluoresce. The probability of being in the initial state is $0.9999 \cong 1$. However, 10^{16} excited molecules can produce substantial fluorescence. The absorption and emission of radiation for weak radiation fields are treated with time-dependent perturbation theory in Chapter 12.

The condition in equation (11.10) eliminates the summation in equation (11.9), and the system of equations reduces to an approximate expression,

$$\dot{c}_m = -\frac{i}{\hbar}\left\langle\Psi_m^0(q,t)\right|\underline{H}'\left|\Psi_\ell^0(q,t)\right\rangle. \tag{11.11}$$

Equation (11.11) with the condition equation (11.10) defines time-dependent perturbation theory. Equation (11.11) is a differential equation for the probability amplitude of finding the system in the state $\left|\Psi_m^0\right\rangle$, given that the system is initially in the state $\left|\Psi_\ell^0\right\rangle$. The restriction to situations in which the probability of the system being in the initial state is ~ 1 uncouples the differential equations in equation (11.9). Since $\left|\Psi_\ell^0\right\rangle$ is the only state with any significant probability, the probability transfer to $\left|\Psi_m^0\right\rangle$ depends exclusively on the coupling of $\left|\Psi_\ell^0\right\rangle$ to $\left|\Psi_m^0\right\rangle$ by $\underline{H}'$.

B. VIBRATIONAL EXCITATION BY A GRAZING ION–MOLECULE COLLISION

As a example of the application of time-dependent perturbation theory, vibrational excitation of a diatomic molecule bound to a surface caused by a grazing collision with a ion will be investigated. Because the molecule is bound to a surface, it is not rotating as it would be in a gas-phase grazing collision. This greatly simplifies the calculation while still illustrating important features of a grazing collision.

The molecule, M, which has a dipole moment, is bound to the surface as illustrated in Figure 11.1. The bond to the surface is very weak, so the M-surface vibrational frequencies are very low. The internal vibrational frequency of M is high. Therefore, to a large extent, the internal vibration is decoupled from the M-surface modes. In real systems, the internal mode and the M-surface modes can participate in vibrational relaxation of vibrational excitations of M. Here, the coupling will be neglected, and the internal mode of M is taken to be a harmonic oscillator. As shown in Figure 11.1, M has the positive side of its dipole at the surface.

A positively charged ion, I^+, flies by the molecule. The ion velocity is V. The ion starts infinitely far away from M at $t = -\infty$, passes by M, and, at $t = +\infty$, it is again infinitely far away. b is the distance of closest approach. (b is referred to as the impact parameter.) The time of closest approach occurs at $t = 0$ (t_2 in Figure 11.1). At any time, t, the distance from I^+ to M is a:

$$a = \sqrt{b^2 + (Vt)^2}. \tag{11.12}$$

As the ion flies by the molecule, the molecular vibrational states experience a time-dependent perturbation. If the molecular oscillator is viewed classically, the nature of the interaction can be readily discerned. There is a Coulombic interaction (a point charge–dipole interaction) of I^+ with M. The more negative side of M is always closer to I^+ than the more positive side. When the bond expands, the partial negative charge on M is somewhat closer to I^+ and the partial positive charge is somewhat further from I^+. The interaction helps expand the spring, which lowers the energy somewhat compared to the energy in the absence of the ion. When the bond contracts, the partial negative charge moves farther from I^+ and the partial positive charge moves closer to I^+. The bond contraction is inhibited by the interaction with the ion, and the energy is raised. The interaction of the ion with the molecular oscillator has odd symmetry; that is, the energy is lowered when the bond expands and is raised when the bond contracts. The interaction is a perturbation of the symmetric parabolic potential of the harmonic oscillator.

A qualitatively correct model is to take the perturbation to be the addition of a cubic term to the Hamiltonian. This has the correct symmetry. Higher-order odd terms (e.g., a fifth-order term) could also be included. The inclusion of higher-order terms does not change the nature of the results, and the influence of their inclusion on the results will be discussed below. The strength of the cubic interaction is inversely proportional to the square of the M–I^+ distance because this is the distance dependence of a point change–dipole interaction. The orientational factor will be neglected for simplicity and because most of the influence

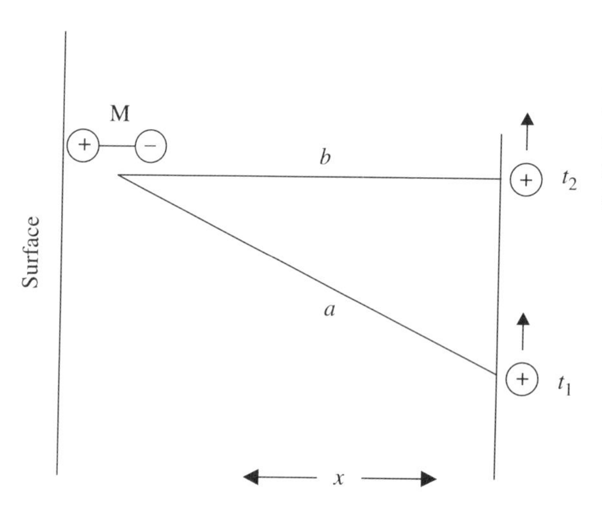

Figure 11.1 A schematic of the molecule–ion grazing collision. The dipolar molecule, M, is bound to the surface. The positively charged ion flies by. The distance of closest approach is b.

of the I^+ on M occurs at the shortest distances where the angle is very small. The interaction is time-dependent because the distance changes as the ion moves by the molecule.

For the model given above, the time-independent piece of the Hamiltonian, $\underline{H}^0$, is the harmonic oscillator Hamiltonian given in terms of raising and lowering operators by equation (6.62). The complete set of time-independent eigenkets, which are the solutions to the time-independent energy eigenvalue problem, are the harmonic oscillator eigenkets, $|n\rangle$. The time-dependent piece of the Hamiltonian is

$$\underline{H}'(t) = A(t)\underline{x}^3 \tag{11.13}$$

$$= \frac{q}{[b^2 + (Vt)^2]}\underline{x}^3. \tag{11.14}$$

$\underline{x}$ is the harmonic oscillator position operator given in equation (6.92), and q is a constant that depends on the size of the dipole moment of M. q determines the strength of the interaction.

If the ion's influence on the molecular vibrations is small, time-dependent perturbation theory can be used to calculate the probability of finding M in excited vibrational states after I^+ flies by. The perturbation will be weak if b is relatively large or q is small. Since the internal vibrational frequency of M is high, initially the molecule will be found in its ground vibrational state, $n = 0$, and the probability of finding the system in other vibrational states is ~ 0.

Equation (11.11) can be employed to calculate the probability that the time-dependent perturbation produced by the ion will cause vibrational states of M to be excited. For this problem, the ket and bra in equation (11.11) are

$$\left|\Psi_\ell^0\right\rangle = |0\rangle\, e^{-iE_0t/\hbar} \tag{11.15}$$

and

$$\left\langle\Psi_m^0\right| = \langle m|\, e^{iE_mt/\hbar}, \tag{11.16}$$

where $|0\rangle$ is the ground state of the harmonic oscillator, the $\langle m|$ are the excited states, and E_i are the energies of the harmonic oscillator eigenstates. $\underline{H}'(t)$ is given in equation (11.14). Substituting equations (11.14)–(11.16) into equation (11.11) gives

$$\dot{c}_m = \frac{-i}{\hbar}\langle m|\frac{q}{(b^2 + (Vt)^2)}\underline{x}^3\,|0\rangle\, e^{i(E_m - E_0)t/\hbar}. \tag{11.17}$$

Since only $\underline{x}^3$ operates on the spatial ket $|0\rangle$, the coefficient can be brought outside of the bracket:

$$\dot{c}_m = \frac{-i}{\hbar}\frac{q}{(b^2 + (Vt)^2)}\langle m|\,\underline{x}^3\,|0\rangle\, e^{i\Delta E_{m0}t/\hbar}, \tag{11.18}$$

where $\Delta E = E_m - E_0$. Multiplying by dt and integrating yields

$$c_m = \frac{-i}{\hbar}q\,\langle m|\,\underline{x}^3\,|0\rangle\int_{-\infty}^{\infty}\frac{e^{i\Delta E_{m0}t/\hbar}}{[b^2 + (Vt)^2]}\,dt. \tag{11.19}$$

c_m is the probability amplitude for finding the molecule in the mth vibrational state after the ion flies by given that the molecule is in the ground vibrational state initially.

To evaluate equation (11.19), first consider the time-independent bracket, $\langle m|\underline{x}^3|0\rangle$ with

$$\underline{x} = \left(\frac{\hbar\omega}{2k}\right)^{1/2} (\underline{a} + \underline{a}^+). \tag{11.20}$$

ω is the vibrational frequency, and k is the Hooke's law force constant (see Chapter 6). Then

$$\underline{x}^3 = \left(\frac{\hbar\omega}{2k}\right)^{3/2} (\underline{a} + \underline{a}^+)^3, \tag{11.21}$$

and

$$(\underline{a} + \underline{a}^+)^3 = \underline{a}^3 + \underline{a}^+\underline{a}^2 + \underline{a}^2\underline{a}^+ + \underline{a}^+\underline{a}\,\underline{a}^+ + \underline{a}\,\underline{a}^+\underline{a} + \underline{a}^{+2}\underline{a} + \underline{a}\,\underline{a}^{+2} + \underline{a}^{+3}. \tag{11.22}$$

Since $\underline{x}^3$ operates on $|0\rangle$, only terms with more raising operators then lowering operators will give nonzero results. Furthermore, terms with a lowering operator on the right will vanish—for example, $\underline{a}^{+2}\underline{a}\,|0\rangle = 0$. Using the above, the bracket becomes

$$\langle m|\,\underline{x}^3\,|0\rangle = \left(\frac{\hbar\omega}{2k}\right)^{3/2} \left[\langle m|\,\underline{a}^+\underline{a}\,\underline{a}^+\,|0\rangle + \langle m|\,\underline{a}\,\underline{a}^+\underline{a}^+\,|0\rangle + \langle m|\,\underline{a}^+\underline{a}^+\underline{a}^+\,|0\rangle\right] \tag{11.23}$$

$$= \left(\frac{\hbar\omega}{2k}\right)^{3/2} \left[1\,\langle m \mid 1\rangle + 2\,\langle m \mid 1\rangle + \sqrt{6}\,\langle m \mid 3\rangle\right]. \tag{11.24}$$

The only values of m for which equation (11.24) is nonzero are $m = 1$ and $m = 3$ because the states are orthogonal. The time-dependent perturbation arising from the ion fly by causes scattering only into the vibrational states 1 and 3; that is, c_1 and c_3 are nonzero. Evaluating the brackets in equation (11.24) for $m = 1$ yields

$$3\left(\frac{\hbar\omega}{2k}\right)^{3/2}, \tag{11.25a}$$

and for $m = 3$ yields

$$\sqrt{6}\left(\frac{\hbar\omega}{2k}\right)^{3/2}. \tag{11.25b}$$

The integral over time in equation (11.19) can be written as

$$\int_{-\infty}^{\infty} \frac{e^{i\Delta E_{m0}t/\hbar}}{[b^2 + (Vt)^2]}\,dt = \int_{-\infty}^{\infty} \frac{\cos \Delta E_{m0}t/\hbar}{[b^2 + (Vt)^2]}\,dt + i\int_{-\infty}^{\infty} \frac{\sin \Delta E_{m0}t/\hbar}{[b^2 + (Vt)^2]}\,dt. \tag{11.26}$$

The second term on the right-hand side is the integral over all time of an odd function times an even function, which equals zero. Writing the argument of the cos term as

$$\Delta E_{m0}t/\hbar = \frac{\Delta E_{m0}Vt}{V\hbar}$$

and substituting $Vt = y$ in equation (11.26), the cos integral can be performed to give

$$\int_{-\infty}^{\infty} \frac{e^{i\Delta E_{m0}t/\hbar}}{[b^2 + (Vt)^2]}\,dt = \frac{\pi}{Vb}e^{-\Delta E_{m0}b/V\hbar}. \tag{11.27}$$

Using the results of equations (11.25) and (11.27) in equation (11.19) provides expressions for the probability amplitudes of finding the molecule in the $m = 1$ and $m = 3$ vibrational states after the ion flies by:

$$c_1 = -i3\frac{q}{\hbar}\left(\frac{\hbar\omega}{2k}\right)^{3/2}\frac{\pi}{Vb}e^{-\Delta E_{10}b/V\hbar} \tag{11.28a}$$

$$c_3 = -i\sqrt{6}\frac{q}{\hbar}\left(\frac{\hbar\omega}{2k}\right)^{3/2}\frac{\pi}{Vb}e^{-\Delta E_{30}b/V\hbar}. \tag{11.28b}$$

The probabilities, $c_m^* c_m$, of finding the molecule in the $m = 1$ and $m = 3$ vibrational states are

$$P_1 = c_1^* c_1 = 9q^2\frac{\hbar\omega^3}{8k^3}\frac{\pi^2}{V^2b^2}e^{-2\omega b/V} \tag{11.29a}$$

$$P_3 = c_3^* c_3 = 6q^2\frac{\hbar\omega^3}{8k^3}\frac{\pi^2}{V^2b^2}e^{-6\omega b/V}, \tag{11.29b}$$

where $\Delta E_{10} = \hbar\omega$ and $\Delta E_{30} = 3\hbar\omega$ have been used.

Equations (11.29) show the manner in which the probabilities depend on the impact parameter, b, the molecule–ion coupling strength, q, the ion velocity, V, and the molecular vibrational parameters, ω and k. The result that probability is only transferred to vibrational states $m = 1$ and 3 arises from the choice of a cubic anharmonic perturbation in equation (11.13). If higher-order odd powers of x, which are consistent with the symmetry of the problem (e.g., $\underline{x}^5$) are included in the time-dependent perturbation, $\underline{H}'$, then similar expressions will be obtained for the probability of scattering to higher odd quantum number vibrational states.

An examination of equations (11.29) reveals that in the limits that $V \to 0$ and $V \to \infty$, both P_1 and P_3 go to zero. In between these limits there are velocities that produce maxima in the probabilities, P_1 and P_3. These velocities can be found by differentiating the expressions for P_1 and P_3 and setting the derivatives equal to zero. For P_1 the derivative is

$$\frac{dP_1}{dV} = 9q^2\frac{\hbar\omega^3}{8k^3}\frac{\pi^2}{b^2}\left[\frac{1}{V^2}e^{-2\omega b/V}(2\omega b/V^2) + e^{-2\omega b/V}(-2/V^3)\right] = 0, \tag{11.30}$$

which yields the condition

$$\frac{\omega b}{V} - 1 = 0. \tag{11.31}$$

Then the velocity that gives the maximum probability, $P_1^{\max}$, of finding the system in the $m = 1$ vibrational level following the grazing ion–molecule collision is

$$V_1^{\max} = \omega b, \tag{11.32}$$

and

$$P_1^{\max} = \frac{9\pi^2 q^2\hbar\omega}{8k^3b^4}e^{-2}. \tag{11.33}$$

The maximum probability depends on the inverse fourth power of the distance of closest approach, b. For the transition to the $m = 3$ level,

$$V_3^{\max} = 3\omega b, \tag{11.34}$$

and

$$P_3^{\max} = \frac{2}{3}\frac{\pi^2 q^2 \hbar\omega}{8k^3 b^4} e^{-2}. \tag{11.35}$$

Therefore, $P_1^{\max} = 27/2\, P_3^{\max}$, and the velocity required to produce $P_3^{\max}$ is three times greater than the velocity required to produce $P_1^{\max}$ for the same impact parameter, b.

To gain insight into the physical reason why the maximum probabilities, $P_1^{\max}$ and $P_3^{\max}$ occur at the velocities given in equations (11.32) and (11.34), consider the angular velocity of the ion relative to the molecule near the point of closest approach.

In Figure 11.2, θ is the angle between a, the line segment from the ion to the molecule at time t, and b, the line segment from the ion to the molecule at $t = 0$. $\sin\theta$ is given by

$$\sin\theta = \frac{d}{a} = \frac{Vt}{\sqrt{b^2 + (Vt)^2}} \tag{11.36}$$

Near the point of closest approach, the angle is small, so $\sin\theta \cong \theta$, and $Vt \ll b$. Then

$$\theta = \frac{Vt}{b}. \tag{11.37}$$

For the $0 \rightarrow 1$ transition, the velocity that gives the maximum probability [equation (11.32)] is $V_1^{\max} = \omega b$. Then

$$\theta = \frac{\omega t b}{b} = \omega t. \tag{11.38}$$

The angular velocity is the change in the angle θ with time, that is,

$$\frac{d\theta}{dt} = \omega. \tag{11.39}$$

For the velocity that produces the maximum probability in the state, $m = 1$, near the point of closest approach, the angular velocity is ω, the resonance frequency of the molecular

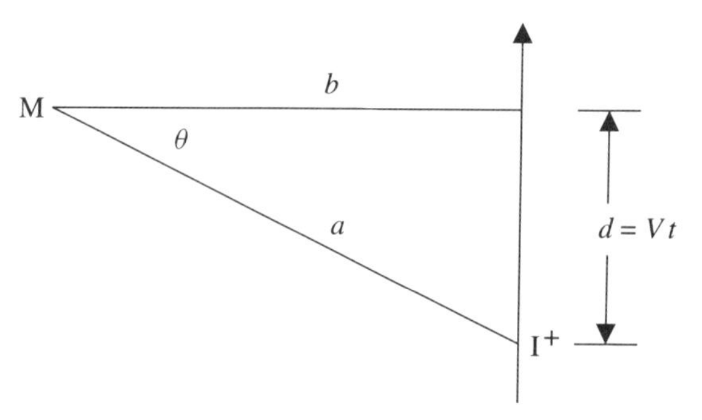

Figure 11.2 A schematic of the ion angular velocity near the point of closest approach.

oscillator. The ion is a charged particle moving by the molecule with angular velocity ω. The motion produces a time-varying electric field that is changing at frequency ω near the point of closest approach. Thus, the molecular oscillator is driven on resonance, and the on resonance perturbation most efficiently induces the transition. For the $0 \rightarrow 3$ transition, $V_3^{\max} = 3\omega b$, and

$$\frac{d\theta}{dt} = 3\omega. \tag{11.40}$$

Again, the transition is driven on resonance, giving rise to the maximum in the transition probability.

CHAPTER 12

Absorption and Emission of Radiation

Time-dependent perturbation theory can be applied to obtain approximate solutions to a wide variety of time-dependent quantum mechanical problems. As discussed in Chapter 11, the principal restriction is that the influence of the time-dependent perturbation on the eigenstates of the system that exist in the absence of the perturbation is small. An important problem that can be successfully treated with time-dependent perturbation is the interaction of an electromagnetic radiation field (light) with atoms and molecules that leads to absorption or emission of light. If the radiation field is not too strong, then the probability of an atom or a molecule making a transition between eigenstates is small, and time-dependent perturbation theory can be applied to calculate transition probabilities. The results of using time-dependent perturbation theory to describe the absorption and emission of light is the basis of many spectroscopic techniques. However, if the radiation field is intense, and the resulting transition probabilities become large, a wide variety of phenomena can occur that cannot be calculated using time-dependent perturbation theory. An example of this type of effect, the transient nutation, is discussed in Chapter 14, Section E.

Since atoms and molecules are composed of charged particles and light can be described classically as an electromagnetic wave, it is necessary to determine the form of the quantum mechanical Hamiltonian for the interaction of charged particles with electric and magnetic fields. To obtain the quantum mechanical Hamiltonian, the classical Hamiltonian must be found. This will be done in general for a charged particle in any combination of electric and magnetic fields. Then the quantum Hamiltonian will be generated by replacing the classical functions with the appropriate quantum mechanical operators. Once the general quantum Hamiltonian is found, it will be made specific so that it will apply to the interaction of a charged particle with a weak electromagnetic wave. This weak field Hamiltonian will be used with time-dependent perturbation theory to describe the absorption and emission of radiation by atoms and molecules. The treatment that is presented below is called semiclassical. The molecular eigenstates are quantized, but the radiation field is treated classically; that is, it is treated as an electromagnetic wave rather than as individual photons. The treatment is sufficient to explain absorption and stimulated emission. However, spontaneous emission (fluorescence) does not emerge naturally from semiclassical theory. The method introduced by Einstein is used to obtain the expression for the rate of spontaneous emission, and a brief discussion of the consequences of quantizing the radiation field, which leads to a correct treatment of spontaneous emission, is presented.

A. THE HAMILTONIAN FOR CHARGED PARTICLES IN ELECTRIC AND MAGNETIC FIELDS

To determine the Hamiltonian for a charged particle in a combination of electric and magnetic fields, it is necessary to obtain an expression for the potential. This can be accomplished by using two expressions for the force on the particle and then solving for the potential. The force, $\vec{F}$, on a particle with charge, e, is

$$\vec{F} = e\left[\vec{E} + \frac{1}{c}\left(\vec{V} \times \vec{B}\right)\right], \tag{12.1}$$

where $\vec{E}$ is the electric field, $\vec{V}$ is the velocity, $\vec{B}$ is the magnetic field, and c is the speed of light. The generalized form for the jth component of the force in Lagrangian Mechanics is

$$F_j = -\frac{\partial U}{\partial q_j} + \frac{d}{dt}\left(\frac{\partial U}{\partial \dot{q}_j}\right), \tag{12.2}$$

where U is the potential, and q, for Cartesian coordinates, is one of x, y, or z. For example, the x component of the force is

$$F_x = -\frac{\partial U}{\partial x} + \frac{d}{dt}\left(\frac{\partial U}{\partial V_x}\right), \tag{12.3}$$

where V_x is the x component of the velocity because $V_x = \dot{x}$.

Using the standard expressions from Maxwell's equations

$$\vec{B} = \vec{\nabla}\, x\, \vec{A} \tag{12.4a}$$

$$\vec{E} = -\frac{1}{c}\frac{\partial}{\partial t}\vec{A} - \vec{\nabla}\phi, \tag{12.4b}$$

where $\vec{A}$ is the vector potential, and ϕ is the scalar potential. Substituting equation (12.4) into equation (12.1) gives

$$\vec{F} = e\left[-\vec{\nabla}\,\phi - \frac{1}{c}\frac{\partial \vec{A}}{\partial t} + \frac{1}{c}(\vec{V} \times \vec{\nabla} \times \vec{A})\right]. \tag{12.5}$$

From equation (12.5), the x component of F, F_x, can be determined:

$$F_x = e\left[-\frac{\partial \phi}{\partial x} - \frac{1}{c}\frac{\partial A_x}{\partial t} + \frac{1}{c}(\vec{V} \times \vec{\nabla} \times \vec{A})_x\right]. \tag{12.6}$$

Evaluating the last term in equation (12.6) gives

$$(\vec{V} \times \vec{\nabla} \times \vec{A})_x = V_y\left(\frac{\partial A_y}{\partial x} - \frac{\partial A_x}{\partial y}\right) - V_z\left(\frac{\partial A_x}{\partial z} - \frac{\partial A_z}{\partial x}\right). \tag{12.7}$$

Adding and subtracting

$$V_x \frac{\partial A_x}{\partial x}$$

in equation (12.7) yields

$$(\vec{V} \times \vec{\nabla} \times \vec{A})_x = V_y \frac{\partial A_y}{\partial x} + V_z \frac{\partial A_z}{\partial x} + V_x \frac{\partial A_x}{\partial x} - V_y \frac{\partial A_x}{\partial y} - V_z \frac{\partial A_x}{\partial z} - V_x \frac{\partial A_x}{\partial x}. \tag{12.8}$$

The total time derivative of A_x is

$$\frac{dA_x}{dt} = \frac{\partial A_x}{\partial t} + \left(V_x \frac{\partial A_x}{\partial x} + V_y \frac{\partial A_x}{\partial y} + V_z \frac{\partial A_x}{\partial z} \right) \tag{12.9}$$

The first term on the right-hand side of equation (12.9) arises from the explicit variation of A_x with t. The term in brackets is due to the motion of the particle. The vector potential is evaluated at the location of the charged particle. If a particle is moving through a nonuniform potential, then the point at which A_x is evaluated will change with time; therefore, A_x will acquire a time dependence determined by the particle velocity and the change in the potential with location. Rearranging equation (12.9) gives

$$\frac{\partial A_x}{\partial t} - \frac{dA_x}{dt} = -V_y \frac{\partial A_x}{\partial y} - V_z \frac{\partial A_x}{\partial z} - V_x \frac{\partial A_x}{\partial x}. \tag{12.10}$$

Using equation (12.10) in equation (12.8) yields

$$(\vec{V} \times \vec{\nabla} \times \vec{A})_x = \frac{\partial}{\partial x}(\vec{V} \cdot \vec{A}) - \frac{dA_x}{dt} + \frac{\partial A_x}{\partial t}, \tag{12.11}$$

since

$$\frac{\partial}{\partial x}(\vec{V} \cdot \vec{A}) = \vec{V} \frac{\partial \vec{A}}{\partial x} + \frac{\vec{A} \partial \vec{V}}{\partial x} = V_x \frac{\partial A_x}{\partial x} + V_y \frac{\partial A_y}{\partial x} + V_z \frac{\partial A_z}{\partial x}, \tag{12.12}$$

and $\vec{V}$ is not an explicit function of the position, so the partial derivative of $\vec{V}$ with respect to x vanishes.

Equation (12.11) can be substituted into equation (12.6) to give

$$F_x = e \left[-\frac{\partial}{\partial x} \left(\phi - \frac{1}{c} \vec{A} \cdot \vec{V} \right) - \frac{1}{c} \frac{dA_x}{dt} \right]. \tag{12.13}$$

Equation (12.13) can be written in the form of equation (12.3) by noting that

$$A_x = \frac{\partial}{\partial V_x} (\vec{A} \cdot \vec{V}), \tag{12.14}$$

since the partial derivatives of the components of $\vec{A}$ with respect to V_x are zero because $\vec{A}$ depends on time and position but is not a function of velocity, and the partial derivatives of the components of $\vec{V}$ with respect to V_x are 1 for the V_x component and 0 for the other two components. Then equation (12.13) becomes

$$F_x = e\left[-\frac{\partial}{\partial x}\left(\phi - \frac{1}{c}\vec{A}\cdot\vec{V}\right) - \frac{1}{c}\frac{d}{dt}\left(\frac{\partial}{\partial V_x}(\vec{A}\cdot\vec{V})\right)\right]. \tag{12.15}$$

Furthermore, since the scalar potential does not depend on the velocity, it can be added to the last term on the right-hand side of equation (12.15) inside the parentheses to give

$$F_x = \left[-\frac{\partial}{\partial x}\left(e\phi - \frac{e}{c}\vec{A}\cdot\vec{V}\right) + \frac{d}{dt}\left(\frac{\partial}{\partial V_x}\left(e\phi - \frac{e}{c}\vec{A}\cdot\vec{V}\right)\right)\right]. \tag{12.16}$$

Equation (12.16) is in the general form of equation (12.3). Therefore, the potential, U, for a charged particle in any combination of electric and magnetic fields is

$$U = e\,\phi - \frac{e}{c}\vec{A}\cdot\vec{V}. \tag{12.17}$$

With knowledge of the potential, the classical Hamiltonian can be found and then converted to the quantum mechanical Hamiltonian. The Lagrangian, L, is

$$L = T - U, \tag{12.18}$$

where T is the kinetic energy. For a charged particle

$$L = T - e\,\phi + \frac{e}{c}\vec{A}\cdot\vec{V}, \tag{12.19}$$

and

$$T = \frac{1}{2}m(\dot{x}^2 + \dot{y}^2 + \dot{z}^2). \tag{12.20}$$

The ith component of the momentum is given by

$$P_i = \frac{\partial L}{\partial \dot{q}_i}. \tag{12.21}$$

Using equation (12.21), the components of the momentum can be found; for example,

$$P_x = m\dot{x} + \frac{e}{c}A_x. \tag{12.22}$$

The classical Hamiltonian is

$$H = P_x\dot{x} + P_y\dot{y} + P_z\dot{z} - L. \tag{12.23}$$

Thus,

$$H = \left(m\dot{x}^2 + \frac{e}{c}A_x\dot{x}\right) + \left(m\dot{y}^2 + \frac{e}{c}\dot{y}A_y\right) + \left(m\dot{z}^2 + \frac{e}{c}\dot{z}A_z\right) - \frac{1}{2}m(\dot{x}^2 + \dot{y}^2 + \dot{z}^2) - \frac{e}{c}(\dot{x}A_x + \dot{y}A_y + \dot{z}A_z) + e\,\phi, \tag{12.24}$$

which reduces to

$$H = \frac{1}{2}m(\dot{x}^2 + \dot{y}^2 + \dot{z}^2) + e\,\phi. \tag{12.25}$$

To convert the classical Hamilton into the quantum mechanical Hamiltonian, it is necessary to rewrite H in terms of the components of the momentum $p_x,\ p_y,\ p_z$ instead of $\dot{x},\ \dot{y},\ \dot{z}$ so that the classical momentum can be replaced with the corresponding quantum mechanical operator. Multiplying equation (12.26) by m/m gives

$$H = \frac{1}{2m}\left((m\dot{x})^2 + (m\dot{y})^2 + (m\dot{z})^2\right) + e\,\phi, \tag{12.26}$$

and using the equation for the components of the momentum (see equation 12.22)

$$P_i = m\dot{q}_i + \frac{e}{c}A_i,$$

the classical Hamiltonian becomes

$$H = \frac{1}{2m}\left[\left(p_x - \frac{e}{c}A_x\right)^2 + \left(p_y - \frac{e}{c}A_y\right)^2 + \left(p_z - \frac{e}{c}A_z\right)^2\right] + e\phi. \tag{12.27}$$

The quantum mechanical Hamiltonian is obtained by making the substitutions of the Schrödinger representation operators for the classical components of the momentum, for example,

$$p_x \Rightarrow -i\hbar\frac{\partial}{\partial x}.$$

Then the term

$$\left(p_x - \frac{e}{c}A_x\right)^2 \Rightarrow -\hbar^2\frac{\partial^2}{\partial x^2} + \frac{e^2}{c^2}|A_x|^2 + i\frac{\hbar e}{c}\frac{\partial}{\partial x}A_x + i\frac{\hbar e}{c}A_x\frac{\partial}{\partial x}. \tag{12.28}$$

The right-hand side of (12.28) can be simplified by recognizing the operators will operate on functions. Take the function to be $\psi(x)$. Operating the third term on the right-hand side of (12.28) on $\psi(x)$ and using the chain rule of differentiation gives

$$i\frac{\hbar e}{c}\frac{\partial}{\partial x}A_x\psi(x) = i\frac{\hbar e}{c}\frac{\partial A_x}{\partial x}\psi(x) + i\frac{\hbar e}{c}A_x\frac{\partial\psi(x)}{\partial x}. \tag{12.29}$$

The second term on the right-hand side of equation (12.29) is the same as the fourth term on the right-hand side of (12.28). Substituting equation (12.29) into (12.28) yields

$$\left(p_x - \frac{e}{c}A_x\right)^2 \Rightarrow -\hbar^2\frac{\partial^2}{\partial x^2} + \frac{e^2}{c^2}|A_x|^2 + i\frac{\hbar e}{c}\frac{\partial A_x}{\partial x} + 2i\frac{\hbar e}{c}A_x\frac{\partial}{\partial x}. \tag{12.30}$$

Making the same substitutions for the terms containing the other components of the momentum in equation (12.27), the quantum mechanical Hamiltonian for a charged particle in any combination of electric and magnetic fields is

$$\underline{H} = \frac{1}{2m}\left(-\hbar^2\nabla^2 + \frac{e^2}{c^2}|A|^2 + \frac{i\hbar e}{c}\vec{\nabla}\cdot\vec{A} + 2\frac{i\hbar e}{c}\vec{A}\cdot\vec{\nabla}\right) + e\,\phi. \tag{12.31}$$

To use the Hamiltonian in equation (12.31) in addressing the problem of the absorption and emission of light by atoms and molecules, it is necessary to make it specific for electromagnetic light waves. An electromagnetic plane wave is represented only by a vector potential. For light (electromagnetic waves), there is no scalar potential, that is,

$$\phi = 0. \tag{12.32}$$

As is discussed below, the light will be taken to be a plane wave, and the relationship of a plane wave to the vector potential is given. Since

$$\vec{\nabla} \cdot \vec{A} + \frac{1}{c^2}\frac{\partial \phi}{\partial t} = 0,$$

which is the Lorentz gauge condition, then,

$$\vec{\nabla} \cdot \vec{A} = 0. \tag{12.33}$$

Therefore, for a charged particle in an electromagnetic radiation field, the quantum mechanical Hamiltonian is

$$\underline{H} = \frac{1}{2m}\left(-\hbar^2\nabla^2 + \frac{e^2}{c^2}\,|A|^2 + 2\frac{i\,\hbar\,e}{c}\vec{A}\cdot\vec{\nabla}\right). \tag{12.34}$$

To this point there are no approximations. However, to use time-dependent perturbation theory, it is necessary for the perturbation to be weak. The treatment will describe the interaction of an atom or molecule with a weak radiation field, that is, low-intensity light. For low-intensity light, the $|A|^2$ term in equation (12.34) is negligible and can be dropped. Then for a weak radiation field, the Hamiltonian becomes

$$\underline{H} = \frac{1}{2m}\left(-\hbar^2\nabla^2 + \frac{2\,i\,\hbar\,e}{c}\vec{A}\cdot\vec{\nabla}\right). \tag{12.35}$$

The Hamiltonian has two terms. The first represents the kinetic energy of the charged particle; the second represents the radiation field. This weak-field Hamiltonian is valid for many situations of experimental interest. Most absorption and emission experiments occur under weak-field conditions, and the results obtained below using this Hamiltonian will provide an accurate context in which to analyze experiments. However, many experiments that utilize very high intensity lasers involve effects that cannot be described in the weak-field limit using time-dependent perturbation theory. Examples of high-field experiments are discussed in Chapter 14.

The Hamiltonian in equation (12.35) is for a single charged particle in a weak radiation field. A molecule or atom is composed of a number of charged particles. For a system of many particles, which interact with each other through a potential V, there will be one kinetic energy term for each particle plus the potential, in addition to the term representing the radiation field. To apply time-dependent perturbation theory, it is necessary to divide the Hamiltonian into time-dependent and time-independent parts, that is,

$$\underline{H} = \underline{H}^0 + \underline{H}'. \tag{12.36}$$

$$\underline{H}^0 = -\sum_j \frac{\hbar^2}{2m_j}\,\nabla_j^2 + V, \tag{12.37}$$

where the first term is the sum of the kinetic energy terms, with m_j the mass of the jth particle. This is the normal many-particle Hamiltonian. It is the time-independent atomic or molecular Hamiltonian, and it has a set of time-independent energy eigenvalues and corresponding eigenvectors. The remaining terms, one for the interaction of the radiation field with each of the charged particles, comprise the time-dependent piece of the Hamiltonian, $\underline{H}'$:

$$\underline{H}' = \sum_j \frac{e}{m_j c} i\hbar \underline{\vec{A}}_j \cdot \underline{\vec{\nabla}}_j. \tag{12.38}$$

Substituting $-i\hbar\underline{\vec{\nabla}} = \underline{\vec{P}}$ gives

$$\underline{H}' = -\sum_j \frac{e}{m_j c} \underline{\vec{A}}_j \cdot \underline{\vec{P}}_j. \tag{12.39}$$

B. APPLICATION OF TIME-DEPENDENT PERTURBATION THEORY

1. The Transition Dipole Bracket

To apply time-dependent perturbation theory to the absorption and induced emission of light, it is necessary to specify the vector potential in equation (12.39). Without loss of generality, the electromagnetic field can be taken to be a plane wave propagating in the z direction, since any other form of the field can be described as a superposition of plane waves. Then

$$\vec{A} = \vec{i} A_x \tag{12.40a}$$

where $\vec{i}$ is unit vector in x direction, $\vec{j}$ and $\vec{k}$ are unit vectors in the y and z directions, respectively, and

$$A_x = A_x^0 \cos\left[2\pi\nu\left(t - \frac{z}{c}\right)\right]. \tag{12.40b}$$

A_x^0 is a constant reflecting the amplitude of the field, and ν is the frequency of the light.

To see that equation (12.40) is a plane wave, consider the expressions for the electric and magnetic fields from Maxwell's equations (12.4):

$$\vec{E} = -\frac{1}{c}\frac{\partial}{\partial t}\vec{A} = \vec{i}\frac{2\pi\nu}{c} A_x^0 \sin 2\pi\nu\left(t - \frac{z}{c}\right) \tag{12.41a}$$

$$\vec{B} = \vec{\nabla} x \vec{A} = \vec{j}\frac{2\pi\nu}{c} A_x^0 \sin 2\pi\nu\left(t - \frac{z}{c}\right). \tag{12.41b}$$

Equations (12.41) show that the vector potential defined in equation (12.40) corresponds to oscillating electric and magnetic waves. The $\vec{E}$ and $\vec{B}$ fields are oscillating in phase at frequency, ν; they are propagating along the z direction at the speed of light, c; they are perpendicular, pointing along the x and y directions, respectively; and they have equal amplitudes, A_x^0.

Time-dependent perturbation theory can be used to calculate the transition probability from a state $\left|\Psi_m^0(q,t)\right\rangle$ to a state $\left|\Psi_n^0(q,t)\right\rangle$. The kets are eigenkets of $\underline{H}^0$. They are functions of coordinates, q, and include time-dependent phase factors. It is necessary to evaluate the bracket

$$\left\langle\Psi_m^0\right|\underline{H}'\left|\Psi_n^0\right\rangle=\left\langle\Psi_m^0\right|-\sum_j\frac{e}{m_jc}\underline{A}_{xj}\underline{P}_{xj}\left|\Psi_n^0\right\rangle. \tag{12.42}$$

For most cases of interest, the wavelength of light, λ, is much greater than size of an atom or molecule. Typically, $\lambda > 2\times10^3$ Å, while the size of an atom or molecule is ~1–10 Å. This is true even for very large molecules, such as proteins, because the group that absorbs light (e.g., the amino acid tryptophan) is small. When λ is large compared to the size of the molecule, A_x is essentially a constant; that is, two particles in different locations in a molecule will experience the same A_x at a given instant of time. Therefore, A_x can be taken out of the bracket, and the bracket becomes

$$\left\langle\Psi_m^0\right|\underline{H}'\left|\Psi_n^0\right\rangle=-\frac{e}{c}A_x\sum_j\frac{1}{m_j}\left\langle\Psi_m^0\right|\underline{P}_{xj}\left|\Psi_n^0\right\rangle \tag{12.43a}$$

$$=i\frac{e\hbar}{c}A_x\sum_j\frac{1}{m_j}\left\langle\Psi_m^0\right|\frac{\partial}{\partial x_j}\left|\Psi_n^0\right\rangle. \tag{12.43b}$$

Since $\frac{\partial}{\partial x_j}$ does not operate on time-dependent part of eigenket, the time-dependent phase factors can be brought outside of the bracket.

$$\left\langle\Psi_m^0(q,t)\right|\underline{H}'\left|\Psi_n^0(q,t)\right\rangle=i\frac{e\hbar}{c}A_x\,e^{i(E_m-E_n)t/\hbar}\sum_j\frac{1}{m_j}\left\langle\psi_m^0(q)\right|\frac{\partial}{\partial x_j}\left|\psi_n^0(q)\right\rangle, \tag{12.44}$$

where on the right-hand side the ket and bra are now time-independent, and E_m and E_n are the energy eigenvalues of the two zeroth-order states.

The bracket on the right-hand side of equation (12.44), which depends on the operator $\partial/\partial x_j$, can be reduced to a bracket which depends on the operator x_j. To see this, consider the ket and bra to be one-dimensional one particle wavefunctions, $\psi_m^0(x)$ and $\psi_n^0(x)$. These wavefunctions satisfy the equations

$$\frac{d^2\psi_m^{0*}}{dx^2}+\frac{2m}{\hbar^2}\left[E_m-V(x)\right]\psi_m^{0*}=0 \tag{12.45}$$

and

$$\frac{d^2\psi_n^0}{dx^2}+\frac{2m}{\hbar^2}\left[E_n-V(x)\right]\psi_n^0=0, \tag{12.46}$$

where equation (12.45) is the Schrödinger equation for the complex conjugate of the wavefunction, ψ_m^0, and equation (12.46) is the Schrödinger equation for the wavefunction ψ_n^0.

Left-multiplying equations (12.45) and (12.46) by $x\psi_n^0$ and $x\psi_m^{0*}$, respectively, subtracting the second resulting equation from the first, and integrating gives

$$\int_{-\infty}^{\infty}\left(x\psi_n^0\frac{d^2\psi_m^{0*}}{dx^2}-x\psi_m^{0*}\frac{d^2\psi_n^0}{dx^2}\right)dx+\frac{2m}{\hbar^2}(E_m-E_n)\int_{-\infty}^{\infty}\psi_m^{0*}x\psi_n^0\,dx$$
$$-\frac{2m}{\hbar^2}\int_{-\infty}^{\infty}\psi_n^0 xV\psi_m^{0*}\,dx+\frac{2m}{\hbar^2}\int_{-\infty}^{\infty}\psi_m^{0*}xV\psi_n^0\,dx=0. \tag{12.47}$$

The last two integrals are the identical brackets $\langle\psi_m^0|\,xV\,|\psi_n^0\rangle$, but in the first bracket the operator operates to the left and in the second bracket the operator operates to the right. For the Hermitian operator, xV, the value of the bracket is independent of the direction of the operation, so the last two terms in equation (12.47) sum to zero. Transposing the second integral, equation (12.47) becomes

$$\frac{2m}{\hbar^2}(E_n-E_m)\int_{-\infty}^{\infty}\psi_m^{0*}x\psi_n^0\,dx=\int_{-\infty}^{\infty}\left(x\psi_n^0\frac{d^2\psi_m^{0*}}{dx^2}-x\psi_m^{0*}\frac{d^2\psi_n^0}{dx^2}\right)dx. \tag{12.48}$$

The left-hand side is a bracket with the operator x rather than $\partial/\partial x_j$. The right-hand side can be integrated by parts, with, for the first term,

$$u=x\psi_n^0$$

and

$$dv=\frac{d^2\psi_m^{0*}}{dx^2}\,dx,$$

and similarly for the second term. Employing the fact that the wavefunctions vanish at infinity, the result of the integration by parts of the two terms is

$$\frac{2m}{\hbar^2}(E_n-E_m)\int_{-\infty}^{\infty}\psi_m^{0*}x\psi_n^0\,dx$$
$$=\int_{-\infty}^{\infty}\left[-\frac{d}{dx}(x\psi_n^0)\frac{d\psi_m^{0*}}{dx}+\frac{d}{dx}(x\psi_m^{0*})\frac{d\psi_n^0}{dx}\right]dx. \tag{12.49}$$

Using the chain rule to perform the derivatives on the right-hand side and collecting terms gives

$$\frac{2m}{\hbar^2}(E_n-E_m)\int_{-\infty}^{\infty}\psi_m^{0*}x\psi_n^0\,dx=\int_{-\infty}^{\infty}\left(-\psi_n^0\frac{d\psi_m^{0*}}{dx}+\psi_m^{0*}\frac{d\psi_n^0}{dx}\right)dx. \tag{12.50}$$

Integrating the first term on the right-hand side by parts and again using the fact that the wavefunctions vanish at infinity, it can be seen that the first term is equal to second term. Then

$$\frac{2m}{\hbar^2}(E_n-E_m)\int_{-\infty}^{\infty}\psi_m^{0*}x\psi_n^0\,dx=2\int_{-\infty}^{\infty}\psi_m^{0*}\frac{d}{dx}\psi_n^0\,dx. \tag{12.51}$$

Equation (12.51) gives the desired result, that is,

$$\left\langle \psi_m^0 \right| \frac{\partial}{\partial x} \left| \psi_n^0 \right\rangle = -\frac{m}{\hbar^2}(E_m - E_n)\left\langle \psi_m^0 \right| \underline{x} \left| \psi_n^0 \right\rangle . \tag{12.52}$$

Equation (12.52) can be generalized to three-dimensional wavefunctions and more than one particle.

Using the equation (12.52) in equation (12.44) gives

$$\left\langle \Psi_m^0(q,t) \right| \underline{H}' \left| \Psi_n^0(q,t) \right\rangle = -i\frac{1}{c\hbar} A_x (E_m - E_n) x_{mn} e^{i(E_m - E_n)t/\hbar} \tag{12.53}$$

with

$$x_{mn} = \left\langle \psi_m^0 \right| e \sum_j \underline{x}_j \left| \psi_n^0 \right\rangle . \tag{12.54}$$

x_{mn} is the bracket on the right-hand side of equation (12.52). It has the dimensions of a dipole, charge times length. However, the bracket does not represent a permanent dipole moment because the operator couples two different states of the zeroth-order Hamiltonian. It is called the transition dipole or transition dipole moment. The operator is the transition dipole moment operator for the radiation field with the electric field of the light along the x axis. If the light is polarized with the field along the y axis, then operator would be $e\underline{y}$. For the radiation field polarized in an arbitrary direction, there is an x, y, and z component of the transition dipole moment operator. The dipole approximation, the simplification that occurred when it was assumed that the vector potential was constant on a molecular distance scale [see equation (12.43)], led to the form of the bracket in equation (12.54), the transition dipole bracket.

2. Absorption and Stimulated Emission Transition Probabilities

The radiation field, a time-dependent perturbation, $\underline{H}'(t)$, generates a probability amplitude, C_m, for finding a molecule or atom (the system) in a state, $\left| \Psi_m^0(q,t) \right\rangle$, given that the system is initially in a state $\left| \Psi_n^0(q,t) \right\rangle$. $\left| \Psi_m^0(q,t) \right\rangle$ and $\left| \Psi_n^0(q,t) \right\rangle$ are eigenstates of the molecular Hamiltonian in the absence of the radiation field. If $\underline{H}'(t)$ is a weak perturbation, then C_m will remain very small, and the initial unit probability of finding the system in $\left| \Psi_n^0(q,t) \right\rangle$ will be essentially unchanged. For this situation, which is commonly encountered in a laboratory spectroscopy experiment, time-dependent perturbation theory (Chapter 11) can be used to calculate C_m, that is,

$$\frac{dC_m}{dt} = -\frac{i}{\hbar} \left\langle \Psi_m^0(q,t) \right| \underline{H}' \left| \Psi_n^0(q,t) \right\rangle . \tag{12.55}$$

The bracket is given in equation (12.53). Thus,

$$\frac{dC_m}{dt} = -\frac{1}{c\hbar^2} A_x (E_m - E_n) x_{mn} e^{i(E_m - E_n)t/\hbar}, \tag{12.56}$$

with x_{mn} defined by equation (12.54). For light of frequency ν, the vector potential is

$$A_x = A_x^0 \cos(2\pi\nu t),$$

which can be written as

$$A_x = \frac{1}{2} A_x^0 (e^{i2\pi \nu t} + e^{-i2\pi \nu t}). \tag{12.57}$$

Then, equation (12.56) becomes

$$\frac{dC_m}{dt} = -\frac{1}{2c\hbar^2} A_x^0 x_{mn} (E_m - E_n) \left(e^{i(E_m - E_n + h\nu)t/\hbar} + e^{i(E_m - E_n - h\nu)t/\hbar}\right). \tag{12.58}$$

Multiplying through by dt, integrating, and choosing a constant of integration such that $C_m = 0$ at $t = 0$ (i.e., the system is initially in the state $|\Psi_n^0\rangle$) gives

$$C_m = \frac{i}{2c\hbar} A_x^0 x_{mn} (E_m - E_n) \left[\frac{e^{i(E_m - E_n + h\nu)t/\hbar} - 1}{(E_m - E_n + h\nu)} + \frac{e^{i(E_m - E_n - h\nu)t/\hbar} - 1}{(E_m - E_n - h\nu)} \right]. \tag{12.59}$$

First, consider the case in which $E_m > E_n$ (Figure 12.1). This situation corresponds to absorption of radiation. Since $E_m > E_n$, the denominator of first term in the square brackets in equation (12.59) goes to $2h\nu$ as $h\nu \to (E_m - E_n)$, while the denominator of the second term goes to 0. Therefore, for the case of absorption, the second term will be large relative to the first term when the frequency of the light is at or near the transition frequency, and the first term can be dropped. This is referred to as the rotating wave approximation. The term that is kept is on or near resonance. The term that is dropped is effectively 2ν off resonance. The oscillating field 2ν off resonance produces a high-frequency Stark effect, which can be observed in some circumstances. For most situations, the effect is insignificant and can be neglected. The term that is kept represents the component of the light field rotating in the sense required to couple to the atomic or molecular transition moment and induce absorption.

The probability of finding the system in the state $|\Psi_m^0\rangle$ at time t is $C_m^* C_m$. Forming the product $C_m^* C_m$ and using the trigonometric identities

$$(e^{ix} - 1)(e^{-ix} - 1) = 2(1 - \cos x) = 4 \sin^2 x/2$$

gives

$$C_m^* C_m = \frac{1}{c^2\hbar^2} \left|A_x^0\right|^2 |x_{mn}|^2 (E_m - E_n)^2 \frac{\sin^2 \left[\dfrac{(E_m - E_n - h\nu)t}{2\hbar}\right]}{[(E_m - E_n - h\nu)]^2}. \tag{12.60}$$

Take $E_m - E_n = E$; E is the energy difference between the two eigenstates of the time independent Hamiltonian. $\Delta E = E - h\nu$; ΔE is the amount the radiation field is off

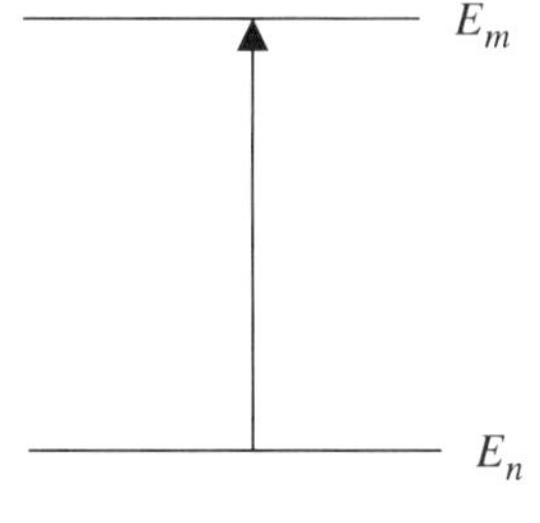

Figure 12.1 A Transition is made from the initial lower energy state, with energy E_n, to the higher energy state, with energy E_m by absorption of radiation.

resonance from the transition energy; that is, it is the difference between E and the energy of the applied radiation field. A plot of equation (12.60) versus ΔE is given in Figure 12.2. The maximum transition probability occurs on resonance, $\Delta E = 0$. As ΔE goes to zero, $\sin^2(\Delta E t/2\hbar)$ goes to $(\Delta E t/2\hbar)^2$, and the $(\Delta E)^2$ in the numerator and denominator of equation (12.60) divides out. Then, the maximum probability is

$$|C_m|^2_{\max} = Q\frac{t^2}{(2\hbar)^2}, \tag{12.61a}$$

where

$$Q = \frac{1}{c^2\hbar^2}\left|A_x^0\right|^2 |x_{mn}|^2 (E_m - E_n)^2. \tag{12.61b}$$

The maximum probability is proportional to t^2. The transition probability is only significant in an interval around $\Delta E = 0$ of width $\sim 4\pi\hbar/t$. This width is essentially determined by the uncertainty principle. For example, if the radiation field is turned on at $t = 0$, for $t = 1$ ps, the width is 67 cm^{-1} between the points of the first minima. This 1-ps time interval corresponds to a 1-ps square pulse. For a square pulse, the uncertainty relation is $\Delta\nu\Delta t = 0.886$, and, for 1-ps it yields an energy uncertainty of 30 cm^{-1} for the full width at half-maximum. The plot of the probability, $C_m^* C_m$ versus ΔE, in Figure 12.2 has the form of the square of a zeroth-order spherical Bessel function. It has a large central lobe and oscillations that die out rapidly as ΔE becomes large. Away from the region around $\Delta E = 0$, the probability oscillates very rapidly between zero and a very small value.

As t increases, the central lobe of the probability curve increases in height and decreases in width. An increasingly large fraction of the probability is contained in the central lobe. Even for a relatively short application of the radiation field of 10 ns, the central lobe is just $\sim$0.0067 cm^{-1} in width, and it contains virtually all of the probability. As $t \to \infty$, equation (12.60) goes to a Dirac delta function, $\delta(\Delta E)$ (see Chapter 3, Section B), and $C_m^* C_m$ is only nonzero exactly on resonance, $h\nu = (E_m - E_n)$. The total probability of finding the system in the final state, $\left|\Psi_m^0\right\rangle$, by making a transition from the initial state, $\left|\Psi_n^0\right\rangle$, is the integral of equation (12.60) over ν (the area under the graph in Figure 12.2), that is,

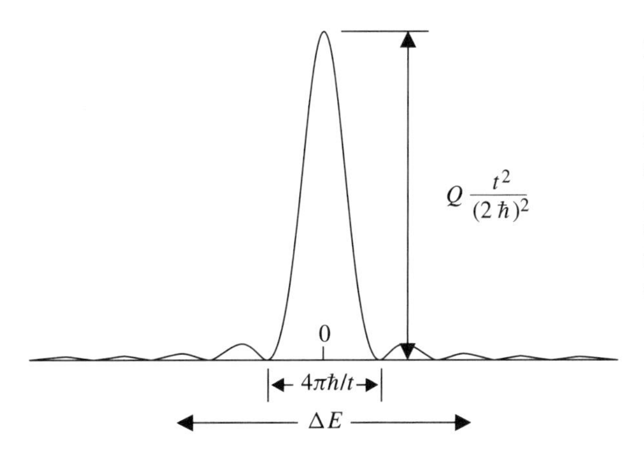

Figure 12.2 A plot of equation (12.60) versus ΔE showing the probability of the system undergoing a transition to higher energy by absorbing radiation at ΔE, where ΔE is the amount the radiation field is off resonance from the transition energy. Q is defined in the text.

$$\int_{-\infty}^{\infty} C_m^* C_m \, d\Delta\nu = \frac{1}{h}\int_{-\infty}^{\infty} C_m^* C_m \, d\Delta E = \frac{Q}{h}\int_{-\infty}^{\infty} \frac{\sin^2(\Delta E t/2\hbar)}{(\Delta E)^2} \, d\Delta E. \quad (12.62)$$

The integral is done over $\Delta\nu = \Delta E/h$, which is the difference between the frequency of the radiation field and $(E_m - E_n)/h$. The integral may be performed from $-\infty$ to ∞ because it is only large very close to $\Delta E = 0$. Making the substitution $x = \Delta E t/2\hbar$, the integral can be written as

$$= \frac{Q}{h}\frac{t}{2\hbar}\int_{-\infty}^{\infty} \frac{\sin^2 x}{x^2} \, dx. \quad (12.63)$$

The result is

$$= \frac{Qt}{4\hbar^2}, \quad (12.64)$$

since the integral of the square of a zeroth-order spherical Bessel function is equal to π. In the integral in equation (12.62), $\left|A_x^0\right|^2$ (contained in Q), which is proportional to the intensity of the light and is frequency-dependent, is taken to be $\left|A_x^0(\nu_{mn})\right|^2$, the value at the transition frequency, $(E_m - E_n)/h$. Even for relatively short pulses (10 ns), the transition probability is basically a delta function, $\delta(\Delta\nu_{mn})$ with $\Delta\nu_{mn} = \Delta E/h$. Then the probability of finding the system in the state $\left|\Psi_m^0\right\rangle$ is

$$C_m^* C_m = \frac{\pi^2 \nu_{mn}^2}{c^2 \hbar^2} \left|A_x^0(\nu_{mn})\right|^2 \left|x_{mn}\right|^2 t, \quad (12.65)$$

where $E_m - E_n$ has been replaced by $h\nu_{mn}$ in the definition of Q.

Equation (12.65) shows that the probability, $C_m^* C_m$, increases linearly with the time duration of the applied radiation field. To use time-dependent perturbation theory, $C_m^* C_m \ll 1$. If t is allowed to increase without limit, this condition would be violated. Atoms or molecules have excited state lifetimes. The lifetime can be caused by spontaneous emission (see below) or by radiationless relaxation, which is frequently dominant in condensed phases, in high density gases, or in sufficiently large isolated molecules. Spontaneous emission or radiationless relaxation returns the system to the lower energy state. In most circumstance in which the radiation field is weak, the lifetime effectively sets a limit on t, and keeps $C_m^* C_m \ll 1$. However, for strong radiation fields, particularly if the transition dipole bracket, x_{mn}, is large, $C_m^* C_m$ can become large. In such situations, time-dependent perturbation theory cannot be used. An example of "coherent coupling" of a radiation field to a two-level molecular system is given in Chapter 14, Sections E, F, and H. It is shown that it is possible for $C_m^* C_m = 1$; that is, the radiation field takes the system from being initially in the lower energy state to having a 100% probability of being found in the excited state.

Equation (12.65) gives $C_m^* C_m$ in terms of the vector potential [equation (12.40)]. The vector potential can be related to the intensity of light through the Poynting vector, $\vec{S}$, given by

$$\vec{S} = \frac{c}{4\pi}\vec{E} \times \vec{B}. \quad (12.66)$$

$\vec{E}$ and $\vec{B}$ are given in equation (12.41). Then

$$\vec{S} = \vec{k}\frac{c}{4\pi}\frac{4\pi^2\nu^2}{c^2}\left|A_x^0\right|^2 \sin^2 2\pi\nu(t - z/c). \tag{12.67}$$

The intensity is the time average magnitude of the Poynting vector. The time average is obtained by integrating the $\sin^2$ term over t from 0 to 2π, which yields 1/2. Therefore, the intensity for x polarized light, I_x, is

$$I_x = \frac{\pi\nu^2}{2c}\left|A_x^0\right|^2. \tag{12.68}$$

Using this relation in equation (12.65) gives

$$C_m^* C_m = \frac{2\pi}{c\hbar^2} I_x \left|x_{mn}\right|^2 t. \tag{12.69}$$

The probability of absorption per unit time depends linearly on the intensity of light.

Equation (12.69) is for light polarized in the x direction and a transition dipole bracket which couples the molecular transition to the light field only for the electric field oscillating along x. In general, the applied radiation field may have components polarized along x, y, and z, and the molecule may have transition dipole brackets that couple the molecular transition to the $\vec{E}$ fields along x, y, and z. Then,

$$C_m^* C_m = \frac{2\pi}{c\hbar^2}\left[I_x\left|x_{mn}\right|^2 + I_y\left|y_{mn}\right|^2 + I_z\left|z_{mn}\right|^2\right]t, \tag{12.70}$$

where I_x, I_y, and I_z are the intensities of light with polarizations along x, y, and z, respectively, and x_{mn}, y_{mn}, and z_{mn} are the transition dipole brackets for light polarized along x, y, and z, respectively.

A related expression describing the magnitude of the radiation field is the radiation density, ρ. ρ is defined as

$$\rho(\nu_{mn}) = \frac{1}{4\pi}\overline{E^2(\nu_{mn})}, \tag{12.71}$$

where $\overline{E^2(\nu_{mn})}$ is the time-averaged magnitude of the electric field.

$$\overline{E^2(\nu_{mn})} = \frac{2\pi^2\nu_{mn}^2}{c^2}\left|A^0(\nu_{mn})\right|^2. \tag{12.72}$$

Then,

$$\left|A^0(\nu_{mn})\right|^2 = \frac{2c^2}{\pi\nu_{mn}^2}\rho(\nu_{mn}). \tag{12.73}$$

For isotropic radiation (i.e., the radiation field has components with equal amplitudes polarized along x, y, and z)

$$\left|A_x^0(\nu_{mn})\right|^2 = \left|A_y^0(\nu_{mn})\right|^2 = \left|A_z^0(\nu_{mn})\right|^2 = \frac{1}{3}\left|A^0(\nu_{mn})\right|^2, \tag{12.74}$$

and $C_m^* C_m$ for isotropic radiation in terms of the radiation density is

$$C_m^* C_m = \frac{2\pi}{3\hbar^2}\left\{|x_{mn}|^2 + |y_{mn}|^2 + |z_{mn}|^2\right\}\rho(\nu_{mn})t. \tag{12.75}$$

The probability that a transition from the lower energy state to the higher energy state ($n \rightarrow m$) will take place in unit time with isotropic radiation is

$$B_{n\rightarrow m}\rho(\nu_{mn}) = \frac{2\pi}{3\hbar^2}|\mu_{mn}|^2\rho(\nu_{mn}), \tag{12.76}$$

where

$$|\mu_{mn}|^2 = |x_{mn}|^2 + |y_{mn}|^2 + |z_{mn}|^2. \tag{12.77}$$

μ_{mn} is the transition dipole bracket,

$$\mu_{mn} = \langle\psi_m^0|\, e\sum_j \underline{\mu}_j\, |\psi_n^0\rangle. \tag{12.78}$$

Equation (12.78) is analogous to equation (12.54) except that the $\underline{\mu}_j$ operator has x, y, and z components.

Following equation (12.59), the problem of absorption of energy from the radiation field is treated. The system begins in the lower energy state labeled by n; and the probability that a transition has occurred to the higher energy state, labeled m, is calculated. If the system is initially in the higher energy state, the identical treatment, except retention of the first exponential term in equation (12.59), gives the results for stimulated (induced) emission. Stimulated emission is the process in which the radiation field acts on the system (which is initially in the higher energy state) and causes a downward transition (i.e., energy is added to the field). The radiation field stimulates or induces a downward transition. [In equation (12.59), n is taken to be higher in energy than m so that the first exponential is the dominate term. To preserve the energy levels as depicted in Figure 12.1, in the following absorption is taken to be an $n \rightarrow m$ transition, and emission is taken to be an $m \rightarrow n$ transition.] Since the results are symmetrical for absorption and stimulated emission,

$$B_{m\rightarrow n}\rho(\nu_{mn}) = B_{n\rightarrow m}\rho(\nu_{mn}), \tag{12.79}$$

where $B_{n\rightarrow m}$ is defined by equation (12.76). $B_{n\rightarrow m}$ and $B_{m\rightarrow n}$ are called the Einstein B coefficients for absorption and stimulated emission, respectively. Equation (12.79) shows that the stimulated emission probability per unit time is equal to the absorption probability per unit time.

C. SPONTANEOUS EMISSION

The treatment of absorption and stimulated emission given above is semiclassical. This means that the molecules or atoms are treated with quantum mechanics, but the radiation field is described classically with Maxwell's equations. However, in addition to absorption and stimulated emission, there is another processes, spontaneous emission, that does not emerge naturally from the semiclassical treatment. Spontaneous emission is the emission of light from an excited state of a molecule or atom in the absence of a radiation field that can cause stimulated emission. If a molecule is placed in an excited state by absorption from a short pulse of light, via the result of a chemical reaction, via bombardment by an electron, or through some other means that subsequently leaves it in the dark, it can, nonetheless, emit a photon. The emission is referred to as fluorescence or phosphorescence. Fluorescence is spin-allowed spontaneous emission; phosphorescence is spin-forbidden spontaneous emission. The important point here is that these processes occur in the absence of a radiation field that induces emission.

The relationship between spontaneous emission and absorption and stimulated emission can be derived using an approach that treats both the molecule and the radiation field quantum mechanically. This will discussed very briefly below. It is also possible to derive the identical results using a thermodynamic argument put forward by Einstein. $A_{m\to n}$ is called the Einstein A coefficient for spontaneous emission. $A_{m\to n}$ is obtained by considering an ensemble of ground-state and excited-state molecules in the absence of a radiation field and requiring that the ensemble have a thermal equilibrium population distribution.

In a sample containing a large number of molecules, if N_m is the number of molecules in upper state with energy E_m, and N_n is the number of molecules in lower state with energy E_n, then at temperature T the Boltzmann distribution law gives

$$\frac{N_m}{N_n} = \frac{e^{-E_m/k_B T}}{e^{-E_n/k_{\mathbf{B}} T}} = e^{-h\nu_{mn}/k_{\mathbf{B}} T}, \tag{12.80}$$

where k_B is Boltzmann's constant. For the system to be in equilibrium, the number of upward transitions must equal the number of downward transitions. Therefore,

$$N_m \left\{A_{m\to n} + B_{m\to n}\, \rho(\nu_{mn})\right\} = N_n B_{n\to m} \rho(\nu_{mn}). \tag{12.81}$$

Note that $B_{n\to m}$ and $B_{m\to n}$ are multiplied by the radiation density, while $A_{m\to n}$ is not, since spontaneous emission will occur in the absence of a radiation field. The presence of $A_{m\to n}$ is necessary to cause the higher energy state to have less population than the lower energy state at equilibrium. Combining equations (12.80) and (12.81) gives

$$e^{-h\nu_{mn}/k_B T} = \frac{B_{n\to m}\rho(\nu_{mn})}{A_{m\to n} + B_{m\to n}\, \rho(\nu_{mn})}. \tag{12.82}$$

Equation (12.82) can be solved for $\rho(\nu_{mn})$ to give

$$\rho(\nu_{mn}) = \frac{A_{m\to n} e^{-h\nu_{mn}/k_B T}}{-B_{m\to n} e^{-h\nu_{mn}/k_B T} + B_{n\to m}}, \tag{12.83}$$

and since $B_{n\to m} = B_{m\to n}$, the equation reduces to

$$\rho(\nu_{mn}) = \frac{\dfrac{A_{m\to n}}{B_{m\to n}}}{e^{h\nu_{mn}/k_B T} - 1}. \tag{12.84}$$

To proceed further, the sample is assumed to be a black body, which has a blackbody radiation density given by

$$\rho(\nu_{mn}) = \frac{8\pi\, h\nu_{mn}^3}{c^3} \frac{1}{e^{h\nu_{mn}/k_B T} - 1}. \tag{12.85}$$

Substituting equation (12.85) into equation (12.84) and solving for $A_{m\to n}$ yields

$$A_{m\to n} = \frac{8\pi\, h\nu_{mn}^3}{c^3} B_{m\to n}. \tag{12.86}$$

Using equation (12.76) for $B_{m\to n}$, the Einstein A coefficient for spontaneous emission—that is, the probability per unit time for an excited atom or molecule to undergo spontaneous emission—is

$$A_{m\to n} = \frac{32\pi^3 \nu_{mn}^3}{3c^3\hbar} |\mu_{mn}|^2. \tag{12.87}$$

This result is within the context of the dipole approximation because the dipole approximation was used in the derivation of the B coefficient. Like absorption or stimulated emission, spontaneous emission depends on the absolute value squared of the transition dipole bracket. However, absorption and stimulated emission depend on the intensity of light [equation (12.69)] or the radiation density (12.76). These processes have zero probability of occurring if the intensity is zero. In contrast, the probability per unit time of a spontaneous emission event is independent of the intensity.

Note that the A coefficient does depend on the cube of the frequency difference between the higher and lower energy states in the transition. Thus, dipole transitions between electronic states, which occur in the visible or ultraviolet regions of the electromagnetic spectrum, have spontaneous emission lifetimes on the order of a few nanoseconds to 100 ns. Dipole transitions between vibrational states, which occur in the infrared, have spontaneous emission lifetimes of tens of microseconds to many milliseconds. The very low frequencies associated with magnetic resonance transitions result in no spontaneous emission occurring. (The time scale for spontaneous emission in NMR, which would be measured in millennia, is made even longer because the transitions depend on very small magnetic transition dipole brackets rather than the much stronger electric transition dipole brackets considered above.)

D. SELECTION RULES

For x polarized light, a dipole transition will occur only if

$$\langle m|\, e \sum_j x_j \,|n\rangle \neq 0. \tag{12.88}$$

In some cases, it is possible to determine if this bracket will be zero or nonzero without fully evaluating the magnitude of the bracket. For the harmonic oscillator, this bracket was evaluated in Chapter 6, equations (6.95)–(6.96). It was determined that the bracket would be zero unless $m = n \pm 1$. This type of result is referred to as a selection rule. Unless the selection rule is met, absorption or emission of light will not occur. If the selection rule is obeyed, the bracket is not necessarily zero, but it still can be zero for other reasons. For example, a homonuclear diatomic molecule, such as H_2, will not absorb or emit light even for $m = n \pm 1$ because the value of the bracket is zero.

In the Schrödinger representation, the dipole bracket takes the form

$$\int \psi_F^*(q)\, \vec{\mu}\, \psi_I(q)\, dq, \tag{12.89}$$

where $\psi_I(q)$ is the initial state of the system, $\psi_F^*(q)$ is the final state of the system, and q are all of the relevant coordinates. It is frequently possible to determine if this bracket will be zero using symmetry arguments. If the wavefunctions are one-dimensional (e.g., functions of x only), then it may be possible to characterize them as odd or even. The dipole operator is x, which is an odd function of x. If $\psi_I(x)$ is even, then the product of $\psi_I(x)$ and x is an odd function. Therefore, if $\psi_F^*(x)$ is even, the integral will vanish because the product of an odd and an even function is odd, and the integral of an odd function over all space is zero. Thus, if the initial state is even, the final state must be odd for a dipole transition to occur. Classifying the functions as odd and even is classifying them by their symmetry with respect to reflection through the line $x = 0$. The result that the product of the three functions in the expression (12.89) must be even or the integral will vanish is a symmetry selection rule.

Molecules, such as anthracene shown in Figure 5.4, are three-dimensional, and their wavefunctions cannot be classified simply as odd or even. However, group theory provides a method for assigning a symmetry to a molecular wavefunction as well as determining the symmetry associated with the product of any number of functions. The symmetry associated with the product must be "totally symmetric", the multidimensional group theory equivalent of even, or the integral (bracket) is identically equal to zero. Considering the symmetry of wavefunctions using group theory is an important method for analyzing transition dipole brackets. It must be emphasized that symmetry properties can determine if a transition is allowed, but they cannot determine the magnitude of an allowed bracket. Even if allowed, a bracket can be virtually zero, resulting in a weak or possibly unobservable optical transition.

E. LIMITATIONS OF THE TIME-DEPENDENT PERTURBATION THEORY TREATMENT

The results derived above are appropriate for many situation commonly encountered in problems dealing with the absorption and emission of light. The treatment is appropriate for weak radiation fields because time-dependent perturbation theory was used in the derivation. Time-dependent perturbation theory will be applicable if the probability of finding the system in the final state, $|m\rangle$, which is higher in energy than the initial state for absorption and lower in energy for emission, is always small; that is, $C_m^* C_m \ll 1$. This will be true for weak radiation fields or if the transition dipole bracket is small. Radiation fields produced by standard laboratory absorption and fluorescence spectrometers are so weak that, even for very highly absorbing molecules, the $C_m^* C_m \ll 1$ condition will be met. This condition does not imply that only a small fraction of the incident light is absorbed by the sample. If at the peak wavelength for absorption, the incident light contains 10^{12} photons per second, 90% of them can be absorbed by a sample with 10^{20} molecules in the optical path, and the probability of finding a molecule in its excited state is at most $\sim 10^{-8}$, one in a hundred million. Since the lifetimes of molecules are generally much shorter than a second, the probability is actually much smaller. If the lifetime of a molecule is 10 ns, then the number of photons per 10 ns is only 10^4. Then the probability of finding a molecule excited is $\sim 10^{-16}$. Even if there are many fewer molecules in the optical path and the light source is significantly more intense, the basic condition for the use of time-dependent perturbation theory can still be met.

In equation (12.43), the dipole approximation was made. This restricted the treatment to a subset of all possible types of transition brackets. Only electric dipole transitions were considered following equation (12.43). Electric dipole transitions give rise to the strongest transitions and are responsible for most absorption and emission of light by atoms and molecules. However, there are circumstances for which the electric dipole transition bracket is identically zero for all polarizations of light, that is,

$$\mu_{mn} = 0.$$

This can occur because of symmetry selection rules or angular momentum selection rules, or because the transition inherently involves the magnetic properties of the system, as in magnetic resonance, rather than the electronic properties of the system. Even if the μ_{mn} is identically zero, absorption or emission of radiation may still occur because of terms that were dropped when the dipole approximation was made. In equation (12.43), the vector potential was taken out of the bracket. A more general approach is to expand the vector potential with a multipole expansion. The first term in such an expansion is the electric dipole term. The higher-order terms are magnetic dipole, electric quadrupole, magnetic quadrupole, electric octapole, and so on, which give rise to magnetic dipole transitions, electric quadrupole transition, magnetic quadrupole transition, and electric octapole transitions, respectively. These higher-order transitions will only be important if the electric dipole transition is either identically or virtually zero.

For electric dipole allowed transitions and strong radiation fields, such as those produced by high-intensity lasers, time-dependent perturbation theory may not be applicable because $C_m^* C_m$ can become substantially greater than zero. As discussed briefly in Chapter 14, Section E, it is even possible for $C_m^* C_m \cong 1$. In such situations, it is necessary to solve the time-dependent Schrödinger equation or an equivalent formulation—for example, a density matrix treatment (Chapter 14). It may be necessary to solve coupled Maxwell and Schrödinger equations.

The time-dependent perturbation theory treatment given above or the solution of coupled Maxwell and Schrödinger equations are referred to as semiclassical treatments. The molecule is treated with quantum mechanics, but the radiation field is treated classically. In the time-dependent perturbation theory treatment presented here, energy is not explicitly conserved. In the absorption process, a molecule becomes excited, but the radiation field does not expressly lose energy. The radiation field is treated classically with Maxwell's equations. In a fully quantum mechanical treatment of the absorption and emission of radiation by atoms and molecules, the light as well as the atoms or molecules are treated quantum mechanically. Light is described as discrete photons, as in the discussion of photon wave packets in Chapter 3, Sections C and D, rather than as a classical wave. In Chapter 1, it was pointed out that problems can arise when light is not described as photons. One of the problems associated with the semiclassical treatment of absorption and emission of radiation is the failure of the theory to describe spontaneous emission on the same footing as absorption and stimulated emission.

In a quantum mechanical treatment of the radiation field, the part of the Hamiltonian associated with the radiation field is written in terms of raising and lowering operators, which are identical in nature to those developed in the Dirac treatment of the harmonic oscillator (Chapter 6, Section B). The operators a^+ and a are usually referred to as creation and annihilation operators, respectively. The ket $|n\rangle$ is the state of the radiation field containing n photons. $|n\rangle$ is an eigenket of the number operator. The application of the number operator to $|n\rangle$ yields n,

$$a^+a\,|n\rangle = n\,|n\rangle\,, \tag{12.90}$$

the number of photons in the radiation field. The application of the creation operator to $|n\rangle$ yields the state $|n+1\rangle$,

$$a^+\,|n\rangle = \sqrt{n+1}\,|n+1\rangle\,, \tag{12.91}$$

that is, it increases the number of photons in the radiation field by 1. The application of the annihilation operator gives the state $|n-1\rangle$,

$$a\,|n\rangle = \sqrt{n}\,|n-1\rangle\,, \tag{12.92}$$

that is, a state with one photon less in the radiation field.

Absorption and emission of radiation by an atom or molecule still depend on the square of a transition bracket, but now the bracket contains operators that operate on the states of the molecule and on the states of the radiation field. Absorption of a photon takes a molecule from a lower energy state to a higher energy state, and the annihilation operator operates on the radiation field state $|n\rangle$ to give a new state in which the radiation field has lost one photon. Thus energy is conserved. The annihilation operator brings out the factor $\sqrt{n}$; and when the bracket is squared, the result is a factor of n. Since the intensity of light is proportional to the number of photons, the probability of absorption is proportional to the intensity of light as in equation (12.69). If $n = 0$, the factor $\sqrt{n}$ is zero, and the probability of absorption is zero. This is the same as the semiclassical result. If the intensity is zero, absorption cannot occur.

Emission of a photon takes a molecule from a higher energy state to a lower energy state. The creation operator operates on the radiation field state $|n\rangle$ to give a new state in which the radiation field has gained one photon. Energy is conserved. The creation operator brings out the factor $\sqrt{n+1}$. When the bracket is squared, the result is a factor of $n+1$. This result is very different from the semiclassical result. When the intensity of the radiation field is high, then n is a very large number, and $n \gg 1$. Thus, the 1 can be dropped, and the probability of emission is proportional to n; that is, it is proportional to the intensity. This is stimulated emission, which is successfully treated by the semiclassical approach. However, in the quantum treatment, when $n = 0$, the intensity is zero, but the transition probability is not zero. The factor $\sqrt{n+1}$ squared gives 1. Thus, even when there are no photons present (the intensity is zero), emission of a photon by an molecule or atom in an excited state is still possible. This is spontaneous emission. Spontaneous emission arises naturally out of a quantum mechanical treatment of the absorption and emission of light because the radiation field is described in terms of photon creation and annihilation operators. It is possible to create a photon when $n = 0$, making possible spontaneous emission. Detailed analysis of the fully quantum mechanical problem yields the result given in equation (12.87), the Einstein A coefficient. However, the quantum mechanical treatment does not require grafting spontaneous emission onto the treatment of absorption and stimulated emission as is necessary in the semiclassical approach presented above.

It is possible to see qualitatively how spontaneous emission arises. Since the states of the radiation field are described quantum mechanically by the kets $|n_\omega\rangle$, where n_ω is the number of photons with frequency ω in the field, the energy in the field for a particular frequency is

$$E = \left(n_\omega + {}^1\!/_2\right)\hbar\omega, \tag{12.93}$$

just like the harmonic oscillator. Thus, even when $n = 0$ for all ω, the radiation field does not have zero energy. The state of the radiation field with all $n = 0$ is called the vacuum state. The vacuum state still has the zero point energy of each mode. This energy is not available for absorption because it is not possible to lower the state below $n = 0$. For a cavity with a volume V, the quantum mechanical electric field operator is

$$\underline{E}_{\vec{k}} = i(\hbar\omega_{\vec{k}}/2\varepsilon_o V)^{1/2}\varepsilon_{\vec{k}}\left\{\underline{a}_{\vec{k}}\exp(-i\omega_{\vec{k}}t + i\vec{k}\cdot\vec{r}) - \underline{a}^+_{\vec{k}}\exp(i\omega_{\vec{k}}t + i\vec{k}\cdot\vec{r})\right\}, \tag{12.94}$$

where $\vec{k}$ is the photon wave vector, $\omega_{\vec{k}}$ is the frequency of a photon with wave vector $\vec{k}$, ε_0 is the permittivity of vacuum, $\varepsilon_{\vec{k}}$ is the photon polarization, and $\vec{r}$ is the position. $\underline{a}_{\vec{k}}$ and $\underline{a}^+_{\vec{k}}$ are the annihilation and creation operators, respectively, for photons with wave vector $\vec{k}$. The exponential terms are the time- and position-dependent phase factors. Since $\underline{E}_{\vec{k}}$ contains $\underline{a}^+_{\vec{k}}$, the electric field does not vanish even when $n = 0$. Therefore, the vacuum state has electric fields at all frequencies. These electric fields are frequently referred to as fluctuations of the vacuum state. For a transition at frequency ω, there is a time-dependent electric field. Absorption cannot occur because the annihilation operator cannot lower the state below $n = 0$. However, since the creation operator brings out a factor $\sqrt{n+1}$, which is squared when the probability is calculated, emission of a photon by an excited atom or molecule can be driven by the fluctuations of the vacuum state. This emission is called spontaneous emission because it occurs although the state of the radiation field initially has $n = 0$. Spontaneous emission is inherently a quantum mechanical effect.

CHAPTER

13 THE MATRIX REPRESENTATION

In previous chapters, states of a quantum mechanical system have been represented by ket vectors and bra vectors—that is, $| \rangle$ and $\langle |$—in an abstract vector space. The matrix representation of quantum mechanics changes none of the ideas that have been developed previously. Rather, the matrix representation provides a convenient way of working with the vectors. The superposition principle is central to quantum theory. A ket vector can be expressed in terms of a complete set of other ket vectors. From a complete set of N vectors, it is possible to form other complete sets of N vectors. In particular, for a Hermitian operator, $\underline{A}$, representing a real dynamical variable, a set of vectors can be found which satisfy the eigenvalue equation,

$$\underline{A}\,|U_i\rangle = \alpha_i\,|U_i\rangle\,,$$

where the $|U_i\rangle$ are the eigenvectors, and the α_i are the eigenvalues (observables). The matrix representation provides a method to find the eigenstates of a given operator, and the associated eigenvalues, in terms of particular superpositions of a complete set of ket vectors (the basis vectors).

A. MATRICES AND OPERATORS

Consider an orthonormal basis in an N-dimensional vector space,

$$\left\{\left|e^j\right\rangle\right\}.$$

Any ket vector in the space can be written as a superposition of the basis vectors,

$$|x\rangle = \sum_{j=1}^{N} x_j \left|e^j\right\rangle, \tag{13.1}$$

with

$$x_j = \left\langle e^j \mid x\right\rangle. \tag{13.2}$$

In terms of the basis $\left\{\left|e^j\right\rangle\right\}$, the operator equation

$$|y\rangle = \underline{A}\,|x\rangle \tag{13.3}$$

becomes

$$\sum_{j=1}^{N} y_j \left| e^j \right\rangle = \underline{A} \sum_{j=1}^{N} x_j \left| e^j \right\rangle \tag{13.4a}$$

$$= \sum_{j=1}^{N} x_j \underline{A} \left| e^j \right\rangle . \tag{13.4b}$$

Left-multiplying by $\left\langle e^i \right|$ gives

$$y_i = \sum_{j=1}^{N} \left\langle e^i \right| \underline{A} \left| e^j \right\rangle x_j . \tag{13.5}$$

The N different y_i are the vector representatives of the vector $|y\rangle$ in terms of the basis set $\{|e^j\rangle\}$. The N^2 scalar products $\langle e^i | \underline{A} | e^j \rangle$ are completely determined by $\underline{A}$ and basis $\{|e^j\rangle\}$. Writing the brackets as

$$a_{ij} = \left\langle e^i \right| \underline{A} \left| e^j \right\rangle , \tag{13.6}$$

the linear transformation [operator equation (13.3)] becomes

$$y_i = \sum_{j=1}^{N} a_{ij} x_j , \qquad i = 1, \ 2, \ \ldots , N. \tag{13.7}$$

Equation (13.7) can be written in terms of the vector representatives of $|x\rangle$ and $|y\rangle$ in the basis $\{|e^j\rangle\}$:

$$x = [x_1, x_2, \ \ldots , x_N] \tag{13.8a}$$

$$y = [y_1, y_2, \ \ldots , y_N] . \tag{13.8b}$$

x and y are sets of numbers, not vectors. They are only meaningful when the basis has been specified. When multiplied by the basis, they give the corresponding vectors. Equation (13.7) is a set of N linear algebraic equations. In terms of the vector representatives, x and y, it can be written as

$$y = \underline{\underline{A}} \, x , \tag{13.9}$$

where $\underline{\underline{A}}$ denotes array of coefficients,

$$\underline{\underline{A}} = (a_{ij}) = \begin{bmatrix} a_{11} & a_{12} & \cdots & a_{1N} \\ a_{21} & a_{22} & \cdots & a_{2N} \\ \vdots & & & \\ a_{N1} & a_{N2} & \cdots & a_{NN} \end{bmatrix} . \tag{13.10}$$

The double line beneath a letter is used to indicate a matrix. The a_{ij}, the brackets given in equation (13.6), are the elements of the matrix $\underline{\underline{A}}$. These are generally referred to as matrix elements rather than brackets. Frequently, a closed bracket is called a matrix element, even if the matrix formulation of quantum mechanics is not being used.

Equation (13.9) is a central result. It shows that the operator equation (13.3) can be expressed as the product of a matrix and a vector representative to give a new vector representative. Each of these is in terms of a specified basis set. The matrix elements, a_{ij}, can be calculated from equation (13.6), since the basis vectors, the $\{|e^j\rangle\}$, and the operator, $\underline{A}$, are known. Thus the linear transformation of a ket into another ket by a linear operator is reduced to multiplication of a vector representative by a matrix.

The necessary mathematical machinery to use the matrix formulation of quantum mechanics is developed in linear algebra. A number of definitions, relationships, and theorems are presented here because they are necessary for the development and application of the matrix representation. The material is not exhaustive, and a full treatment of the necessary linear algebra can be found in the large number of books on the subject.

Two matrices, $\underline{\underline{A}}$ and $\underline{\underline{B}}$, are equal,

$$\underline{\underline{A}} = \underline{\underline{B}} \tag{13.11a}$$

if

$$a_{ij} = b_{ij}, \tag{13.11b}$$

that is, the corresponding matrix elements are equal.

The unit matrix is defined as

$$\underline{\underline{1}} = \underline{\underline{\delta}}_{ij} = \begin{bmatrix} 1 & 0 & \cdots & 0 \\ 0 & 1 & \cdots & 0 \\ \vdots & \vdots & & \vdots \\ 0 & 0 & \cdots & 1 \end{bmatrix}. \tag{13.13}$$

The unit matrix has ones down the principal diagonal (the line from the upper left-hand corner to the lower right-hand corner of the matrix). The unit matrix is the identity in matrix transformations. Its application produces the identity transformation

$$y_i = \sum_{j=1}^{N} \delta_{ij} x_j = x_i,$$

where the ith element of y_i is x_i, which corresponds to

$$|y\rangle = \underline{1}\,|x\rangle = |x\rangle\,.$$

The zero matrix is defined as

$$\underline{\underline{0}} = \begin{bmatrix} 0 & 0 & \cdots & 0 \\ 0 & 0 & \cdots & 0 \\ \vdots & & & \vdots \\ 0 & 0 & \cdots & 0 \end{bmatrix}, \tag{13.12a}$$

and

$$\underline{\underline{0}}x = 0. \tag{13.12b}$$

Two matrices can be multiplied to give a new matrix. Consider the operator equations

$$|y\rangle = \underline{A}\,|x\rangle \tag{13.13a}$$

and

$$|z\rangle = \underline{B}\,|y\rangle\,. \tag{13.13b}$$

These are the equivalent of

$$|z\rangle = \underline{B}\,\underline{A}\,|x\rangle\,. \tag{13.14}$$

Using the same basis for both transformations, equation (13.13b) becomes

$$z_k = \sum_{i=1}^{N} b_{ki}\, y_i, \tag{13.15a}$$

or

$$z = \underline{\underline{B}}\, y, \tag{13.15b}$$

where $\underline{\underline{B}}$ is the matrix (b_{ki}). There are the analogous equations for equation (13.13a). Then the transformation in equation (13.14) is

$$z = \underline{\underline{B}}\,\underline{\underline{A}}\, x = \underline{\underline{C}}x, \tag{13.16}$$

where

$$\underline{\underline{C}} = \underline{\underline{B}}\,\underline{\underline{A}}, \tag{13.17}$$

which has matrix elements

$$c_{kj} = \sum_{i=1}^{N} b_{ki}a_{ij}. \tag{13.18}$$

Equation (13.18) is the law of matrix multiplication. It defines the elements of the matrix $\underline{\underline{C}}$ in terms of the elements of the matrices $\underline{\underline{A}}$ and $\underline{\underline{B}}$. The matrix $\underline{\underline{C}}$ transforms the vector representative x into the vector representative z.

Matrix multiplication is associative; that is,

$$(\underline{\underline{A}}\,\underline{\underline{B}})\underline{\underline{C}} = \underline{\underline{A}}(\underline{\underline{B}}\,\underline{\underline{C}}). \tag{13.19}$$

In general, matrix multiplication is not commutative.

$$\underline{\underline{A}}\,\underline{\underline{B}} \neq \underline{\underline{B}}\,\underline{\underline{A}}. \tag{13.20}$$

The fact that matrix multiplication is not commutative plays a very important role in the matrix representation. Since operators do not necessarily commute, matrices, which will represent operators, must also have this property.

Matrices can be multiplied by complex numbers and added.

$$\alpha \underline{\underline{A}} + \beta \underline{\underline{B}} = \underline{\underline{C}}. \tag{13.21}$$

The elements of the matrix $\underline{\underline{C}}$ are

$$c_{ij} = \alpha a_{ij} + \beta b_{ij}. \tag{13.22}$$

The inverse of a matrix $\underline{\underline{A}}$ is defined as the matrix $\underline{\underline{A}}^{-1}$ such that

$$\underline{\underline{A}}\,\underline{\underline{A}}^{-1} = \underline{\underline{A}}^{-1}\underline{\underline{A}} = \underline{\underline{1}}. \tag{13.23}$$

The inverse of a product is

$$(\underline{\underline{A}}\,\underline{\underline{B}})^{-1} = \underline{\underline{B}}^{-1}\underline{\underline{A}}^{-1}, \tag{13.24}$$

the product of the inverses in reverse order.

For the matrix defined as $\underline{\underline{A}} = (a_{ij})$, the transpose of a matrix is

$$\tilde{\underline{\underline{A}}} = (a_{ji}); \tag{13.25}$$

the rows and columns are interchanges.

The complex conjugate of a matrix is

$$\underline{\underline{A}}^* = (a_{ij}^*); \tag{13.26}$$

the matrix with the complex conjugate of each element.

The Hermitian conjugate of a matrix is

$$\underline{\underline{A}}^+ = (a_{ji}^*); \tag{13.27}$$

the complex conjugate of the transpose.

The transpose of the product of matrices is the product of the transposes in reverse order:

$$\left(\widetilde{\underline{\underline{A}}\,\underline{\underline{B}}}\right) = \tilde{\underline{\underline{B}}}\,\tilde{\underline{\underline{A}}}. \tag{13.28}$$

The determinant of the transpose of a matrix is the determinant of the matrix.

$$|\tilde{\underline{\underline{A}}}| = |\underline{\underline{A}}|. \tag{13.29}$$

The complex conjugate of the product of two matrices is the product of their complex conjugates:

$$(\underline{\underline{A}}\,\underline{\underline{B}})^* = \underline{\underline{A}}^*\underline{\underline{B}}^* \tag{13.30}$$

The determinant of the complex conjugate of a matrix is the complex conjugate of the determinant:

$$|\underline{\underline{A}}^*| \;=\; |\underline{\underline{A}}|^*. \tag{13.31}$$

The Hermitian conjugate of the product of matrices is the product of the Hermitian conjugates in reverse order:

$$(\underline{\underline{A}}\,\underline{\underline{B}})^{+} = \underline{\underline{B}}^{+}\underline{\underline{A}}^{+}. \tag{13.32}$$

The determinant of the Hermitian conjugate is the complex conjugate of the determinant:

$$|\underline{\underline{A}}^{+}| = |\underline{\underline{A}}|^{*}. \tag{13.33}$$

The inverse of a matrix is the transpose of the cofactor matrix (matrix of the signed minors) divided by the determinant:

$$\underline{\underline{A}}^{-1} = \frac{\tilde{\underline{\underline{A}}}^{C}}{|\underline{\underline{A}}|}. \tag{13.34}$$

A matrix can only have an inverse if $|\underline{\underline{A}}| \neq 0$. If $|\underline{\underline{A}}| = 0$, then $\underline{\underline{A}}$ is said to be singular. For $\underline{\underline{C}} = \underline{\underline{A}}\,\underline{\underline{B}}$, since $|\underline{\underline{C}}| = |\underline{\underline{A}}|\,|\underline{\underline{B}}|$, if either $|\underline{\underline{A}}| = 0$ or $|\underline{\underline{B}}| = 0$, then $\underline{\underline{C}}$ is also singular. The procedure given in equation (13.34) is frequently unnecessary because the matrix of interest is unitary. The unitary property is defined below.

Matrices that appear frequently in quantum theory have been given special names.

Symmetric	$\underline{\underline{A}} = \tilde{\underline{\underline{A}}}$
Hermitian	$\underline{\underline{A}} = \underline{\underline{A}}^{+}$
Real	$\underline{\underline{A}} = \underline{\underline{A}}^{*}$
Imaginary	$\underline{\underline{A}} = -\underline{\underline{A}}^{*}$
Unitary	$\underline{\underline{A}}^{-1} = \underline{\underline{A}}^{+}$
Diagonal	$a_{ij} = a_{ij}\,\delta_{ij}$

As well be seen below, two of these that are particularly important are Hermitian and unitary matrices. Note that to find the inverse of a unitary matrix, it is only necessary to take its Hermitian conjugate.

A matrix raised to the power n is the matrix multiplied by itself n times, that is,

$$\underline{\underline{A}}^{0} = \underline{\underline{1}}, \quad \underline{\underline{A}}^{1} = \underline{\underline{A}}, \qquad \underline{\underline{A}}^{2} = \underline{\underline{A}}\,\underline{\underline{A}}\ldots; \tag{13.35}$$

using this, the exponential of a matrix can be written as

$$e^{\underline{\underline{A}}} = 1 + \underline{\underline{A}} + \frac{\underline{\underline{A}}^{2}}{2!} + \cdots. \tag{13.36}$$

A one-column matrix is a column vector:

$$x = \begin{bmatrix} x_1 \\ x_2 \\ \vdots \\ x_N \end{bmatrix} \tag{13.37}$$

x is actually a vector representative. It defines the vector $|x\rangle$ in terms of the specified basis set. Multiplying each element of the column vector x by its basis vector gives $|x\rangle$. $y = \underline{\underline{A}}\, x$ is

$$\begin{bmatrix} y_1 \\ y_2 \\ \vdots \\ y_N \end{bmatrix} = \begin{bmatrix} a_{11} & a_{12} & \cdots & a_N \\ a_{21} & a_{22} & & \\ \vdots & & & \vdots \\ a_{N1} & a_{N2} & \cdots & a_{NN} \end{bmatrix} \begin{bmatrix} x_1 \\ x_2 \\ \vdots \\ x_N \end{bmatrix}. \tag{13.38}$$

The transpose of a column vector is a row vector:

$$\tilde{x} = (x_1, x_2 \ldots, x_N) \tag{13.39}$$

Taking the transpose of both sides of the equation $y = \underline{\underline{A}}\, x$ gives

$$\tilde{y} = \tilde{x}\tilde{\underline{\underline{A}}}, \tag{13.40}$$

and taking the Hermitian conjugate of $y = \underline{\underline{A}}\, x$ gives

$$y^+ = x^+ \underline{\underline{A}}^+. \tag{13.41}$$

B. CHANGE OF BASIS SET

An important transformation that can be performed using matrix methods is to change from one orthonormal basis set to another. For the orthonormal basis $\{|e^i\rangle\}$, the basis vectors have the property

$$\langle e^i \mid e^j \rangle = \delta_{ij} \qquad (i, j = 1, 2, \ldots, N). \tag{13.42}$$

Any vector in the N dimension vector space spanned by $\{|e^i\rangle\}$ can be expressed as a superposition of the basis vectors. It is also possible to find other basis vectors that span the same space. By superposition, N new linearly independent vectors can be generated which form a basis $\{|e^{i'}\rangle\}$, with the new vectors given by

$$\left|e^{i'}\right\rangle = \sum_{k=1}^{N} u_{ik} \left|e^k\right\rangle, \qquad i = 1, 2, \ldots, N. \tag{13.43}$$

The u_{ik} are suitably chosen complex numbers. The for the appropriate choice of the u_{ik}, the new basis will also be orthonormal, that is,

$$\left\langle e^{j'} \middle| e^{i'} \right\rangle = \delta_{ij}. \tag{13.44}$$

A result from linear algebra defines the nature of the u_{ik} that will result in the new basis being orthonormal. Using equation (13.43) and a corresponding equation for the bra, and substituting into equation (13.44) gives

$$\sum_l \sum_k u_{jl}^* u_{ik} \langle e^l \mid e^k \rangle = \delta_{ij}. \tag{13.45}$$

Since the basis $\{|e^i\rangle\}$ is orthonormal, the bracket is zero unless $l = k$, in which case it is 1. Then,

$$\sum_k u_{jk}^* u_{ik} = \delta_{ij}. \tag{13.46}$$

For this condition to be met, the matrix of the coefficients in equation (13.43),

$$\underline{\underline{U}} = (u_{ik})$$

must satisfy

$$\underline{\underline{U}}^+ \underline{\underline{U}} = \underline{\underline{1}}. \tag{13.47}$$

Thus, $\underline{\underline{U}}$ must be nonsingular, and

$$\underline{\underline{U}}^{-1} = \underline{U}^+. \tag{13.48}$$

The important result is that the new basis, $\{|e^{i'}\rangle\}$, will be orthonormal if the matrix $\underline{\underline{U}}$ is unitary. A unitary transformation takes one orthonormal basis into another orthonormal basis. The unitary property can be written in a more symmetrical form as

$$\underline{\underline{U}}\,\underline{\underline{U}}^+ = \underline{\underline{U}}^+\underline{\underline{U}} = \underline{\underline{1}}. \tag{13.49}$$

The new basis $\{|e'\rangle\}$, will be orthonormal if the transformation matrix is unitary. A unitary transformation substitutes the orthonormal basis $\{|e'\rangle\}$ for orthonormal basis $\{|e\rangle\}$. $|x\rangle$ is a vector that defines a line in a vector space. It can be written in terms of an orthonormal basis as

$$|x\rangle = \sum_i x_i \left|e^i\right\rangle \tag{13.50}$$

or, in terms of a different orthonormal basis, as

$$|x\rangle = \sum_i x_i' \left|e^{i'}\right\rangle. \tag{13.51}$$

$|x\rangle$ is the same vector, defining the same directed line segment, regardless of which basis set is used to represented it. The unitary transformation $\underline{\underline{U}}$ can be used to change from a vector representative of $|x\rangle$ in one basis to a vector representative of $|x\rangle$ in another basis. If x is the representative of $|x\rangle$ in the unprimed basis and x' is the vector representative in the primed basis, then

$$x' = \underline{\underline{U}}\,x \tag{13.52a}$$

and

$$x = \underline{\underline{U}}^+ x'. \tag{13.52b}$$

x and x' are representatives in different basis, but the vector, $|x\rangle$, is unchanged.

As a simple example, consider the basis, $\{\hat{x}, \hat{y}, \hat{z}\}$. The vector $|s\rangle$ is a directed line segment in real space. In terms of the $\{\hat{x}, \hat{y}, \hat{z}\}$ basis,

$$|s\rangle = 7\hat{x} + 7\hat{y} + 1\hat{z}. \tag{13.53}$$

The vector representative of $|s\rangle$ in the $\{\hat{x}, \hat{y}, \hat{z}\}$ basis is

$$s = \begin{bmatrix} 7 \\ 7 \\ 1 \end{bmatrix}. \tag{13.54}$$

The basis used and illustrated in Figure 13.1 is not unique. There are an infinite number of other orthonormal basis sets that will span three dimensional space. Another basis can be obtained by rotating the axis system 45° around the z axis. The new representative, s', of the vector $|s\rangle$ is obtained by performing the operation

$$s' = \underline{\underline{U}}\, s, \tag{13.55}$$

where $\underline{\underline{U}}$ is the rotation matrix

$$\underline{\underline{U}} = \begin{pmatrix} \cos\theta & \sin\theta & 0 \\ -\sin\theta & \cos\theta & 0 \\ 0 & 0 & 1 \end{pmatrix}. \tag{13.56}$$

For a 45° rotation

$$\underline{\underline{U}} = \begin{pmatrix} \sqrt{2}/2 & \sqrt{2}/2 & 0 \\ -\sqrt{2}/2 & \sqrt{2}/2 & 0 \\ 0 & 0 & 1 \end{pmatrix}. \tag{13.57}$$

Then

$$s' = \begin{pmatrix} \sqrt{2}/2 & \sqrt{2}/2 & 0 \\ -\sqrt{2}/2 & \sqrt{2}/2 & 0 \\ 0 & 0 & 1 \end{pmatrix} \begin{pmatrix} 7 \\ 7 \\ 1 \end{pmatrix} = \begin{pmatrix} 7\sqrt{2} \\ 0 \\ 1 \end{pmatrix} \tag{13.58}$$

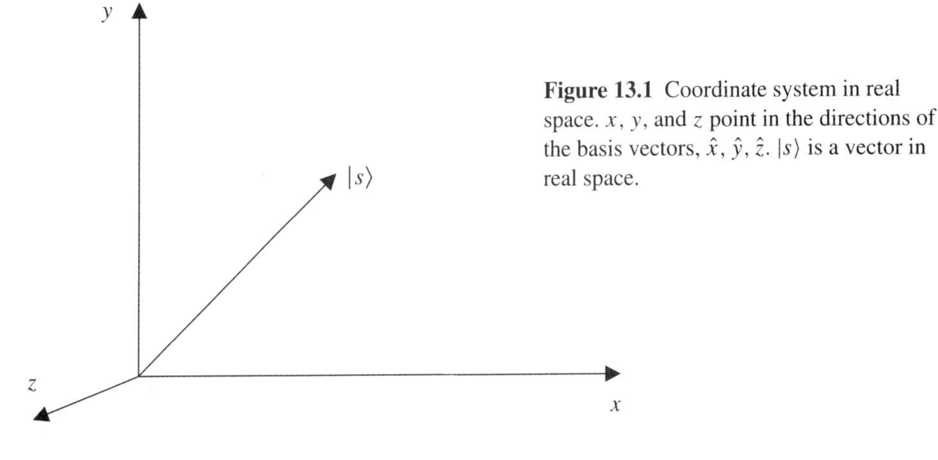

Figure 13.1 Coordinate system in real space. x, y, and z point in the directions of the basis vectors, $\hat{x}$, $\hat{y}$, $\hat{z}$. $|s\rangle$ is a vector in real space.

and

$$s' = \begin{bmatrix} 7\sqrt{2} \\ 0 \\ 1 \end{bmatrix}. \tag{13.59}$$

s' represents the same vector as s but in a different basis. The properties calculated for a vector are independent of the basis used to obtain its vector representative. For example, the length of the vector is $(\langle s \mid s \rangle)^{1/2}$ (see Chapter 2, Section A). In the original basis, the length is calculated as

$$\begin{aligned}[\langle s \mid s \rangle]^{1/2} &= (s^* \cdot s)^{1/2} \\ &= (49 + 49 + 1)^{1/2} = (99)^{1/2}.\end{aligned}$$

In the new basis, the length is

$$\begin{aligned}[\langle s \mid s \rangle]^{1/2} &= (s'^* \cdot s')^{1/2} \\ &= (2 \cdot 49 + 0 + 1)^{1/2} = (99)^{1/2}.\end{aligned}$$

The length and other properties of a vector are independent of the orthonormal basis set used to describe the vector.

Consider the linear transformation of the vector $|x\rangle$ into the vector $|y\rangle$ by the operator $\underline{A}$ in terms of their representatives in the basis $\{|e\rangle\}$:

$$y = \underline{\underline{A}}\, x, \tag{13.60a}$$

that is,

$$y_i = \sum_j a_{ij} x_j. \tag{13.60b}$$

If the basis is changed to a new orthonormal basis $\{|e'\rangle\}$ via with the unitary matrix $\underline{\underline{U}}$, equation (13.60a) can be changed to the new basis as well:

$$y' = \underline{\underline{U}}\, y = \underline{\underline{U}}\,\underline{\underline{A}}\, x = \underline{\underline{U}}\,\underline{\underline{A}}\,\underline{\underline{U}}^{+} x', \tag{13.61}$$

where equation (13.52) has been used. The equation in the $\{|e'\rangle\}$ basis is

$$y' = \underline{\underline{A}}' x', \tag{13.62}$$

with the matrix $\underline{\underline{A}}'$ given by

$$\underline{\underline{A}}' = \underline{\underline{U}}\,\underline{\underline{A}}\,\underline{\underline{U}}^{+}; \tag{13.63}$$

and since $\underline{\underline{U}}$ is unitary,

$$\underline{\underline{A}}' = \underline{\underline{U}}\,\underline{\underline{A}}\,\underline{\underline{U}}^{-1}. \tag{13.64}$$

Equation (13.64) is a very important result. A matrix representing an operator in one orthonormal basis can be transformed into the matrix representing the operator in another orthonormal basis using equation 13.64. This transformation is called a similarity transformation.

Any equation involving vector representatives and matrices can be transformed from one orthonormal basis to another. The equations

$$y = \underline{\underline{A}}\, x, \qquad \underline{\underline{A}}\,\underline{\underline{B}} = \underline{\underline{C}}, \qquad \underline{\underline{A}} + \underline{\underline{B}} = \underline{\underline{C}} \tag{13.65}$$

become, after a change of basis from the unprimed basis to the primed basis,

$$y' = \underline{\underline{A}}'\, x', \qquad \underline{\underline{A}}'\underline{\underline{B}}' = \underline{\underline{C}}', \qquad \underline{\underline{A}}' + \underline{\underline{B}}' = \underline{\underline{C}}'.$$

For example,

$$\underline{\underline{A}}\,\underline{\underline{B}} = \underline{\underline{C}}$$

$$\underline{\underline{U}}\,\underline{\underline{A}}\,\underline{\underline{B}}\,\underline{\underline{U}}^{+} = \underline{\underline{U}}\,\underline{\underline{C}}\,\underline{\underline{U}}^{+}.$$

Since $\underline{\underline{U}}^{+}\underline{\underline{U}}$ equals the identity matrix, it can be inserted between $\underline{\underline{A}}\,\underline{\underline{B}}$ on the left-hand side of the equation to give

$$\underline{\underline{U}}\,\underline{\underline{A}}\,\underline{\underline{U}}^{+}\underline{\underline{U}}\,\underline{\underline{B}}\,\underline{\underline{U}}^{+} = \underline{\underline{U}}\,\underline{\underline{C}}\,\underline{\underline{U}}^{+},$$

which yields

$$\underline{\underline{A}}'\underline{\underline{B}}' = \underline{\underline{C}}'.$$

The developments given above demonstrate that there is an isomorphism between vectors and operators in an abstract vector space and their vector and matrix representatives. Equations (13.65) are representatives of the operator equations

$$|y\rangle = \underline{A}\,|x\rangle\,, \qquad \underline{A}\,\underline{B} = \underline{C}, \qquad \underline{A} + \underline{B} = \underline{C}.$$

Because of isomorphism, it is unnecessary to distinguish abstract vectors and operators from their representatives. Matrices and vector representatives can be used in place of operators and abstract vectors. For example, the scalar product defined in an abstract vector space as $\langle x \mid y\rangle$ goes over to

$$\langle x \mid y\rangle = \sum_i x_i^* y_i = x^* \cdot y,$$

where x^* and y are representatives of vectors $\langle x|$ and $|y\rangle$. The scalar product is independent of particular basis because it is defined in terms of the abstract vectors, and the properties of vectors are independent of the orthonormal basis used to represent them.

$$x^* y = (\underline{\underline{U}}\, x)^* \underline{\underline{U}}\, y = x'^* y'$$

x'^* and y' are also representatives of $\langle x|$ and $|y\rangle$, but in the primed rather than the unprimed orthonormal basis.

C. HERMITIAN OPERATORS AND MATRICES

As discussed in Chapter 2, Sections B and C, a Hermitian operator has the property

$$\langle x|\, \underline{A}\, |y\rangle = \overline{\langle y|\, \underline{A}\, |x\rangle}.$$

In quantum theory, real dynamical variables (observables) are represented by Hermitian operators. Observables are the eigenvalues of Hermitian operators. Observables are calculated by solving the eigenvalue equation for the Hermitian operator corresponding to the observable of interest. In the matrix representation of quantum mechanics, a Hermitian operator is represented by a Hermitian matrix—that is, a matrix with the property $\underline{\underline{A}} = \underline{\underline{A}}^{+}$.

In linear algebra, there is a theorem that relates matrices to the solution of eigenvalues problems. While the eigenvalue problem is a subject of mathematics in the field of linear algebra, the theorem provides the foundation of the matrix representation of quantum mechanics. The proof of the theorem is quite lengthy and will not be given. It can be found in comprehensive books on linear algebra.

❖ **Theorem:** For a Hermitian operator $\underline{A}$ in a linear vector space of N dimensions, there exists an orthonormal basis $|U^1\rangle,\ |U^2\rangle \cdots |U^N\rangle$ relative to which $\underline{A}$ is represented by a diagonal matrix:

$$\underline{\underline{A}}' = \begin{pmatrix} \alpha_1 & 0 & 0 & \cdots \\ 0 & \alpha_2 & & 0 \\ 0 & & & \vdots \\ \vdots & 0 & \cdots & \alpha_N \end{pmatrix}. \tag{13.66}$$

The vectors $|U^i\rangle$ and the corresponding real number α_i are solutions of the eigenvalue equation

$$\underline{A}\, |U\rangle = \alpha\, |U\rangle\,, \tag{13.67}$$

and there are no other eigenvalues.

While the theorem as stated applies to finite matrices, the analogous statements can be made for infinite matrices as well.

The theorem is used in the following manner. An operator $\underline{A}$ can be represented by a matrix $\underline{\underline{A}}$ in some orthonormal basis $\{|e^i\rangle\}$. The basis used to represent the matrix can be selected in any convenient manner. There is another basis, $\{|U^i\rangle\}$, in which the matrix, $\underline{\underline{A}}'$, representing the operator, is diagonal. To go from the arbitrary basis $\{|e^i\rangle\}$ to the eigenvector basis, $\{|U^i\rangle\}$, there is a unitary transformation, that is,

$$\{|U^i\rangle\} = \underline{\underline{U}}\{|e^i\rangle\}. \tag{13.68}$$

Since $\underline{\underline{U}}$ takes the arbitrary basis into the eigenvector basis, the similarity transformation,

$$\underline{\underline{A}}' = \underline{\underline{U}}\,\underline{\underline{A}}\,\underline{\underline{U}}^{-1}, \tag{13.69}$$

takes the matrix $\underline{\underline{A}}$ in the arbitrary basis into the matrix $\underline{\underline{A}}'$ in the eigenvector basis. Therefore, in the matrix representation, solving the eigenvalue problem becomes the problem of matrix diagonalization. The matrix $\underline{\underline{A}}'$ is diagonal and has the eigenvalues as its diagonal elements, and the matrix $\underline{\underline{U}}$ transforms the arbitrary basis into the eigenvectors. If the initial basis that is chosen is the eigenvector basis, then the initial matrix will be diagonal with the eigenvalues as its elements.

Prior to the introduction of the matrix representation, the state of a system is described by a ket vector. Dynamical variables are represented by linear operators. Application of a linear operator to a ket produces a linear transformation

$$|y\rangle = \underline{A}\,|x\rangle\,.$$

Real dynamical variables (observables) are represented by Hermitian operators. Observables are eigenvalues of Hermitian operators

$$\underline{A}\,|S\rangle = \alpha\,|S\rangle\,.$$

Solution of the eigenvalue problem gives eigenvalues and eigenvectors.

In matrix representation, nothing is fundamentally changed. A Hermitian operator is replaced with its corresponding Hermitian matrix.

$$\underline{A} \rightarrow \underline{\underline{A}}.$$

In the proper basis, the matrix $\underline{\underline{A}}'$ is the diagonalized Hermitian matrix, and the diagonal matrix elements are the eigenvalues (observables). A suitable similarity transformation

$$\underline{\underline{A}}' = \underline{\underline{U}}\,\underline{\underline{A}}\,\underline{\underline{U}}^{-1}$$

takes the matrix $\underline{\underline{A}}$, written in an arbitrary basis set, into the matrix $\underline{\underline{A}}'$, which is diagonal, and the unitary matrix $\underline{\underline{U}}$ takes the arbitrary basis into the basis composed of the eigenvectors.

The matrix representation is another way of dealing with operators and solving eigenvalue problems. Since a matrix can replace an operator and a vector representative can replace a ket vector, all of the relationships among operators and ket vectors that have been described previously apply to matrices and vector representatives. For example, the statement that two Hermitian operators will only have simultaneous eigenvectors if and only if they commute becomes a statement about matrix diagonalization. Two Hermitian matrices, $\underline{\underline{A}}$ and $\underline{\underline{B}}$, can be simultaneously diagonalized by the same unitary transformation if and only if they commute.

D. THE HARMONIC OSCILLATOR IN THE MATRIX REPRESENTATION

The harmonic oscillator is treated in Chapter 6 in both the Schrödinger representation and using the Dirac raising and lower operator approach. In some textbooks, the Dirac method is treated as an example of the matrix representation. However, the development of the harmonic oscillator with raising and lowering operators only employed the application of the general ideas of quantum theory and did not require any of the matrix representation developments given above. Here, the results of the Dirac treatment will be rewritten using the matrix representation. Since the problem has already been solved, the eigenkets are known. By treating the problem using the eigenkets as the basis set, the problem of finding a unitary transformation from an arbitrary basis to the eigenvector basis is avoided. In Section E the method for obtaining such a transformation is presented.

Using the development of Chapter 6, Section B, the harmonic oscillator Hamiltonian is written as

$$\underline{H} = \frac{1}{2}(\underline{P}^2 + \underline{x}^2) \tag{13.70a}$$

$$= \frac{1}{2}(\underline{a}\,\underline{a}^+ + \underline{a}^+\underline{a}), \tag{13.70b}$$

where $\underline{P}^2$ is the square of the momentum operator, and $\underline{x}^2$ is the square of the position operator. $\underline{a}$ is the lowering operator, and $\underline{a}^+$ is the raising operator. The eigenkets of the harmonic oscillator energy eigenvalue problem are the kets $|n\rangle$, the occupation number kets. The application of the lowering and raising operators to the kets $|n\rangle$ gives

$$\underline{a}\,|n\rangle = \sqrt{n}\,|n-1\rangle \tag{13.71a}$$

$$\underline{a}^+\,|n\rangle = \sqrt{n+1}\,|n+1\rangle\,. \tag{13.71b}$$

The lowering and raising operators can be written as matrices in the basis of the kets $|n\rangle$. To write the matrix for the lowering operator, it is necessary to determine the matrix elements of $\underline{a}$ for all $|n\rangle$ (i.e., all $\langle m|\,\underline{a}\,|n\rangle$) and then collect all of the matrix elements in a matrix (a_{ij}). The matrix elements are

$$\begin{array}{l}
\langle 0|\,\underline{a}\,|0\rangle = 0 \\
\langle 0|\,\underline{a}\,|1\rangle = \sqrt{1} \\
\langle 0|\,\underline{a}\,|2\rangle = 0 \\
\quad\vdots \\
\langle 1|\,\underline{a}\,|0\rangle = 0 \\
\langle 1|\,\underline{a}\,|1\rangle = 0 \\
\langle 1|\,\underline{a}\,|2\rangle = \sqrt{2} \\
\langle 1|\,\underline{a}\,|3\rangle = 0 \\
\quad\vdots
\end{array}\quad. \tag{13.72}$$

The matrix for the lowering operator a has the form

$$\underline{\underline{a}} = \begin{array}{c} \\ \langle 0| \\ \langle 1| \\ \langle 2| \\ \langle 3| \\ \cdot \\ \cdot \\ \cdot \\ \cdot \end{array}
\begin{array}{c}
\begin{array}{ccccc} |0\rangle & |1\rangle & |2\rangle & |3\rangle & \cdots \end{array} \\
\begin{pmatrix}
0 & \sqrt{1} & 0 & 0 & 0 & \cdots \\
0 & 0 & \sqrt{2} & 0 & 0 & \\
0 & 0 & 0 & \sqrt{3} & 0 & \\
0 & 0 & 0 & 0 & \sqrt{4} & \\
\vdots & & & & & \ddots
\end{pmatrix}
\end{array}. \tag{13.73}$$

To place these matrix elements into the matrix, it is useful to write the basis kets across the top of the matrix and to write the corresponding bras down the left-hand side of the matrix. The position of any matrix element in the array is clear. The matrix element is the ket on the top, which is operated on by the operator, and the matrix element is closed by the bra in the column on the left. The lowering operator matrix has the elements $\sqrt{n}$ on a diagonal one above the principal diagonal.

In the same manner, the raising operator matrix, $\underline{\underline{a}}^{+}$, is obtained.

$$\underline{\underline{a}}^{+} = \begin{pmatrix} 0 & 0 & 0 & 0 & \cdot & \cdot \\ \sqrt{1} & 0 & 0 & 0 & \cdot & \cdot \\ 0 & \sqrt{2} & 0 & 0 & \cdot & \cdot \\ 0 & 0 & \sqrt{3} & 0 & \cdot & \cdot \\ 0 & 0 & 0 & \sqrt{4} & \cdot & \cdot \\ \cdot & \cdot & \cdot & \cdot & \cdot & \cdot \end{pmatrix}. \tag{13.74}$$

The raising operator has the elements $\sqrt{n+1}$ on a diagonal one below the principal diagonal.

The Hamiltonian matrix $\underline{\underline{H}}$ is

$$\underline{\underline{H}} = \frac{1}{2}(\underline{\underline{a}}\,\underline{\underline{a}}^{+} + \underline{\underline{a}}^{+}\underline{\underline{a}}). \tag{13.75}$$

$\underline{\underline{H}}$ is obtained from the matrices $\underline{\underline{a}}$ and $\underline{\underline{a}}^{+}$ by matrix multiplication [equation (13.18)] and then addition of the resulting matrices:

$$\underline{\underline{a}}\,\underline{\underline{a}}^{+} = \begin{pmatrix} 0 & \sqrt{1} & 0 & 0 & \cdot & \cdot \\ 0 & 0 & \sqrt{2} & 0 & \cdot & \cdot \\ 0 & 0 & 0 & \sqrt{3} & \cdot & \cdot \\ 0 & 0 & 0 & 0 & \sqrt{4} & \cdot \\ \cdot & \cdot & \cdot & \cdot & \cdot & \cdot \\ \cdot & \cdot & \cdot & \cdot & \cdot & \cdot \end{pmatrix} \begin{pmatrix} 0 & 0 & 0 & 0 & \cdot & \cdot \\ \sqrt{1} & 0 & 0 & 0 & \cdot & \cdot \\ 0 & \sqrt{2} & 0 & 0 & \cdot & \cdot \\ 0 & 0 & \sqrt{3} & 0 & \cdot & \cdot \\ 0 & 0 & 0 & \sqrt{4} & \cdot & \cdot \\ \cdot & \cdot & \cdot & \cdot & \cdot & \cdot \end{pmatrix} \tag{13.76a}$$

$$= \begin{pmatrix} 1 & 0 & 0 & 0 & \cdot \\ 0 & 2 & 0 & 0 & \cdot \\ 0 & 0 & 3 & 0 & \cdot \\ 0 & 0 & 0 & 4 & \cdot \\ \cdot & \cdot & \cdot & \cdot & \cdot \end{pmatrix} \tag{13.76b}$$

and

$$\underline{\underline{a}}^{+}\underline{\underline{a}} = \begin{pmatrix} 0 & 0 & 0 & 0 & \cdot & \cdot & \cdot \\ \sqrt{1} & 0 & 0 & 0 & \cdot & \cdot & \cdot \\ 0 & \sqrt{2} & 0 & 0 & \cdot & \cdot & \cdot \\ 0 & 0 & \sqrt{3} & 0 & \cdot & \cdot & \cdot \\ 0 & 0 & 0 & \sqrt{4} & \cdot & \cdot & \cdot \\ \cdot & \cdot & \cdot & \cdot & \cdot & \cdot & \cdot \end{pmatrix} \begin{pmatrix} 0 & \sqrt{1} & 0 & 0 & \cdot & \cdot \\ 0 & 0 & \sqrt{2} & 0 & \cdot & \cdot \\ 0 & 0 & 0 & \sqrt{3} & \cdot & \cdot \\ 0 & 0 & 0 & 0 & \sqrt{4} & \cdot \\ \cdot & \cdot & \cdot & \cdot & \cdot & \cdot \\ \cdot & \cdot & \cdot & \cdot & \cdot & \cdot \end{pmatrix} \tag{13.77a}$$

$$= \begin{pmatrix} 0 & 0 & 0 & 0 & \cdot \\ 0 & 1 & 0 & 0 & \cdot \\ 0 & 0 & 2 & 0 & \cdot \\ 0 & 0 & 0 & 3 & \cdot \\ \cdot & \cdot & \cdot & \cdot & \cdot \end{pmatrix}. \tag{13.77b}$$

Adding the matrices $\underline{\underline{a}}\,\underline{\underline{a}}^{+}$ and $\underline{\underline{a}}^{+}\underline{\underline{a}}$ and multiplying by 1/2 gives $\underline{\underline{H}}$:

$$\underline{\underline{H}} = \frac{1}{2}\begin{pmatrix} 1 & 0 & 0 & 0 & \cdot \\ 0 & 3 & 0 & 0 & \cdot \\ 0 & 0 & 5 & 0 & \cdot \\ 0 & 0 & 0 & 7 & \cdot \\ \cdot & \cdot & \cdot & \cdot & \cdot \end{pmatrix} = \begin{pmatrix} 1/2 & 0 & 0 & 0 & \cdot \\ 0 & 3/2 & 0 & 0 & \cdot \\ 0 & 0 & 5/2 & 0 & \cdot \\ 0 & 0 & 0 & 7/2 & \cdot \\ \cdot & \cdot & \cdot & \cdot & \cdot \end{pmatrix}. \tag{13.78}$$

In the basis $|n\rangle$, $\underline{\underline{H}}$ is diagonal. The diagonal elements are the eigenvalues, $n + 1/2$, and the vectors $|n\rangle$ are the eigenvectors. In conventional units (Chapter 6, Section B) the matrix will be multiplied by $h\nu$, and the energies of the harmonic oscillator eigenstates are $E = h\nu(n + 1/2)$, the result found previously.

Other operators associated with the harmonic oscillator problem can also be written in matrix form. For example, $\underline{x}$ and $\underline{P}$ can be written in terms of lowering and raising operators, so the matrices for $\underline{x}$ and $\underline{P}$ can be found using the matrices $\underline{\underline{a}}$ and $\underline{\underline{a}}^{+}$, equations (13.73) and (13.74).

E. SOLVING THE EIGENVALUE PROBLEM BY MATRIX DIAGONALIZATION

In Section D, the basis is composed of the eigenvectors, so the problem of finding the transformation into the eigenvector basis was avoided. The utility of the matrix representation is that the matrix representing a Hermitian operator can be obtained using any basis. The nondiagonal matrix written using some initial basis is then diagonalized to give the eigenvalues and eigenvectors of the operator.

In the matrix representation, the eigenvalue equation is

$$\underline{\underline{A}}\, u = \alpha u, \tag{13.79}$$

where $\underline{\underline{A}}$ is the matrix representing the Hermitian operator (observable), the u are the vector representatives of the eigenkets, and the α are the eigenvalues. Equation (13.79) can be written in terms of the matrix elements and the components of the vector representatives as

$$\sum_{j=1}^{N} (a_{ij} - \alpha\,\delta_{ij})\, u_j = 0 \qquad (i = 1, 2, \ldots, N). \tag{13.80}$$

This is a system of N equations in the N unknown components of the eigenvectors, the u_j.

$$
\begin{aligned}
(a_{11} - \alpha)u_1 + a_{12}u_2 + a_{13}u_3 + \cdots &= 0 \\
a_{21}u_1 + (a_{22} - \alpha)u_2 + a_{23}u_3 + \cdots &= 0 \\
a_{31}u_1 + a_{32}u_2 + (a_{33} - \alpha)u_3 + \cdots &= 0 \\
&\vdots
\end{aligned}
\tag{13.81}
$$

(A similar situation was encountered in the treatment of degenerate perturbation theory, Chapter 9, Section C. Here, the results will be an exact solution of the eigenvalue problem, but the procedure is the same.) In the system of equations, the u_j, the representatives of the eigenvectors in the initially chosen basis set, are not known, and the α, the eigenvalues, are not known. However, besides the trivial solution

$$u_1 = u_2 = \cdots u_N = 0,$$

such a system of equations will only have a solution if the determinant of the coefficients of the u_j vanishes, that is,

$$
\begin{vmatrix}
(a_{11} - \alpha) & a_{12} & a_{13} & \cdots \\
a_{21} & (a_{22} - \alpha) & a_{23} & \cdots \\
a_{31} & a_{32} & (a_{33} - \alpha) & \\
\vdots & & & \ddots
\end{vmatrix} = 0. \tag{13.82}
$$

Expanding this determinant gives an Nth-degree equation for the unknown α's, the eigenvalues. Then, substituting one eigenvalue at a time into the system of equations gives N equations in the N unknown u_i, the eigenvector representative associated with the particular eigenvalue α that was substituted. However, because the equations can be multiplied by a constant without changing the relationships, there are only $N - 1$ conditions. The necessary additional condition is obtained by requiring that the eigenvector representative is normalized:

$$u_1^* u_1 + u_2^* u_2 + \cdots + u_N^* u_N = 1. \tag{13.83}$$

The set of equations is solved in turn for each eigenvalue, giving the complete set of eigenvector representatives.

The simplest example is the degenerate two state problem that was discussed in Chapter 8. In Section C of that chapter, two specific superpositions of the basis states were shown to be the eigenstates, and the eigenvalues were obtained. In this example, the procedure outlined above will be used to obtain the eigenvalues and eigenvectors given in Chapter 8.

In the problem discussed in Chapter 8, the Hamiltonian, $\underline{H}$ operates on a basis consisting of two time-independent orthonormal kets $|\alpha\rangle$ and $|\beta\rangle$:

$$\underline{H}\,|\alpha\rangle = E_0\,|\alpha\rangle + \gamma\,|\beta\rangle \tag{13.84a}$$

$$\underline{H}\,|\beta\rangle = E_0\,|\beta\rangle + \gamma\,|\alpha\rangle\,. \tag{13.84b}$$

Equations (13.84) define $\underline{H}$. The four matrix elements of $\underline{H}$ are

$$\begin{aligned} \langle\alpha|\,\underline{H}\,|\alpha\rangle &= E_0 \\ \langle\beta|\,\underline{H}\,|\alpha\rangle &= \gamma \\ \langle\alpha|\,\underline{H}\,|\beta\rangle &= \gamma \\ \langle\beta|\,\underline{H}\,|\beta\rangle &= E_0. \end{aligned} \tag{13.85}$$

These are found by multiplying both sides of equations (13.84) by $\langle\alpha|$ and $\langle\beta|$. The Hamiltonian matrix, $\underline{\underline{H}}$, is

$$\begin{array}{cc} & \begin{array}{cc} |\alpha\rangle & |\beta\rangle \end{array} \\ \underline{\underline{H}} = \begin{array}{c} \langle\alpha| \\ \langle\beta| \end{array} & \begin{pmatrix} E_0 & \gamma \\ \gamma & E_0 \end{pmatrix}. \end{array} \tag{13.86}$$

The eigenvalues, λ, are obtained by forming the determinant

$$\begin{vmatrix} E_0 - \lambda & \gamma \\ \gamma & E_0 - \lambda \end{vmatrix} = 0 \tag{13.87}$$

from $\underline{\underline{H}}$, setting it equal to zero, and, then, expanding the determinant. Expanding the determinant gives the quadratic equation in λ

$$(E_0 - \lambda)^2 - \gamma^2 = 0. \tag{13.88}$$

Solving the quadratic equation yields the two eigenvalues

$$\lambda_+ = E_0 + \gamma \tag{13.88a}$$

$$\lambda_- = E_0 - \gamma. \tag{13.88b}$$

The eigenkets are superpositions of basis vectors,

$$\begin{aligned} |+\rangle &= a_+\,|\alpha\rangle + b_+\,|\beta\rangle \\ |-\rangle &= a_-\,|\alpha\rangle + b_-\,|\beta\rangle\,, \end{aligned} \tag{13.89}$$

where the ket $|+\rangle$ and $|-\rangle$ are the eigenvectors associated with the eigenvalues λ_+ and λ_-, respectively. The vector representatives are $[a_+, b_+]$ and $[a_-, b_-]$. To find the vector representative for $|+\rangle$, λ_+ is substituted into equations (13.81) to give the equations

$$(H_{11} - \lambda_+)a_+ + H_{12}b_+ = 0 \tag{13.90a}$$

$$H_{21}a_+ + (H_{22} - \lambda_+)b_+ = 0, \tag{13.90b}$$

where H_{ij} are the matrix elements of $\underline{\underline{H}}$. With $\lambda_+ = E_0 + \gamma$, the equations are

$$-\gamma a_+ + \gamma b_+ = 0 \tag{13.91a}$$

$$\gamma a_+ - \gamma b_+ = 0. \tag{13.91b}$$

The two equations are identical. As noted above, the system of N equations only gives $N-1$ conditions. These equations yield

$$a_+ = b_+. \tag{13.92}$$

The coefficients can be taken to be real. Then the normalization condition is

$$a_+^2 + b_+^2 = 1, \tag{13.93}$$

and, therefore,

$$a_+ = b_+ = \frac{1}{\sqrt{2}}. \tag{13.94}$$

Then, in terms of the basis $\{|\alpha\rangle, |\beta\rangle\}$, the eigenvector $|+\rangle$ is

$$|+\rangle = \frac{1}{\sqrt{2}}|\alpha\rangle + \frac{1}{\sqrt{2}}|\beta\rangle. \tag{13.95}$$

To find the other eigenvector, λ_- is substituted into the system of equations. The condition

$$a_- = -b_- \tag{13.96}$$

is obtained. With the normalization condition,

$$a_- = \frac{1}{\sqrt{2}}, \qquad b_- = -\frac{1}{\sqrt{2}}, \tag{13.97}$$

and, in terms of the basis $\{|\alpha\rangle, |\beta\rangle\}$, the eigenvector $|-\rangle$ is

$$|-\rangle = \frac{1}{\sqrt{2}}|\alpha\rangle - \frac{1}{\sqrt{2}}|\beta\rangle. \tag{13.98}$$

In the formal development, a similarity transformation takes the initially nondiagonal matrix into the diagonal matrix with the eigenvalues on the diagonal. The necessary unitary matrix to perform the similarity transformation is the matrix of the vector representatives of the eigenvectors. However, to find these, it is necessary to first find the eigenvalues. Therefore, in practice, finding the eigenvalues does not involve performing the similarity transformation.

The Hamiltonian matrix, equation (13.86), can be diagonalized by performing the similarity transformation using the eigenvector representatives as the columns of the unitary matrix in the similarity transformation. Calling the transformed Hamiltonian matrix, $\underline{\underline{H}}'$, it is given by

$$\underline{\underline{H}}' = \begin{pmatrix} 1/\sqrt{2} & 1/\sqrt{2} \\ 1/\sqrt{2} & -1/\sqrt{2} \end{pmatrix} \begin{pmatrix} E_0 & \gamma \\ \gamma & E_0 \end{pmatrix} \begin{pmatrix} 1/\sqrt{2} & 1/\sqrt{2} \\ 1/\sqrt{2} & -1/\sqrt{2} \end{pmatrix}. \tag{13.99}$$

Factoring out a $1/\sqrt{2}$ from both matrices,

$$\underline{\underline{H}}' = \frac{1}{2} \begin{pmatrix} 1 & 1 \\ 1 & -1 \end{pmatrix} \begin{pmatrix} E_0 & \gamma \\ \gamma & E_0 \end{pmatrix} \begin{pmatrix} 1 & 1 \\ 1 & -1 \end{pmatrix}. \tag{13.100}$$

Performing the multiplication of the Hamiltonian matrix and $\underline{\underline{U}}^{-1}$, the matrix on the right,

$$\underline{\underline{H}}' = \frac{1}{2}\begin{pmatrix} 1 & 1 \\ 1 & -1 \end{pmatrix} \begin{pmatrix} E_0 + \gamma & E_0 - \gamma \\ E_0 + \gamma & -E_0 + \gamma \end{pmatrix}. \quad (13.101)$$

Performing the remaining matrix multiplication and multiplying each element by 1/2 yields

$$\underline{\underline{H}}' = \begin{pmatrix} E_0 + \gamma & 0 \\ 0 & E_0 - \gamma \end{pmatrix}. \quad (13.102)$$

The Hamiltonian matrix is diagonal, and the diagonal matrix elements are the eigenvalues.

Finding the eigenvalues and eigenvectors analytically becomes increasingly difficult as the size of the matrix increases. A 10×10 matrix gives a tenth-order equation for the eigenvalues. In some cases, the Hamiltonian matrix may be block diagonal; that is, a large matrix is composed of smaller blocks of matrix elements that are not connected by off-diagonal matrix elements. The smaller blocks may be amenable to analytical solution. However, the great power of the matrix representation comes from the ability of computers to numerically find the eigenvalues and eigenvectors of very large matrices. Matrix diagonalization routines are part of most mathematical packages for computers. Therefore, large basis sets and very large matrices are not an obstacle to the use of the matrix representation.

CHAPTER

14 The Density Matrix and Coherent Coupling of Molecules to Light

In Chapter 13, the matrix representation of quantum mechanics was used to solve the eigenvalue problem. Matrices can also be used to solve time-dependent problems. The density matrix formalism permits the direct calculation of time-dependent and time-independent probabilities and observables without the intermediate step of calculating probability amplitudes. Density matrices also play an important role in quantum statistical mechanics. Here, some aspects of the density matrix formalism are developed and illustrated.

A. THE DENSITY OPERATOR AND THE DENSITY MATRIX

The state of a system at an instant of time, t, can be described by the ket

$$|t\rangle = \sum_n C_n(t)\,|n\rangle\,, \tag{14.1}$$

where the set $\{|n\rangle\}$ is a complete orthonormal basis set, and

$$\sum_n |C_n(t)|^2 = 1, \tag{14.2}$$

that is, $|t\rangle$ is normalized. The density operator is

$$\underline{\rho}(t) = |t\rangle\langle t|\,. \tag{14.3}$$

The density operator can be represented in terms of the density matrix, $\underline{\underline{\rho}}(t)$, using the basis set $\{|n\rangle\}$. The matrix elements of $\underline{\underline{\rho}}(t)$ are

$$\rho_{ij}(t) = \langle i|\,\underline{\rho}(t)\,|j\rangle\,. \tag{14.4}$$

Consider a two-state system. Then

$$|t\rangle = C_1(t)\,|1\rangle + C_2(t)\,|2\rangle\,, \tag{14.5}$$

and

$$\rho_{11} = \langle 1 \mid t\rangle\langle t \mid 1\rangle$$
$$= \langle 1|\ [C_1\,|1\rangle + C_2\,|2\rangle]\,[C_1^*\,\langle 1| + C_2^*\,\langle 2|]\ |1\rangle$$
$$\rho_{11} = C_1 C_1^* \tag{14.6}$$
$$\rho_{12} = \langle 1 \mid t\rangle\langle t \mid 2\rangle$$
$$\rho_{12} = C_1 C_2^* \tag{14.7}$$
$$\rho_{21} = \langle 2 \mid t\rangle\langle t \mid 1\rangle$$
$$\rho_{21} = C_2 C_1^* \tag{14.8}$$
$$\rho_{22} = \langle 2 \mid t\rangle\langle t \mid 2\rangle$$
$$\rho_{22} = C_2 C_2^*. \tag{14.9}$$

The 2×2 density matrix is

$$\underline{\underline{\rho}}(t) = \begin{bmatrix} C_1C_1^* & C_1C_2^* \\ C_2C_1^* & C_2C_2^* \end{bmatrix}. \tag{14.10}$$

Given that the system is in the state $|t\rangle$, the diagonal matrix elements, $C_1C_1^*$ and $C_2C_2^*$, are the probabilities of finding the system in the states $|1\rangle$ and $|2\rangle$, respectively. Since $\sum_n |C_n(t)|^2 = 1$, the trace of the density matrix (the sum of the diagonal elements) is one,

$$Tr\,\underline{\underline{\rho}}(t) = 1, \tag{14.11}$$

independent of the dimension of the matrix. Also, in general,

$$\rho_{ij} = \rho_{ji}^*. \tag{14.12}$$

B. THE TIME DEPENDENCE OF THE DENSITY MATRIX

The time dependence of the density operator can be found by employing the time dependent Schrödinger equation. The time derivative of the density operator is

$$\dot{\underline{\rho}} = \frac{d\,\underline{\rho}(t)}{dt}; \tag{14.13}$$

and employing the chain rule for differentiation and the definition of the density operator, the derivative is

$$\frac{d\underline{\rho}(t)}{dt} = \left(\frac{d}{dt}\,|t\rangle\right)\langle t| + |t\rangle\left(\frac{d}{dt}\,\langle t|\right). \tag{14.14}$$

Using the time-dependent Schrödinger equation (Chapter 5, Section A) to substitute for the time derivatives of the ket and bra gives

$$\frac{d\underline{\rho}(t)}{dt} = \frac{1}{i\hbar}\underline{H}(t)\,|t\rangle\langle t| + \frac{1}{-i\hbar}\,|t\rangle\langle t|\,\underline{H}(t), \tag{14.15}$$

where the $\underline{H}(t)$ on the extreme right of the equation operates to the left on the bra. Then

$$\begin{aligned}\frac{d\underline{\rho}(t)}{dt} &= \frac{1}{i\hbar}[\underline{H}(t)\,|t\rangle\langle t| - |t\rangle\langle t|\,\underline{H}(t)] \\ &= \frac{1}{i\hbar}[\underline{H}(t), \underline{\rho}(t)].\end{aligned} \tag{14.16}$$

Therefore,

$$i\hbar\,\dot{\underline{\rho}}(t) = [\underline{H}(t), \underline{\rho}(t)]; \tag{14.17}$$

$i\hbar$ times the time derivative of the density operator equals the commutator of the Hamiltonian with the density operator.

Equation (14.17) can be written in matrix form in terms of an orthonormal basis set. The density matrix equation of motion is

$$\dot{\underline{\underline{\rho}}}(t) = -\frac{i}{\hbar}\left[\underline{\underline{H}}(t), \underline{\underline{\rho}}(t)\right]. \tag{14.18}$$

In general, since $\rho_{ij} = C_i C_j^*$

$$\begin{aligned}\dot{\rho}_{ij} &= C_i\left(\frac{dC_j^*}{dt}\right) + C_j^*\left(\frac{dC_i}{dt}\right) \\ &= C_i\dot{C}_j^* + C_j^*\dot{C}_i.\end{aligned} \tag{14.19}$$

Equation (14.19) defines the time derivative of the density matrix elements.

For a 2×2 density matrix, the equation of motion is explicitly

$$\begin{bmatrix}\dot{\rho}_{11} & \dot{\rho}_{12} \\ \dot{\rho}_{21} & \dot{\rho}_{22}\end{bmatrix} = -\frac{i}{\hbar}\left\{\begin{bmatrix}H_{11} & H_{12} \\ H_{21} & H_{22}\end{bmatrix}\begin{bmatrix}\rho_{11} & \rho_{12} \\ \rho_{21} & \rho_{22}\end{bmatrix} - \begin{bmatrix}\rho_{11} & \rho_{12} \\ \rho_{21} & \rho_{22}\end{bmatrix}\begin{bmatrix}H_{11} & H_{12} \\ H_{21} & H_{22}\end{bmatrix}\right\}. \tag{14.20a}$$

Performing the matrix multiplications, the equations of motion of the density matrix elements are

$$\dot{\rho}_{11} = -\dot{\rho}_{22} = -\frac{i}{\hbar}(H_{12}\rho_{21} - H_{21}\rho_{12}) \tag{14.20b}$$

$$\dot{\rho}_{12} = \dot{\rho}_{21}^* = -\frac{i}{\hbar}[(H_{11} - H_{22})\rho_{12} + (\rho_{22} - \rho_{11})H_{12}]. \tag{14.20c}$$

In many problems, the Hamiltonian is composed of a piece that is time-dependent and a piece that is time-independent. This is the situation in studies of molecules subjected to an outside influence—for example, a radiation field in a spectroscopic experiment or a thermal bath, which is responsible for time-dependent solute–solvent interactions. In such situations it is natural to express the density matrix in terms of the basis set composed of the eigenkets of the time-independent piece of the Hamiltonian. In this basis set, the equation of motion of the density matrix elements only depends on the time-dependent piece of the Hamiltonian. To see this, the time derivative of the density matrix operator is evaluated in two ways. It is evaluated using the Schrödinger equation as in equations (14.14)–(14.16), and it is evaluated by explicitly taking the time derivative after expressing the density matrix operator in terms of the eigenket basis set. Cancellation of terms and finding the matrix elements give the expression for $C_i \dot{C}_j^* + C_j^* \dot{C}_i$, the time derivative of the density matrix elements [equation (14.19)].

The total Hamiltonian is

$$\underline{H} = \underline{H}_0 + \underline{H}_I(t), \tag{14.21}$$

where $\underline{H}_0$ is time-independent, and $\underline{H}_I(t)$ is time-dependent and is responsible for interactions among the eigenstates of $\underline{H}_0$. The basis set $\{|n\rangle\}$ are the eigenstates of $\underline{H}_0$,

$$\underline{H}_0 |n\rangle = E_n |n\rangle .$$

The state $|t\rangle$ in terms of this basis $\{|n\rangle\}$ is

$$|t\rangle = \sum_n C_n(t) |n\rangle . \tag{14.22}$$

The time derivative of the density operator is

$$\left(\frac{d}{dt} |t\rangle\right) \langle t| + |t\rangle \left(\frac{d}{dt} \langle t|\right), \tag{14.23a}$$

and, using the Schrödinger equation, it is

$$= \frac{1}{i\hbar} \underline{H}(t) |t\rangle\langle t| + \frac{1}{-i\hbar} |t\rangle\langle t| \underline{H}(t) \tag{14.23b}$$

$$= \frac{1}{i\hbar} \underline{H}_0 |t\rangle\langle t| + \frac{1}{i\hbar} \underline{H}_I |t\rangle\langle t| + \frac{1}{-i\hbar} |t\rangle\langle t| \underline{H}_0 + \frac{1}{-i\hbar} |t\rangle\langle t| \underline{H}_I . \tag{14.23c}$$

Substituting the expansion for $|t\rangle$, equation (14.22), into the derivative terms of equation (14.23a) gives

$$\left(\frac{d}{dt} \sum_n C_n |n\rangle\right) \langle t| + |t\rangle \left(\frac{d}{dt} \sum_n C_n^* \langle n|\right) \tag{14.24}$$

$$= \left(\sum_n \dot{C}_n |n\rangle\right) \langle t| + \left(\sum_n C_n \frac{d}{dt} |n\rangle\right) \langle t| + |t\rangle \left(\sum_n \dot{C}_n^* \langle n|\right) + |t\rangle \left(\sum_n C_n^* \frac{d}{dt} \langle n|\right)$$

The right-hand side of equation (14.23c) equals the right hand-side of equation (14.24). The Schrödinger equation can be written as

$$\left(\sum_n C_n \frac{d}{dt} |n\rangle\right) = \frac{1}{i\hbar} \underline{H}_0 |t\rangle \tag{14.25a}$$

and

$$\left(\sum_n C_n^* \frac{d}{dt} \langle n|\right) = \frac{1}{-i\hbar} \langle t| \underline{H}_0. \tag{14.25b}$$

Multiplying both sides of equation (14.25a) on the right by $\langle t|$ and both sides of equation (14.25b) on the left by $|t\rangle$ gives expressions that show that the first and third terms on the right-hand side of equation (14.23c) cancel the second and fourth terms on the right-hand side of in equation (14.24). After canceling terms, the equation is

$$\left(\sum_n \dot{C}_n |n\rangle\right) \langle t| + |t\rangle \left(\sum_n \dot{C}_n^* \langle n|\right) = \frac{1}{i\hbar} [\underline{H}_I, \underline{\rho}]. \tag{14.26}$$

Now, consider the matrix elements, ij, of equation (14.26). The matrix elements of the left-hand side are

$$\begin{aligned} &\sum_n \dot{C}_n \langle i \mid n\rangle \langle t \mid j\rangle + \langle i \mid t\rangle \sum_n \dot{C}_n^* \langle n \mid j\rangle \\ &= \dot{C}_i C_j^* + C_i \dot{C}_j^* \\ &= \dot{\rho}_{ij}. \end{aligned} \tag{14.27}$$

The last equality is from equation (14.19). Therefore, equation (14.27) becomes

$$\underline{\underline{\dot{\rho}}}(t) = -\frac{i}{\hbar} \left[\underline{\underline{H}}_I(t), \underline{\underline{\rho}}(t)\right]. \tag{14.28}$$

In the basis set of $\underline{H}_0$, $\underline{H}_0$ cancels out of the calculation of the equations of motions of the density matrix elements. This is equivalent to the initial steps, made prior to the introduction of approximations, in the development of time-dependent perturbation theory (Chapter 11, Section A). The time dependence of the coefficients of the expansion of $|t\rangle$ in the eigenkets of $\underline{H}_0$ depends only on $\underline{H}_I(t)$. As will be shown explicitly below, in the absence of $\underline{H}_I(t)$, the coefficients are time-independent, and the only time dependence comes from the time-dependent phase factors contained in the eigenkets of $\underline{H}_0$.

C. THE TIME-DEPENDENT TWO-STATE PROBLEM

In this section, the time-dependent degenerate two-state problem, which was examined in detail using the time-dependent Schrödinger equation in Chapter 8, will be treated with the density matrix formalism. The problem consists of two states coupled by a time-independent interaction. The eigenstates of this system were found in Chapter 13, Section E. As discussed in Chapter 8, the two-state problem is a prototype for physical phenomena such as electronic excitation transfer, electron transfer, and vibrational excitation transfer between different molecules or, in the case of vibrational excitations, different states of the same molecule.

The basis states, $\{|1\rangle, |2\rangle\}$, are degenerate eigenkets of $\underline{H}_0$ with energy E:

$$\underline{H}_0 |1\rangle = E |1\rangle = \hbar\omega_0 |1\rangle \tag{14.29a}$$

$$\underline{H}_0 |2\rangle = E |2\rangle = \hbar\omega_0 |2\rangle . \tag{14.29b}$$

The eigenstates of $\underline{H}_0$ interact through $\underline{H}_I$:

$$\underline{H}_I |1\rangle = \hbar\beta |2\rangle \tag{14.30a}$$

$$\underline{H}_I |2\rangle = \hbar\beta |1\rangle \tag{14.30b}$$

($\hbar\beta = \gamma$ in Chapter 8.) The general state of the system is given by equation (14.5). If the system is initially prepared in $|1\rangle$, which is not an eigenket of the full Hamiltonian, $\underline{H} = \underline{H}_0 + \underline{H}_I$, the time evolution of the system can be calculated with the density matrix equation of motion, equation (14.28) with

$$\underline{\underline{H}}_I = \hbar \begin{bmatrix} 0 & \beta \\ \beta & 0 \end{bmatrix}. \tag{14.31}$$

Because the basis states are degenerate, time-dependent phase factors do not occur as part of the off-diagonal matrix elements. This is a special case. Evaluating the right-hand side of equation (14.28) gives

$$\dot{\underline{\underline{\rho}}} = i\beta \begin{bmatrix} (\rho_{12} - \rho_{21}) & (\rho_{11} - \rho_{22}) \\ -(\rho_{11} - \rho_{22}) & -(\rho_{12} - \rho_{21}) \end{bmatrix}. \tag{14.32}$$

The equations of motion of the density matrix elements are

$$\dot{\rho}_{11} = i\beta(\rho_{12} - \rho_{21}) \tag{14.33a}$$

$$\dot{\rho}_{22} = -i\beta(\rho_{12} - \rho_{21}) \tag{14.33b}$$

$$\dot{\rho}_{12} = i\beta(\rho_{11} - \rho_{22}) \tag{14.33c}$$

$$\dot{\rho}_{21} = -i\beta(\rho_{11} - \rho_{22}). \tag{14.33d}$$

The solutions to equations (14.33) can be found by taking the derivative of both sides of equation (14.33a):

$$\ddot{\rho}_{11} = i\beta(\dot{\rho}_{12} - \dot{\rho}_{21}) \tag{14.34}$$

Substituting equations (14.33c) and (14.33d) into the right-hand side gives

$$\ddot{\rho}_{11} = -2\beta^2(\rho_{11} - \rho_{22}). \tag{14.35}$$

Since $\mathrm{Tr}\,\rho = 1$,

$$\rho_{11} + \rho_{22} = 1$$

and

$$\rho_{22} = 1 - \rho_{11}. \tag{14.36}$$

Substituting gives an equation only in ρ_{11}:

$$\ddot{\rho}_{11} = 2\beta^2 - 4\beta^2 \rho_{11}. \tag{14.37}$$

For the initial condition that $\rho_{11} = 1$ at $t = 0$, which implies $\rho_{22} = 0$, $\rho_{12} = 0$, $\rho_{21} = 0$ at $t = 0$, and using equation (14.33a), the solution is

$$\rho_{11} = \cos^2(\beta t). \tag{14.38}$$

Then from equation (14.36) we

$$\rho_{22} = \sin^2(\beta t). \tag{14.39}$$

ρ_{11} and ρ_{22} are the probabilities of finding the system in the states $|1\rangle$ and $|2\rangle$, respectively. The probability oscillates between the two states at frequency β. This is identical to the results obtained for the probabilities in Chapter 8 [equations (8.26)] with $\beta = \gamma/\hbar$ using the time-dependent Schrödinger equation. However, the density matrix treatment gives the probabilities directly without first having to calculate probability amplitudes.

The off-diagonal density matrix elements, ρ_{12} and ρ_{21}, can also be calculated. Substituting ρ_{11} and ρ_{22} into equation (14.33c) yields

$$\dot{\rho}_{12} = i\beta(\cos^2 \beta t \; - \sin^2 \beta t). \tag{14.40}$$

Then

$$\rho_{12} = i\beta \int (\cos^2 \beta t - \sin^2 \beta t)\, dt \tag{14.41}$$

and

$$\rho_{12} = \frac{i}{2} \sin(2\beta t). \tag{14.42}$$

Since $\rho_{ij} = \rho_{ji}^*$,

$$\rho_{21} = -\frac{i}{2} \sin(2\beta t). \tag{14.43}$$

The importance of the off-diagonal density matrix elements will be illustrated below in conjunction with the interaction of two nondegenerate states with a radiation field.

D. EXPECTATION VALUE OF AN OPERATOR

The expectation value or average value of an operator representing an observable (Chapter 4, Section C) at an instant of time is

$$\langle \underline{A} \rangle = \langle t | \, \underline{A} \, | t \rangle \, . \tag{14.44}$$

Writing $|t\rangle$ in terms of the complete orthonormal basis set $\{|j\rangle\}$,

$$|t\rangle = \sum_j C_j(t)\,|j\rangle\,,$$

the matrix elements of $\underline{A}$, A_{ij} in the basis $\{|j\rangle\}$ are

$$A_{ij} = \langle i|\,\underline{A}\,|j\rangle\,. \tag{14.45}$$

Then

$$\begin{aligned}\langle t|\,\underline{A}\,|t\rangle &= \left(\sum_i C_i^*\,\langle i|\right)\underline{A}\left(\sum_j C_j\,|j\rangle\right)\\ &= \sum_{i,j} C_i^*(t)C_j(t)\,\langle i|\,\underline{A}\,|j\rangle\,.\end{aligned} \tag{14.46}$$

The matrix elements of the density operator, ρ_{ji}, (note the order of the indices), are

$$\begin{aligned}\rho_{ji} &= \langle j|\,\underline{\rho}(t)\,|i\rangle\\ &= \langle j\mid t\rangle\,\langle t\mid i\rangle\\ &= C_i^*(t)C_j(t).\end{aligned} \tag{14.47}$$

Equation (14.46) becomes

$$\langle t|\,\underline{A}\,|t\rangle = \sum_{i,j}\langle j|\,\underline{\rho}(t)\,|i\rangle\langle i|\,\underline{A}\,|j\rangle\,. \tag{14.48}$$

Equation (14.48) has the form of matrix multiplication [see equation (13.18)] except that only the diagonal matrix elements are calculated, and they are summed.

$$\langle t|\,\underline{A}\,|t\rangle = \sum_j \langle j|\,\underline{\rho}(t)\,\underline{A}\,|j\rangle\,. \tag{14.49}$$

Therefore,

$$\left\langle \underline{A}\right\rangle = Tr\left(\underline{\underline{\rho}}(t)\,\underline{\underline{A}}\right)\,. \tag{14.50}$$

The expectation value of the operator $\underline{A}$ is the trace of the matrix product of the density matrix $\underline{\underline{\rho}}$ and the matrix $\underline{\underline{A}}$. Both are calculated in the basis $\{|j\rangle\}$. The density matrix carries the time dependence. In calculating the matrix elements, A_{ij}, time-dependent phase factors arising from the basis vectors are not included as part of the matrix elements.

As an example, the average value of the energy, $\overline{E} = \langle \underline{H}\rangle$, for the time-dependent two-state problem treated in the last section is calculated.

$$\overline{E} = \langle \underline{H}\rangle = Tr\,\underline{\underline{\rho}}\,\underline{\underline{H}} \tag{14.51}$$

$$\begin{aligned}Tr\,\underline{\underline{\rho}}\,\underline{\underline{H}} &= Tr\begin{bmatrix}\rho_{11} & \rho_{12}\\ \rho_{21} & \rho_{22}\end{bmatrix}\begin{bmatrix}E & \hbar\beta\\ \hbar\beta & E\end{bmatrix}\\ &= E(\rho_{11}+\rho_{22}) + \hbar\beta(\rho_{12}+\rho_{21})\end{aligned} \tag{14.52}$$

Substituting the density matrix elements from equations (14.38), (14.39), (14.42), and (14.43) gives

$$\langle \underline{H} \rangle = E\,(\cos^2 \beta t + \sin^2 \beta t) + \frac{i\hbar\beta}{2}(\sin 2\beta t - \sin 2\beta t)$$
$$= E. \tag{14.53}$$

This is the same result as obtained previously. From equations (13.88), the eigenvalues are $E + \hbar\beta$ and $E - \hbar\beta$, and, as shown in Chapter 8, Section C, the time-evolving state considered here is an equal superposition of the two eigenkets. Therefore, the average value of the energy is the average of the two eigenvalues, which is E, the result obtained from $Tr \underline{\underline{\rho}}\, \underline{\underline{H}}$.

E. COHERENT COUPLING OF A TWO-STATE SYSTEM BY AN OPTICAL FIELD

In Chapter 12, time-dependent perturbation theory was used to study the problem of optical absorption and emission of light by atoms or molecules. The use of time-dependent perturbation theory requires that the probability of finding the system in a state other than the initial state is small. This condition frequently occurs for very weak optical fields or very weak coupling of the states of the system to the field (small transition dipole bracket). In this section, the density matrix treatment will be used to describe the results of applying an optical field that couples strongly to the states of a system. As in the time-dependent perturbation theory treatment, the treatment is semiclassical (i.e., the molecular states are treated quantum mechanically, but the radiation field is treated classically). However, the treatment is nonperturbative. Therefore, the changes in probability are not restricted to be small. If the frequency of light is on or near resonance (i.e., equal to or approximately equal to the energy difference between two molecular states), these states will be strongly coupled by the light and other states can be neglected. For this reason, it is sufficient to treat a two level system. The system is illustrated in Figure 14.1.

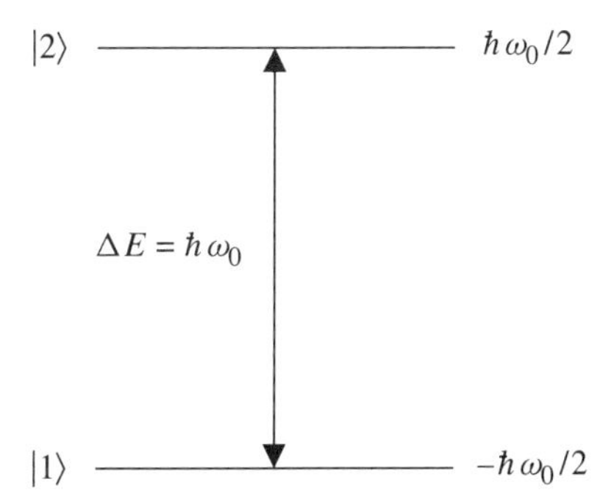

Figure 14.1 The states $|1\rangle$ and $|2\rangle$ have energies $-\hbar\omega_0/2$ and $\hbar\omega/2$, respectively. They are coupled by an optical field on or near the resonance energy, $\Delta E = \hbar\omega_0$.

The zero of energy is taken to be midway between the two states, so the energies of states $|1\rangle$ and $|2\rangle$ are $-\hbar\omega/2$ and $\hbar\omega/2$, respectively. The kets $|1\rangle$ and $|2\rangle$ are eigenkets of the molecular Hamiltonian, $\underline{H}_0$, that is,

$$\underline{H}_0 |1\rangle = -\hbar\omega_0/2 |1\rangle \tag{14.54a}$$

$$\underline{H}_0 |2\rangle = \hbar\omega_0/2 |2\rangle . \tag{14.54b}$$

The difference in energy of the two states is $\Delta E = \hbar\omega_0$.

The states are coupled by optical radiation of frequency. The transition is taken to be dipole-allowed (see Chapter 12, Section B). Dipole allowed transitions can occur between two vibrational states of a molecule or two electronic states of an atom or molecule. In the semiclassical treatment, the interaction term in the Hamiltonian that represents the coupling of the two states by the optical field can be written as (Chapter 12, Sections B)

$$\underline{H}_I(t) = \hbar e \, \underline{x}_{12} \, E_0 \, \cos(\omega t), \tag{14.55}$$

where $e\underline{x}_{12}$ is the transition dipole operator for x polarized light with amplitude E_0, $\underline{x}_{12}$ is the x position operator, and e is the charge on the electron. The frequency, ω, is near or on resonance with the transition frequency, ω_0. Initially, the development is general. Then, the initial condition will be chosen such that the system is in its lower energy state, $|1\rangle$, at $t = 0$, that is, $\rho_{11}(0) = 1$.

$\underline{H}_I(t)$ couples the states $|1\rangle$ and $|2\rangle$. It does not have diagonal matrix elements. The matrix elements of $\underline{H}_I(t)$ are

$$\begin{aligned} \langle 1| \underline{H}_I(t) |2\rangle &= \hbar e E_0 \, \cos(\omega t) \, \langle 1| \underline{x}_{12} |2\rangle \\ &= \hbar e E_0 \, \cos(\omega t) \, e^{-i\omega_0 t/2} \left\langle 1' \right| \underline{x}_{12} \left| 2' \right\rangle e^{-i\omega_0 t/2} \\ &= \hbar \mu E_0 \, \cos(\omega t) \, e^{-i\omega_0 t}, \end{aligned} \tag{14.56}$$

where the primed ket and bra indicate the spatial part of the eigenstates of $\underline{H}_0$, and the time-dependent phase factors are written out explicitly in the second line. μ is the value of the bracket times the charge, e. It is the magnitude of the transition dipole operator. The matrix element $\langle 2| \underline{H}_I(t) |1\rangle$ is the complex conjugate of $\langle 1| \underline{H}_I(t) |2\rangle$, that is,

$$\langle 2| \underline{H}_I(t) |1\rangle = \hbar \mu^* E_0 \, \cos(\omega t) e^{i\omega_0 t}. \tag{14.57}$$

μ^* is the complex conjugate of μ. In the following, μ is taken to be real, $\mu^* = \mu$, which does not change the final results. The Rabi frequency, ω_1, is defined as

$$\omega_1 = \mu E_0. \tag{14.58}$$

Then,

$$\langle 1| \underline{H}_I(t) |2\rangle = \hbar\omega_1 \, \cos(\omega t) e^{-i\omega_0 t} \tag{14.59}$$

$$\langle 2| \underline{H}_I(t) |1\rangle = \hbar\omega_1 \, \cos(\omega t) e^{i\omega_0 t}. \tag{14.60}$$

The matrix $\underline{\underline{H}}_I(t)$ is

$$\underline{\underline{H}}_I(t) = \hbar \begin{bmatrix} 0 & \omega_1 \cos(\omega t) e^{-i\omega_0 t} \\ \omega_1 \cos(\omega t) e^{i\omega_0 t} & 0 \end{bmatrix}. \tag{14.61}$$

The general state of the system in terms of the eigenkets of $\underline{H}_0$ is

$$|t\rangle = C_1(t)\,|1\rangle + C_2(t)\,|2\rangle\,. \tag{14.62}$$

The equation of motion of the density matrix elements is given by equation (14.28). As discussed in Section B, since the basis set is composed of the eigenkets of $\underline{H}_0$, $\underline{H}_0$ does not enter into the equation of motion for the product of the coefficients $C_i C_j^*$, which are the density matrix elements. Substituting $\underline{\underline{H}}_I(t)$ [equation (14.61)] into equation (14.28), and multiplying out the commutator gives

$$\begin{bmatrix} \dot{\rho}_{11} & \dot{\rho}_{12} \\ \dot{\rho}_{21} & \dot{\rho}_{22} \end{bmatrix} =$$

$$\begin{bmatrix} i\omega_1 \cos(\omega t)(e^{i\omega_0 t}\rho_{12} - e^{-i\omega_0 t}\rho_{21}) & i\omega_1 \cos(\omega t) e^{-i\omega_0 t}(\rho_{11} - \rho_{22}) \\ -i\omega_1 \cos(\omega t) e^{i\omega_0 t}(\rho_{11} - \rho_{22}) & -i\omega_1 \cos(\omega t)(e^{i\omega_0 t}\rho_{12} - e^{-i\omega_0 t}\rho_{21}) \end{bmatrix}. \tag{14.63}$$

The equations of motion of the density matrix elements are

$$\dot{\rho}_{11} = i\omega_1 \cos(\omega t)(e^{i\omega_0 t}\rho_{12} - e^{-i\omega_0 t}\rho_{21}) \tag{14.64a}$$

$$\dot{\rho}_{22} = -i\omega_1 \cos(\omega t)(e^{i\omega_0 t}\rho_{12} - e^{-i\omega_0 t}\rho_{21}) \tag{14.64b}$$

$$\dot{\rho}_{12} = i\omega_1 \cos(\omega t) e^{-i\omega_0 t}(\rho_{11} - \rho_{22}) \tag{14.64c}$$

$$\dot{\rho}_{21} = -i\omega_1 \cos(\omega t) e^{i\omega_0 t}(\rho_{11} - \rho_{22}), \tag{14.64d}$$

since $\rho_{11} + \rho_{22} = 1$, $\dot{\rho}_{11} = -\dot{\rho}_{22}$ and since $\rho_{12} = \rho_{21}^*$, $\dot{\rho}_{12} = \dot{\rho}_{21}^*$.

To this point the treatment is exact. The $\cos(\omega t)$ term can be written as

$$\cos(\omega t) = \frac{1}{2}(e^{i\omega t} + e^{-i\omega t}). \tag{14.65}$$

To simplify the solutions to equations (14.64), the rotating wave approximation is made as in the time-dependent perturbation treatment (Chapter 12, Section B.2). The radiation field frequency, ω, is close to the transition frequency, ω_0. The exponential terms in equations (14.64) will combine with the exponentials of equation (14.65) to giving terms of the form

$$e^{\pm i(\omega_0 - \omega)t} \quad \text{and} \quad e^{\pm i(\omega_0 + \omega)t}.$$

Because $\omega \approx \omega_0$, the exponential terms with argument $(\omega_0 - \omega)$ are near resonance, while the exponential terms with argument $(\omega_0 + \omega)$ are $\sim 2\omega_0$ off resonance. In the rotating wave approximating, the terms that are far off resonance are dropped. The terms with $(\omega_0 + \omega)$ act like a very high frequency Stark effect and give rise to a small change in the transition frequency known as the Bloch–Siegert frequency shift.

With the rotating wave approximation, the equations of motion of the density matrix elements become

$$\dot{\rho}_{11} = i\frac{\omega_1}{2}(e^{i(\omega_0-\omega)t}\rho_{12} - e^{-i(\omega_0-\omega)t}\rho_{21}) \tag{14.66a}$$

$$\dot{\rho}_{22} = -i\frac{\omega_1}{2}(e^{i(\omega_0-\omega)t}\rho_{12} - e^{-i(\omega_0-\omega)t}\rho_{21}) \tag{14.66b}$$

$$\dot{\rho}_{12} = i\frac{\omega_1}{2}e^{-i(\omega_0-\omega)t}(\rho_{11} - \rho_{22}) \tag{14.66c}$$

$$\dot{\rho}_{21} = -i\frac{\omega_1}{2}e^{i(\omega_0-\omega)t}(\rho_{11} - \rho_{22}). \tag{14.66d}$$

These equations are called the optical Bloch equations. They can be used to obtain the time-dependent populations of the two levels (ρ_{11} and ρ_{22}), as well as the "coherences" (ρ_{12} and ρ_{21}).

First, consider the case of an on-resonance optical field, that is, $\omega = \omega_0$. Then, the optical Bloch equations become

$$\dot{\rho}_{11} = i\frac{\omega_1}{2}(\rho_{12} - \rho_{21}) \tag{14.67a}$$

$$\dot{\rho}_{22} = -i\frac{\omega_1}{2}(\rho_{12} - \rho_{21}) \tag{14.67b}$$

$$\dot{\rho}_{12} = i\frac{\omega_1}{2}(\rho_{11} - \rho_{22}) \tag{14.67c}$$

$$\dot{\rho}_{21} = -i\frac{\omega_1}{2}(\rho_{11} - \rho_{22}). \tag{14.67d}$$

These equations are identical to equations (14.33) for the time-dependent two-state problem with $\beta = \omega_1/2$. On-resonance, the coupling of the radiation field to the two levels with energy difference $\Delta E = \hbar\omega_0$ produces the identical equations of motion as the time-independent coupling of two degenerate states by a time-independent interaction. In effect, the resonant interaction of the radiation field with the pair of molecular eigenstates removes their energy difference and removes the time dependence from the field.

If the system is initially in the lower energy state, $|1\rangle$—that is, $\rho_{11} = 1$, $\rho_{22} = 0$, $\rho_{12} = 0$, and $\rho_{21} = 0$—the time-dependent density matrix elements are

$$\rho_{11} = \cos^2(\omega_1 t/2) \tag{14.68a}$$

$$\rho_{22} = \sin^2(\omega_1 t/2) \tag{14.68b}$$

$$\rho_{12} = \frac{i}{2}\sin(\omega_1 t) \tag{14.68c}$$

$$\rho_{21} = -\frac{i}{2}\sin(\omega_1 t) \tag{14.68d}$$

The system begins in state $|1\rangle$ and oscillates coherently between $|1\rangle$ and $|2\rangle$. The oscillation of the population between the two states is called a transient nutation. The frequency at which the population oscillates between the two states is ω_1, the Rabi frequency.

If a pulse of light is applied on-resonance with

$$\omega_1 t = \pi,$$

then $\rho_{11} = 0$ and $\rho_{22} = 1$. The system is in the excited state. For many identical molecules or atoms, subject to the same radiation field, the population will be transferred completely from the ground state, $|1\rangle$ to the excited state, $|2\rangle$. The population is said to be inverted. The pulse that is required to invert the population is called a π pulse. If

$$\omega_1 t = \pi/2,$$

then $\rho_{11} = 0.5$ and $\rho_{22} = 0.5$. This pulse is called a π-over-two pulse. It equalizes the populations of the two states. A $\pi/2$ pulse maximizes ρ_{12} and ρ_{21}. These density matrix elements are referred to as the "coherent components."

When the field is not on resonance (i.e., $\omega \neq \omega_0$), the solutions are more complicated. For the same initial conditions—$\rho_{11} = 1$, $\rho_{22} = 0$, $\rho_{12} = 0$, and $\rho_{21} = 0$—the solutions can be written in terms of a frequency called the "effective field," ω_e. Defining

$$\Delta\omega = \omega_0 - \omega, \tag{14.69}$$

then

$$\omega_e = (\Delta\omega^2 + \omega_1^2)^{1/2}. \tag{14.70}$$

The solutions to the optical Bloch equations are

$$\rho_{11} = 1 - \frac{\omega_1^2}{\omega_e^2}\sin^2(\omega_e t/2) \tag{14.71a}$$

$$\rho_{22} = \frac{\omega_1^2}{\omega_e^2}\sin^2(\omega_e t/2) \tag{14.71b}$$

$$\rho_{12} = \frac{\omega_1}{\omega_e^2}\left[\frac{i\omega_e}{2}\sin(\omega_e t) - \Delta\omega\sin^2(\omega_e t/2)\right]e^{-i\Delta\omega t} \tag{14.71c}$$

$$\rho_{12} = \frac{\omega_1}{\omega_e^2}\left[-\frac{i\omega_e}{2}\sin(\omega_e t) - \Delta\omega\sin^2(\omega_e t/2)\right]e^{i\Delta\omega t}. \tag{14.71d}$$

The populations, ρ_{11} and ρ_{22}, still oscillate as in the on-resonance case. However, the oscillation frequency is ω_e rather than the Rabi frequency, ω_1. The maximum probability in state $|2\rangle$ is no longer 1. The maximum value is

$$\rho_{22}^{\max} = \frac{\omega_1^2}{\omega_e^2}. \tag{14.72}$$

As $\Delta\omega$ increases for fixed ω_1, the frequency of the population oscillation increases and $\rho_{22}^{\max}$ decreases. The results for ρ_{11} and ρ_{22}, except for the notation and the form in which they are written, are identical to the results [equations (8.40)] for the time-dependent two-state problem treated in Chapter 8 when the two states are not degenerate. In that problem, the coupling is time-independent, and the two states differ in energy by ΔE. Here, $\Delta E = \hbar\Delta\omega$ is the difference between the radiation field energy and the transition energy. The time-dependent coupling at ω acting on states with frequency difference ω_0 behaves like a time-independent coupling between states with frequency difference $\Delta\omega$.

The condition of near resonance occurs when

$$\omega_1 > \Delta\omega,$$

which results in

$$\omega_e \cong \omega_1.$$

In this case, ρ_{11} and ρ_{22} reduce to the on-resonance equations [equations (14.68)], and ρ_{12} and ρ_{21} are given by

$$\rho_{12} = \frac{i}{2}\sin(\omega_1 t)e^{-i\Delta\omega t} \tag{14.73a}$$

$$\rho_{21} = -\frac{i}{2}\sin(\omega_1 t)e^{i\Delta\omega t}. \tag{14.73b}$$

These equations are identical to the on-resonance equations [equations (14.68)] except for an additional time-dependent phase factor. For a $\pi/2$ pulse, which maximizes the coherent density matrix elements, ρ_{12} and ρ_{21}, $\omega_1 t = \pi/2$, and the arguments of the exponential factors $\Delta\omega t \ll \pi/2$. Therefore, ρ_{12} and ρ_{21} will be virtually identical to the on-resonance case. Thus, the behavior of a system driven near resonance is essentially identical to a system driven on resonance.

F. FREE PRECESSION

Following a pulse of angle $\theta = \omega_1 t$ (referred to as the flip angle), for an on-resonant or near-resonant ensemble of identical two-state systems with initial condition $\rho_{11} = 1$, the density matrix elements are

$$\rho_{11} = \cos^2(\theta/2) \tag{14.74a}$$

$$\rho_{22} = \sin^2(\theta/2) \tag{14.74b}$$

$$\rho_{12} = \frac{i}{2}\sin\theta \tag{14.74c}$$

$$\rho_{21} = -\frac{i}{2}\sin\theta. \tag{14.74d}$$

Following the pulse (no radiation field), the Hamiltonian is $\underline{\underline{H}}_0$. The time evolution of the density matrix is determined by $\underline{\underline{H}}_0$, with

$$\dot{\underline{\underline{\rho}}} = -\frac{i}{\hbar}[\underline{\underline{H}}_0, \underline{\underline{\rho}}]. \tag{14.75}$$

The Hamiltonian matrix is

$$\underline{\underline{H}}_0 = \begin{bmatrix} -\omega_0/2 & 0 \\ 0 & \omega_0/2 \end{bmatrix}. \tag{14.76}$$

Multiplying out the commutator gives the equations of motion of the density matrix elements:

$$\dot{\rho}_{11} = 0 \tag{14.77a}$$

$$\dot{\rho}_{22} = 0 \tag{14.77b}$$

$$\dot{\rho}_{12} = i\omega_0\rho_{12} \tag{14.77c}$$

$$\dot{\rho}_{21} = -i\omega_0\rho_{21}. \tag{14.77d}$$

The solutions are

$$\rho_{11} = \text{a constant} = \rho_{11}(0) \tag{14.78a}$$

$$\rho_{22} = \text{a constant} = \rho_{22}(0) \tag{14.78b}$$

$$\rho_{12} = \rho_{12}(0)e^{i\omega_0 t} \tag{14.78c}$$

$$\rho_{21} = \rho_{21}(0)e^{-i\omega_0 t}. \tag{14.78d}$$

$t = 0$ is taken to be the time at the end of the pulse, and the initial conditions, $\rho_{ij}(0)$, are given by equations (14.74). The diagonal density matrix elements are independent of time. In the absence of the radiation field ($\underline{H}_I(t)$), the populations do not change. The off-diagonal density matrix elements do not change in magnitude but have time dependence through the time-dependent phase factors.

To demonstrate the nature of the off-diagonal density matrix elements (the coherences), consider the expectation value of the dipole operator, $e\underline{x} = \underline{\mu}$:

$$\langle \underline{\mu} \rangle = Tr\, \underline{\underline{\rho}}\, \underline{\underline{\mu}}. \tag{14.79}$$

As shown in equations (14.56), $\underline{\mu}$ has off-diagonal matrix elements only (taken to be real):

$$\underline{\underline{\mu}} = \begin{bmatrix} 0 & \mu \\ \mu & 0 \end{bmatrix}. \tag{14.80}$$

Then substituting equation (14.78) for the density matrix elements gives

$$Tr\, \underline{\underline{\rho}}\, \underline{\underline{\mu}} = Tr \begin{bmatrix} \rho_{11}(0) & \rho_{12}(0)e^{i\omega_0 t} \\ \rho_{21}(0)e^{-i\omega_0 t} & \rho_{22}(0) \end{bmatrix} \begin{bmatrix} 0 & \mu \\ \mu & 0 \end{bmatrix} \tag{14.81}$$

and

$$\langle \underline{\mu} \rangle = \mu \left[\rho_{12}(0)e^{i\omega_0 t} + \rho_{21}(0)e^{-i\omega_0 t} \right]. \tag{14.82}$$

For a flip angle of θ,

$$\langle \underline{\mu} \rangle = -\mu\, \sin\theta \sin(\omega_0 t). \tag{14.83}$$

Following a pulse of flip angle θ, the system has an electric dipole oscillating at frequency ω_0. The coherent oscillation in the absence of an applied radiation fielded is called free precession. In the semiclassical treatment being used here, the radiation field is treated classically. An oscillating electric dipole at frequency ω_0 gives rise to an oscillating electric field at this frequency. Therefore, an ensemble of identical atoms or molecules will radiate at the transition frequency. Since the dipole associated with each molecule is radiating in phase with the others, the $\bar{E}$ fields add constructively. The intensity of the emitted light is $I \propto |E|^2$. Therefore, the constructive interference produces a greater intensity than a collection of emitters with random phase relationships. This enhanced intensity is called coherent emission.

G. PURE AND MIXED DENSITY MATRICES

The density matrix developments and examples given above are for pure systems; that is, they are for a single system or an ensemble of identical systems. However, many situations involve mixtures of systems that are not identical. The total system is composed of subensembles of identical systems. Each subensemble is described as before. All of the relationships for the density operator are linear. Therefore, it is straightforward to find the density matrix for a mixed system.

For P_k defined as the probability of having the density matrix $\underline{\underline{\rho}}_k$, which describes the kth subensemble, with

$$0 \leq P_1,\ P_2,\ \ldots,\ P_k,\ \ldots \leq 1 \tag{14.84}$$

and

$$\sum_k P_k = 1, \tag{14.85}$$

then the density matrix for the mixed system is

$$\underline{\underline{\rho}}(t) = \sum_k P_k \underline{\underline{\rho}}_k(t). \tag{14.86}$$

If the distribution of subensembles is continuous, then

$$\underline{\underline{\rho}}(t) = \int P(k)\underline{\underline{\rho}}(k,t)\,dk, \tag{14.87}$$

with

$$\int P(k)\,dk = 1.$$

As an example, consider a radiation field acting on two subensembles of two-state systems. The subensembles, 1 and 2, contain the same number of molecules. The molecules in the subensembles differ only in their transition frequencies, which are ω_{01} and ω_{02}, rather than ω_0. If the frequency of the applied field, ω, is very close to the transition frequencies, and the differences between the applied frequency and the two transition frequencies are small compared to ω_1, both transitions are near resonance. Thus, for a given duration pulse of the radiation field, both subensembles will have the same flip angle, θ. Since there are the same number of molecules in each subensemble, $P_1 = 0.5$ and $P_2 = 0.5$.

The expectation value of the dipole moment operator is

$$\begin{aligned} \langle \underline{\mu} \rangle &= Tr \underline{\underline{\rho}}(t) \underline{\underline{\mu}} \\ &= \sum_k P_k Tr \underline{\underline{\rho}}_k \underline{\underline{\mu}}, \end{aligned} \tag{14.88}$$

and with the conditions given above, using equation (14.83),

$$\begin{aligned} \langle \underline{\mu} \rangle &= -\frac{1}{2} \mu \sin\theta [\sin(\omega_{01} t) + \sin(\omega_{02} t)] \\ &= -\mu \sin\theta \left[\sin \frac{1}{2}(\omega_{01} + \omega_{02}) t \cos \frac{1}{2}(\omega_{01} - \omega_{02}) t \right]. \end{aligned} \tag{14.89}$$

Taking the frequency centered between the two transitions as ω_0 and taking the shift from the center frequency as δ, $\omega_{01} = \omega_0 + \delta$ and $\omega_{02} = \omega_0 - \delta$, then

$$\langle \underline{\mu} \rangle = -\mu \sin\theta \left[\sin(\omega_0 t) \cos(\delta t) \right], \tag{14.90}$$

with $\delta \ll \omega_0$. The dipole, which gives rise to emitted radiation, is oscillating at the high-frequency ω_0, but the magnitude is modulated at the frequency δ. The emitted radiation has a beat frequency δ. The observation of such a beat is the basis for time-domain Fourier transform NMR. In NMR, the transitions are magnetic dipole, and $\langle \underline{\mu} \rangle$ is an oscillating magnetic dipole. The oscillating magnetic dipole induces a signal in a pick-up coil. The Fourier transform of the beats gives the frequencies of the subensemble transitions relative to the frequency of the applied field. In NMR, ω_1 can be large enough to make all of the transitions in a spectrum near resonance. The Fourier transform of the complicated beat pattern that results from many subensembles having different transition frequencies is the spectrum.

H. THE FREE INDUCTION DECAY

Frequently, even a single optical or magnetic resonance spectral line is not composed of molecules that have identical transition frequencies. This can arise because identical molecules have different local environments in media such as liquids, glasses, crystals, or proteins. The absorption line is said to be inhomogeneously broadened because the line is composed of a spread of frequencies associated with the molecules in the different environments. In many cases, the distribution of transition frequencies about the center frequency, ω_0, is a Gaussian. The system will be described by a mixed density matrix, and the density matrix is given by

$$\underline{\underline{\rho}}(t) = \frac{1}{\sqrt{2\pi\sigma^2}} \int_{-\infty}^{\infty} e^{-(\omega_h - \omega_0)^2 / 2\sigma^2} \underline{\underline{\rho}}(\omega_h, t)\, d\omega_h, \tag{14.91}$$

where ω_h is the transition frequency of a subensemble in the inhomogeneous line, and σ is the standard deviation. The term multiplying the integral is the normalization constant. The frequency of the radiation field is at line center; that is, $\omega = \omega_0$. If σ is small compared to ω_1, $\sigma \ll \omega_1$, then the entire inhomogeneous line is near resonance, and application of a pulse will cause each subensemble to have the same flip angle, θ.

Following a pulse, each subensemble with transition frequency ω_h will undergo free precession at ω_h. The ensemble averaged expectation value of the dipole moment operator, $\langle \underline{\mu} \rangle$, is given by

$$\begin{aligned} \langle \underline{\mu} \rangle &= Tr\underline{\underline{\rho}}(t)\underline{\underline{\mu}} \\ &= \frac{1}{\sqrt{2\pi\sigma^2}} \int_{-\infty}^{\infty} e^{-(\omega_h-\omega_0)^2/2\sigma^2} Tr\underline{\underline{\rho}}(\omega_h, t)\underline{\underline{\mu}}\, d\omega_h. \end{aligned} \tag{14.92}$$

Using the result of equation (14.83), for frequency ω_h and flip angle θ, the integral becomes

$$\langle \underline{\mu} \rangle = -\frac{\mu \sin\theta}{\sqrt{2\pi\sigma^2}} \int_{-\infty}^{\infty} e^{-(\omega_h-\omega_0)^2/2\sigma^2} \sin(\omega_h t)\, d\omega_h. \tag{14.93}$$

Substituting $\delta = (\omega_h - \omega_0)$, the integral is

$$\begin{aligned} \langle \underline{\mu} \rangle = &-\frac{\mu \sin\theta}{\sqrt{2\pi\sigma^2}} \\ &\left[\cos(\omega_0 t) \int_{-\infty}^{\infty} e^{-\delta^2/2\sigma^2} \sin\delta\, d\delta + \sin(\omega_0 t) \int_{-\infty}^{\infty} e^{-\delta^2/2\sigma^2} \cos\delta\, d\delta\right]. \end{aligned} \tag{14.94}$$

The first integral is zero because it is the product of an even function and an odd function integrated over a symmetrical interval. Evaluating the second integral yields

$$\langle \underline{\mu} \rangle = -\mu \sin\theta \sin(\omega_0 t) \exp\left[-\frac{t^2}{2(1/\sigma)^2}\right]. \tag{14.95}$$

$t = 0$ is taken at the end of the excitation pulse.

Equation (14.95) describes an oscillating electric dipole which will generate an electric field emitted from the ensemble of molecules. The dipole oscillates at the center frequency, ω_0, but its magnitude decreases with time. The magnitude decays as a Gaussian in time. The standard deviation of the time decay is $1/\sigma$, the inverse of the spectral width of the absorption line. The oscillating dipole produces radiation with a magnitude that decays in time. This is called a free induction decay. Its Fourier transform is the spectral line shape. The spread in transition frequencies causes each subensemble to precess at a different frequency. The free induction decay occurs because the freely processing subensembles lose the phase relationship established at $t = 0$. The dephasing causes the coherent emission of light to decay.

As discussed above in Section F, free precession does not change the diagonal density matrix elements. The excited state population, ρ_{22}, is not changed by the free induction decay. However, following the free induction decay, only incoherent spontaneous emission occurs. The $\overline{E}$ fields of the individually emitting oscillators do not add in phase. On a time scale short compared to the free induction decay, the emission is coherent; the $\overline{E}$ fields add in phase, producing enhanced emission. As can be seen from equation (14.82), free precession is a property of the off-diagonal density matrix elements. The diagonal density matrix elements determine the populations of the levels in a system. The free induction decay and the coherent emission embodied in equation (14.95) involve the off-diagonal density matrix elements, which describe the coherences in a system.

Equations (14.83) or (14.95) suggest that the maximum coherent emission will occur for a $\pi/2$ pulse and that no coherent emission will occur for a π pulse because $\langle \underline{\mu} \rangle$ is zero. These results arise from treating the radiation field classically in the semiclassical approach used above. When the radiation field is treated quantum mechanically, it is found that the maximum intensity of coherent optical emission occurs for a π pulse. The creation operator (see Chapter 12, Section E) associated with the $\overline{E}$ field operator carries with it a time-dependent phase factor. On a time scale short compared to the free induction decay, photons created by spontaneous emission have $\overline{E}$ fields with well-defined phase relationships. Thus, the photon $\overline{E}$ fields add in phase, giving rise to coherent emission. The number of photons emitted is proportional to the excited state population, ρ_{22}. Therefore, maximum intensity of coherent optical emission will occur for a π pulse. In NMR, the precessing magnetic dipole is detected. Spontaneous emission, coherent or otherwise, is negligible (see Chapter 12 Section C), and the maximum signal does occur following a $\pi/2$ pulse.

CHAPTER 15 ANGULAR MOMENTUM

In Chapter 7, the energy eigenvalues and wavefunctions for the hydrogen atom were obtained by solving the Schrödinger equation. The three-dimensional differential equation in spherical polar coordinates that arose was separated into three one-dimensional equations. The two equations that involved angles, φ and θ, gave solutions that depend on two quantum numbers, m and ℓ. Together, the solutions to the angular equations describe the orbital angular momentum of the hydrogen atom or any centrally symmetric system. The combined angular solutions are called the spherical harmonics. The spherical harmonics are the product of an exponential term with the quantum number m and the associated Legendre polynomials, with the quantum numbers m and ℓ. It was found that the factor involving ℓ could have values $\ell(\ell+1)$ with $\ell = 0, 1, 2, \ldots$, and the factor involving m could take on values $m = 0, \pm 1, \pm 2, \ldots, \pm \ell$. These solutions to the angular part of the hydrogen atom Schrödinger equation are the solutions to the orbital angular momentum eigenvalue problem.

In this chapter, the problem of angular momentum will be treated in a general manner using a raising and lower operator formalism. In Chapter 6 the harmonic oscillator was treated using the Schrödinger equation and the Dirac raising and lowering operator approach. The results of the two treatments were identical. However, in most situations, it is much easier to work with the raising and lowering operators than with the wavefunctions and differential equations that are inherent in the Schrödinger representation. The treatment of angular momentum with raising and lowering operators provides an alternative formalism to the Schrödinger representation that is very useful in many problems. While it recovers the results that emerged from the treatment of the hydrogen atom, it shows that these results are not complete. In addition to integer values for the angular momentum quantum number (usually called ℓ for orbital angular momentum), the full treatment shows that the angular momentum quantum numbers can also have half integer values.

A. ANGULAR MOMENTUM OPERATORS

Classical angular momentum is a vector, $\vec{J}$, given by the cross product

$$\vec{J} = \vec{r} \times \vec{p}, \tag{15.1}$$

where $\vec{r}$ is the radius vector from the origin, and $\vec{p}$ is the linear momentum. A simple way to obtain the components of the angular momentum is to write the cross product as a determinant.

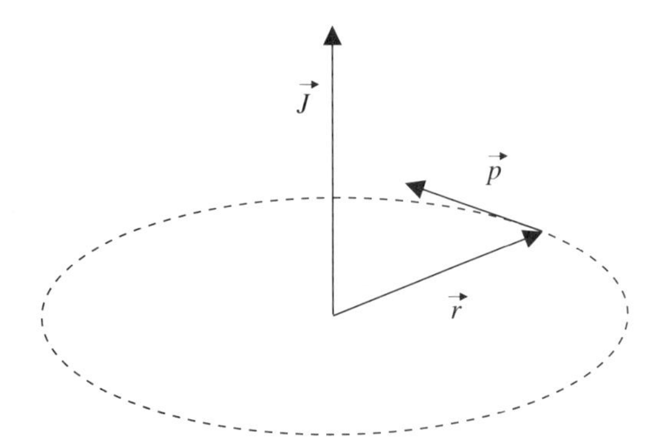

Figure 15.1 $\vec{r}$ is the radius vector. $\vec{p}$ is the linear momentum, which, at any instant, is tangent to the circle traced by the tip of $\vec{r}$. $\vec{J}$ is the angular momentum vector, which is perpendicular to plane containing $\vec{r}$ and $\vec{p}$.

$$\vec{J} = \begin{vmatrix} \hat{i} & \hat{j} & \hat{k} \\ x & y & z \\ p_x & p_y & p_z \end{vmatrix} \tag{15.2}$$

$\hat{i}$, $\hat{j}$, and $\hat{k}$ are unit vectors in the x, y, and z directions, respectively. p_x, p_y, and p_z are the x, y, and z components of the linear momentum. Multiplying out the determinant yields

$$J_x = yp_z - zp_y \tag{15.3a}$$

$$J_y = zp_x - xp_z \tag{15.3b}$$

$$J_z = xp_y - yp_x. \tag{15.3c}$$

The unit vectors have been dropped because the subscript on J_i gives the direction of the particular component of the angular momentum. The scalar product of the angular moment vector with itself is

$$\vec{J} \cdot \vec{J} = J^2 = J_x^2 + J_y^2 + J_z^2, \tag{15.4}$$

which is a scalar quantity. The square root of J^2 is the length of the angular momentum vector.

The quantum mechanical angular momentum operators for the components of the angular momentum are found by replacing the classical positions q with the operator $\underline{q}$ and the classical momentum p_q with the momentum operator $-i\hbar(\partial/\partial q)$ in the determinant in accord with Dirac's quantum condition (Chapter 4, Section A).

$$\underline{\vec{J}} = -i\hbar \begin{vmatrix} \hat{i} & \hat{j} & \hat{k} \\ \underline{x} & \underline{y} & \underline{z} \\ \frac{\partial}{\partial x} & \frac{\partial}{\partial y} & \frac{\partial}{\partial z} \end{vmatrix}. \tag{15.5}$$

Then the quantum mechanical operators for the components of the angular momentum are

$$\underline{J}_x = -i\hbar \left(\underline{y}\frac{\partial}{\partial z} - \underline{z}\frac{\partial}{\partial y} \right) \tag{15.6a}$$

$$\underline{J}_y = -i\hbar \left(\underline{z}\frac{\partial}{\partial x} - \underline{x}\frac{\partial}{\partial z} \right) \tag{15.6b}$$

$$\underline{J}_z = -i\hbar \left(\underline{x}\frac{\partial}{\partial y} - \underline{y}\frac{\partial}{\partial x} \right) \tag{15.6c}$$

and

$$\underline{\vec{J}} \cdot \underline{\vec{J}} = \underline{J}_x^2 + \underline{J}_y^2 + \underline{J}_z^2. \tag{15.7}$$

Consider the commutator

$$\left[\underline{J}_x, \underline{J}_y \right] = \underline{J}_x \underline{J}_y - \underline{J}_y \underline{J}_x. \tag{15.8}$$

Substituting equations (15.6a) and (15.6b), written in units of $\hbar$,

$$\begin{aligned} \underline{J}_x \underline{J}_y &= -\left(\underline{y}\frac{\partial}{\partial z} - \underline{z}\frac{\partial}{\partial y} \right) \left(\underline{z}\frac{\partial}{\partial x} - \underline{x}\frac{\partial}{\partial z} \right) \\ &= -\left(\underline{y}\frac{\partial}{\partial z}\underline{z}\frac{\partial}{\partial x} - \underline{y}\frac{\partial}{\partial z}\underline{x}\frac{\partial}{\partial z} - \underline{z}\frac{\partial}{\partial y}\underline{z}\frac{\partial}{\partial x} + \underline{z}\frac{\partial}{\partial y}\underline{x}\frac{\partial}{\partial z} \right) \end{aligned} \tag{15.9a}$$

and

$$\underline{J}_y \underline{J}_x = -\left(\underline{z}\frac{\partial}{\partial x}\underline{y}\frac{\partial}{\partial z} - \underline{z}\frac{\partial}{\partial x}\underline{z}\frac{\partial}{\partial y} - \underline{x}\frac{\partial}{\partial z}\underline{y}\frac{\partial}{\partial z} + \underline{x}\frac{\partial}{\partial z}\underline{z}\frac{\partial}{\partial y} \right). \tag{15.9b}$$

Subtracting gives

$$\begin{aligned} \left[\underline{J}_x, \underline{J}_y \right] &= -\left[\underline{y}\frac{\partial}{\partial x}\left(\frac{\partial}{\partial z}\underline{z} - \underline{z}\frac{\partial}{\partial z} \right) + \underline{x}\frac{\partial}{\partial y}\left(\underline{z}\frac{\partial}{\partial z} - \frac{\partial}{\partial z}\underline{z} \right) \right] \\ &= -\left(\underline{y}\frac{\partial}{\partial x} - \underline{x}\frac{\partial}{\partial y} \right) \left[\frac{\partial}{\partial z}, \underline{z} \right] \\ &= \left(\underline{x}\frac{\partial}{\partial y} - \underline{y}\frac{\partial}{\partial x} \right) \left[\frac{\partial}{\partial z}, \underline{z} \right] \\ &= i\underline{J}_z \left[\frac{\partial}{\partial z}, \underline{z} \right]. \end{aligned} \tag{15.10}$$

However, it is easily shown that

$$\left[\frac{\partial}{\partial z}, \underline{z} \right] = 1. \tag{15.11}$$

From the definition of the momentum operator,

$$\frac{\partial}{\partial \underline{z}} = \frac{\underline{P}_z}{-i\hbar}. \tag{15.12}$$

Then the left-hand side of equation (15.11) becomes

$$-\frac{1}{i\hbar}\left[\underline{P}_z, \underline{z}\right] = -\frac{1}{i\hbar}(-1)\left[\underline{z}, \underline{P}_z\right], \tag{15.13}$$

and since the commutator $\left[\underline{z}, \underline{P}_z\right] = i\hbar$, the right-hand side of equation (15.13) is

$$= \frac{1}{i\hbar}(i\hbar) = 1. \tag{15.14}$$

Then the term in brackets in equation (15.10) is 1, and

$$\left[\underline{J}_x, \underline{J}_y\right] = i\underline{J}_z, \tag{15.15}$$

in units of $\hbar$. In conventional units, $\left[\underline{J}_x, \underline{J}_y\right] = i\hbar\underline{J}_z$. Similarly, in units of $\hbar$

$$\left[\underline{J}_y, \underline{J}_z\right] = i\underline{J}_x \tag{15.16}$$

$$\left[\underline{J}_z, \underline{J}_x\right] = i\underline{J}_y. \tag{15.17}$$

The operators representing the components of the angular momentum do not commute. Using the commutators of the components of $\underline{\vec{J}}$, it is readily shown that

$$\left[\underline{J}^2, \underline{J}_z\right] = \left[\underline{J}^2, \underline{J}_x\right] = \left[\underline{J}^2, \underline{J}_y\right] = 0; \tag{15.18}$$

that is, the operator for the square of the angular momentum commutes with each of its components.

Equations (15.15)–(15.18) give the fundamental commutator relationships for angular momentum operators. Since the square of the total angular momentum operator commutes with the operators for each component, $\underline{J}^2$ and one component of the angular momentum will have simultaneous eigenvectors and can be simultaneously observed. This is fundamentally different from the classical description of angular momentum. The commutators show that it is possible to know the total length of the angular momentum vector (the square root of the eigenvalue of $\underline{J}^2$) and the projection of the angular momentum vector onto one axis. In classical mechanics it is possible to know all three components of the angular momentum. While the choice of which component of the angular momentum will be measured is arbitrary, the axis system is usually selected such that the component corresponds to the eigenvalue of $\underline{J}_z$. Thus, the matrices representing the operators $\underline{J}^2$ and $\underline{J}_z$ can be simultaneously diagonalized by the same unitary transformation.

It can also be shown that

$$\left[\underline{H}, \underline{\vec{J}}\right] = 0. \tag{15.19}$$

The operator $\underline{\vec{J}}$ acts like a rotation operator. A rotation of an isolated system does not change its energy. The fact that $\underline{H}$ commutes with $\underline{\vec{J}}$ means that

$$\left[\underline{H}, \underline{J}^2\right] = 0. \tag{15.20}$$

Therefore, $\underline{H}$, $\underline{J}^2$, and $\underline{J}_z$ are all simultaneous observables. In the treatment of the hydrogen atom (Chapter 7), it was found that three quantum numbers, n, ℓ, and m, were necessary to define that state of a system. For example, the $2p_0$ orbital (also referred to as the $2p_z$ orbital) has $n = 2$, $\ell = 1$ (p orbital), and $m = 0$. There are three p orbitals, distinguished by three different values of m. The ℓ quantum number differentiates the $2p_0$ from the $2s$ orbital. Both have $m = 0$ and $n = 2$. The n quantum number distinguishes the $2p$ orbitals from the $3p$ orbitals, and so on. The three operators, $\underline{H}$, $\underline{J}^2$, and $\underline{J}_z$, are sufficient to define the states of the hydrogen atom by providing three simultaneous observables; the energy, the total magnitude of the orbital angular momentum, and the projection of the angular momentum on the z axis. There are always enough commuting operators to completely define the state of a system.

B. THE EIGENVALUES OF $\underline{J}^2$ AND $\underline{J}_z$

Since $\underline{J}^2$ and $\underline{J}_z$ commute, there is a set of ket vectors, $|\lambda m\rangle$ (taken to be orthonormal), which are simultaneous eigenvectors of the two operators; that is, the matrices $\underline{\underline{J}}^2$ and $\underline{\underline{J}}_z$ are simultaneously diagonal in basis composed of these ket vectors. In units of $\hbar$,

$$\underline{J}^2 |\lambda m\rangle = \lambda |\lambda m\rangle \tag{15.21}$$

$$\underline{J}_z |\lambda m\rangle = m |\lambda m\rangle . \tag{15.22}$$

The approach used to find the eigenvalues and eigenvectors is analogous to the raising and lowering operator treatment of the harmonic oscillator. The operators

$$\underline{J}_+ = \underline{J}_x + i\underline{J}_y \tag{15.23a}$$

$$\underline{J}_- = \underline{J}_x - i\underline{J}_y \tag{15.23b}$$

are formed. To proceed, a number of commutators and identities are required. Using definitions of $\underline{J}_+$ and $\underline{J}_-$ and the commutators given in equations (15.15)–(15.18), the following commutators can be derived:

$$[\underline{J}_+, \underline{J}_z] = -\underline{J}_+ \tag{15.24a}$$

$$[\underline{J}_-, \underline{J}_z] = \underline{J}_- \tag{15.24b}$$

$$[\underline{J}_+, \underline{J}_-] = 2\underline{J}_z . \tag{15.24c}$$

The following useful identities can be also be obtained:

$$\underline{J}_+\underline{J}_- = \underline{J}^2 - \underline{J}_z^2 + \underline{J}_z \tag{15.25a}$$

$$\underline{J}_-\underline{J}_+ = \underline{J}^2 - \underline{J}_z^2 + \underline{J}_z . \tag{15.25b}$$

The first step is to find a relationship between λ and m. The expectation value of

$$\langle\lambda m| \underline{J}^2 |\lambda m\rangle \geq \langle\lambda m| \underline{J}_z^2 |\lambda m\rangle \tag{15.26}$$

since

$$\langle\lambda m|\, \underline{J}^2 \,|\lambda m\rangle = \langle\lambda m|\, \underline{J}_z^2 \,|\lambda m\rangle + \langle\lambda m|\, \underline{J}_x^2 \,|\lambda m\rangle + \langle\lambda m|\, \underline{J}_y^2 \,|\lambda m\rangle . \tag{15.27}$$

The $\underline{J}_i$'s are Hermitian operators, and therefore the $\underline{J}_i^2$ brackets are the squares of real numbers, which are positive. Then the right-hand side of equation (15.27) is the sum of three positive numbers. $\langle\lambda m|\, \underline{J}^2 \,|\lambda m\rangle$ is equal to the sum of three positive numbers. Therefore, it must be greater than or equal to one of these numbers. Since $|\lambda m\rangle$ is an eigenket of $\underline{J}^2$ with eigenvalue λ,

$$\langle\lambda m|\, \underline{J}^2 \,|\lambda m\rangle = \lambda , \tag{15.28}$$

and $|\lambda m\rangle$ is an eigenket of $\underline{J}_z$ with eigenvalue m,

$$\langle\lambda m|\, \underline{J}_z^2 \,|\lambda m\rangle = m^2 . \tag{15.29}$$

Therefore, using equation (15.26) gives the result

$$\lambda \geq m^2 . \tag{15.30}$$

From equation (15.24a),

$$\underline{J}_z \underline{J}_+ = \underline{J}_+ \underline{J}_z + \underline{J}_+ . \tag{15.31}$$

Consider

$$\begin{aligned} \underline{J}_z \left[\underline{J}_+ \,|\lambda m\rangle\right] &= \underline{J}_+ \underline{J}_z \,|\lambda m\rangle + \underline{J}_+ \,|\lambda m\rangle \\ &= \underline{J}_+ m \,|\lambda m\rangle + \underline{J}_+ \,|\lambda m\rangle \\ &= (m+1)\left[\underline{J}_+ \,|\lambda m\rangle\right] . \end{aligned} \tag{15.32}$$

$\underline{J}_+ \,|\lambda m\rangle$, the term in square brackets on the left-hand side of the first line of equation (15.32), is the same as the term in square brackets on the right-hand side of the last line. $\underline{J}_+ \,|\lambda m\rangle$ is some ket vector because applying an operator to a ket returns a ket. $\underline{J}_z$ operating on $\underline{J}_+ \,|\lambda m\rangle$ gives the same term back times a number, $(m + 1)$. Therefore, $\underline{J}_+ \,|\lambda m\rangle$ is an eigenket of $\underline{J}_z$ with eigenvalue $(m + 1)$. Furthermore,

$$\left[\underline{J}^2, \underline{J}_+\right] = 0, \tag{15.33}$$

since $\underline{J}^2$ commutes with each of its components. Then

$$\begin{aligned} \underline{J}^2 \left[\underline{J}_+ \,|\lambda m\rangle\right] &= \underline{J}_+ \underline{J}^2 \,|\lambda m\rangle \\ &= \lambda \left[\underline{J}_+ \,|\lambda m\rangle\right] . \end{aligned} \tag{15.34}$$

Thus, $\underline{J}_+ \,|\lambda m\rangle$ is an eigenvector of $\underline{J}^2$ with eigenvalue λ.

$\underline{J}_+ \,|\lambda m\rangle$ is an eigenket of both $\underline{J}^2$ and $\underline{J}_z$ with eigenvalues λ and $m + 1$, respectively. Therefore, operating $\underline{J}_+$ on the ket $|\lambda m\rangle$ gives a new eigenket, $|\lambda m + 1\rangle$. $\underline{J}_+$ is a raising operator. Operating it on an eigenket gives a new eigenket with m increased by 1 and λ left unchanged.

Repeated applications of $\underline{J}_+$ to a ket $|\lambda m\rangle$ will give eigenvectors with larger and larger values of m. However, because of the relation equation (15.30), the raising operator cannot be applied indefinitely. There is some maximum value of m, call it j,

$$m_{\max} = j,$$

for which if the raising operator is applied to $|\lambda j\rangle$, the relation $\lambda \geq m^2$ will be violated. For $m = j$,

$$\underline{J}_+ |\lambda j\rangle = 0 \tag{15.35a}$$

with

$$|\lambda j\rangle \neq 0. \tag{15.35b}$$

In a manner analogous to that used in equations (15.31) and (15.32), it can be shown that $\underline{J}_- |\lambda m\rangle$ is an eigenket of $\underline{J}_z$ with eigenvalue $m - 1$, and it is an eigenket of $\underline{J}^2$ with eigenvalue λ. $\underline{J}_-$ is a lowering operator. Operating it on an eigenket gives a new eigenket with m decreased by 1 and λ left unchanged.

Repeatedly operating $\underline{J}_-$ on $|\lambda j\rangle$, where j is the maximum value of m, gives a sequence of eigenkets with eigenvalues of $\underline{J}_z$:

$$m = j, j - 1, j - 2, \ldots.$$

However, the lowering operator cannot be applied indefinitely because $\lambda \geq m^2$. There is some smallest value of m, call it j' such that application of the lowering operator to the ket $|\lambda j'\rangle$ gives

$$\underline{J}_- |\lambda j'\rangle = 0 \tag{15.36a}$$

with

$$|\lambda j'\rangle \neq 0. \tag{15.36b}$$

Since integer steps were taken in going from j, the largest value of m, to j', the smallest value of m, j and j' must differ by an integer, that is,

$$j = j' + \text{an integer}. \tag{15.37}$$

Left-multiplying equation (15.35a) by $\underline{J}_-$ and left-multiplying equation (15.36a) by $\underline{J}_+$ give

$$\underline{J}_-\underline{J}_+ |\lambda j\rangle = 0 \tag{15.38a}$$

$$\underline{J}_+\underline{J}_- |\lambda j'\rangle = 0. \tag{15.38b}$$

Using the identities of equation (15.25), these can be written as

$$\underline{J}_-\underline{J}_+ |\lambda j\rangle = 0 = (\underline{J}^2 - \underline{J}_z^2 - \underline{J}_z) |\lambda j\rangle \tag{15.39a}$$

$$\underline{J}_+\underline{J}_- |\lambda j'\rangle = 0 = (\underline{J}^2 - \underline{J}_z^2 + \underline{J}_z) |\lambda j'\rangle. \tag{15.39b}$$

The kets on the right-hand sides are eigenkets of the operators being applied to them. Performing the operations gives

$$\underline{J}_- \underline{J}_+ |\lambda j\rangle = 0 = (\lambda - j^2 - j)\, |\lambda j\rangle \tag{15.40a}$$

$$\underline{J}_+ \underline{J}_- \left|\lambda j'\right\rangle = 0 = (\lambda - j'^2 + j')\left|\lambda j'\right\rangle. \tag{15.40b}$$

Since $|\lambda j\rangle \neq 0$ and $\left|\lambda j'\right\rangle \neq 0$, the coefficients of these kets on the right-hand sides of equations (15.40) must be zero. Therefore, from equation (15.40a)

$$\lambda = j(j+1) \tag{15.41a}$$

and from equation (15.40b)

$$\lambda = (-j')(-j'+1). \tag{15.41b}$$

Furthermore, since $j > j'$, equations (15.41) give

$$j' = -j. \tag{15.42}$$

The maximum value of m is j. The minimum value of m is j'. Starting with j, application of the lowering operator reduces j in integer steps until j' is reached. The interval from j to j' is $2j$. Thus,

$$2j = \text{an integer}. \tag{15.43}$$

The final result is that the eigenvalues, λ, of $\underline{J}^2$ are $j(j+1)$ with the possible values of j;

$$j = 0, \frac{1}{2}, 1, \frac{3}{2}, \ \ldots . \tag{15.44}$$

j can take on integer and half-integer values, in contrast to the solution of the hydrogen atom orbital angular momentum problem. In the hydrogen atom problem, only integer values of the angular momentum arose from the solution of the Schrödinger equation. The eigenvalues $\underline{J}_z$ are

$$m = j, j-1, \ \ldots, -j+1, -j. \tag{15.45}$$

j is m's largest value, and $-j$ is its smallest value. Labeling the eigenkets by the quantum numbers j and m, the eigenvalue equations are

$$\underline{J}^2 \,|jm\rangle = j(j+1)\,|jm\rangle \tag{15.46}$$

$$\underline{J}_z \,|jm\rangle = m\,|jm\rangle \tag{15.47}$$

For each value of j, there are $2j+1$ different m states. Equations (15.46) and (15.47) are in units of $\hbar$. In conventional units, the right-hand side of equation (15.46) is multiplied by $\hbar^2$, and the right-hand side of equation (15.47) is multiplied by $\hbar$.

In the development given above, the action of the raising and lowering operators, $\underline{J}_+$ and $\underline{J}_-$, was determined. For example, it was shown that $\underline{J}_+ |\lambda m\rangle$ is an eigenket of both $\underline{J}^2$

and $\underline{J}_z$. When $\underline{J}_+$ and $\underline{J}_-$ are applied to an eigenket, they return another eigenket multiplied by a constant. Since the state of a system is determined by the direction of a ket vector, not its length (provided it is not zero; see Chapter 2), the constant factor that occurs on application of $\underline{J}_+$ and $\underline{J}_-$ does not change the arguments made above. However, to use $\underline{J}_+$ and $\underline{J}_-$ in calculations, it is necessary to know the constants. These can be derived and are

$$\underline{J}_+ \left| jm \right\rangle = \sqrt{(j-m)(j+m+1)} \left| jm+1 \right\rangle \tag{15.48}$$

$$\underline{J}_- \left| jm \right\rangle = \sqrt{(j+m)(j-m+1)} \left| jm-1 \right\rangle . \tag{15.49}$$

C. ANGULAR MOMENTUM MATRICES

The total angular momentum quantum number j can take on values $j = 0, 1/2, 1, 3/2, 2, \ldots$. All of the different angular momentum states can be grouped according to the particular j state:

$$\begin{array}{ll} j=0 & m=0 \\ j=1/2 & m=1/2, -1/2 \\ j=1 & m=1, 0, -1 \\ j=3/2 & m=3/2, 1/2, -1/2, -3/2 \\ j=2 & m=2, 1, 0, -1, -2 \\ \vdots & \vdots \end{array}$$

The integer values of j are identical to the results obtained from solving the angular parts of the hydrogen atom Schrödinger equation. In the hydrogen atom, the angular momentum quantum numbers were called ℓ and m; otherwise the results are identical. $j = 0, m = 0$ corresponds to an s orbital. $j = 1, m = 1, 0, -1$ corresponds to the three different p orbitals, and so on. However, the orbital angular momentum states obtained from the solution to the hydrogen atom wave equation have integer values of the total quantum numbers only, but the full solution to the angular momentum problem also includes angular momentum states with half-integer values.

The eigenvalues of the $\underline{J}^2$ operator are the square of the total angular momentum vector. The length of the angular momentum vector for a particular value of j is $\sqrt{j(j+1)}$. In conventional units the length is $\hbar\sqrt{j(j+1)}$. For a given j, eigenvalues of the $\underline{J}_z$ operator, m, are the various projections of the angular momentum vector onto the z axis. In conventional units, the length of the projection is $\hbar m$. The nature of the m states can be visualized as shown in Figure 15.2 for $j = 1, m = 1, 0, -1$. In Figure 15.2, the dashed line is the z axis. The length of the angular momentum vector, in units of $\hbar$, is $\sqrt{2}$. The three possible projections on the z axis, in units of $\hbar$, have lengths 1, 0, and -1. Since only the length of the vector and its projection on the z axis can be known, for each m state, the tip of the vector can be thought of as delocalized; that is, the tip of the vector can be found anywhere on the circle. For $m = 0$, the angular momentum vector is perpendicular to the z axis, so the projection is 0. A similar diagram can be drawn for each value of j.

The matrix elements of $\underline{J}^2$, $\underline{J}_z$, $\underline{J}_+$, and $\underline{J}_-$ are

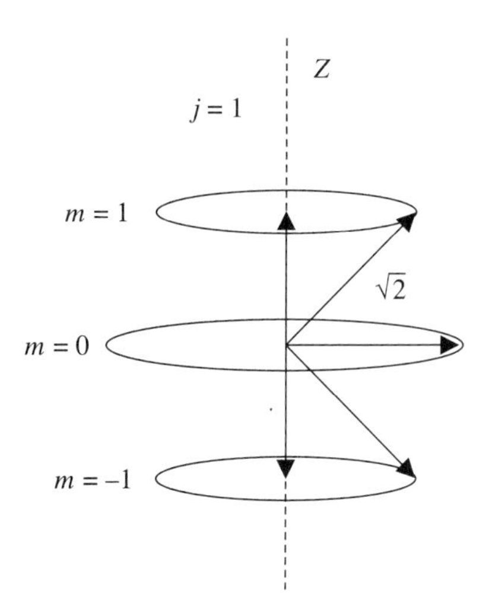

Figure 15.2 Schematic of the angular momentum eigenstates for $j = 1, m = 1, 0, -1$ in units of $\hbar$. The m states are the projections of the total angular momentum vector onto the z axis. The circles illustrate that the tip of the vector for each m state is delocalized about the z axis.

$$\langle j'm' | \underline{J}^2 | jm \rangle = j(j+1) \qquad \delta_{j'j}\delta_{m',m} \tag{15.50}$$

$$\langle j'm' | \underline{J}_z | jm \rangle = m \qquad \delta_{j'j}\delta_{m',m} \tag{15.51}$$

$$\langle j'm' | \underline{J}_+ | jm \rangle = \sqrt{(j-m)(j+m+1)} \qquad \delta_{j'j}\delta_{m',m+1} \tag{15.52}$$

$$\langle j'm' | \underline{J}_- | jm \rangle = \sqrt{(j+m)(j-m+1)} \qquad \delta_{j'j}\delta_{m',m-1}. \tag{15.53}$$

Each of the angular momentum operators could be written as a single matrix as shown:

$$\begin{array}{cc|ccc|}
 & j & 0 & 1/2 & 1 \\
j' & m & 0 & \tfrac{1}{2}\ -\tfrac{1}{2} & 1\ 0\ -1 \\
\hline
0 & 0 & (\ ) & (0) & (0) \cdots \\
1/2 & \begin{array}{c}1/2\\-1/2\end{array} & (0) & (\ ) & (0) \\
1 & \begin{array}{c}1\\0\\-1\end{array} & (0) & (0) & (\ ) \\
 & & \vdots & & \ddots
\end{array}$$

However, none of the operators change the value of j. Therefore, the matrix has blocks on the diagonal for a given j and various values of m, but all of the blocks that are off-diagonal are zero. Nonzero off-diagonal blocks would require an operator that changed j. Therefore, the matrices for the four operators can be grouped by the j value. Each j block is independent of the others and can be considered separately.

Using equations (15.50)–(15.53), the first few sets of matrices for the four operators $\underline{J}^2$, $\underline{J}_z$, $\underline{J}_+$, and $\underline{J}_-$ are:

$j = 0$

$$\underline{\underline{J}}_+ = (0) \qquad \underline{\underline{J}}_- = (0)$$
$$\underline{\underline{J}}_z = (0) \qquad \underline{\underline{J}}^2 = (0) \tag{15.54}$$

$j = \frac{1}{2}$

$$\underline{\underline{J}}_+ = \begin{pmatrix} 0 & 1 \\ 0 & 0 \end{pmatrix} \qquad \underline{\underline{J}}_- = \begin{pmatrix} 0 & 0 \\ 1 & 0 \end{pmatrix}$$
$$\underline{\underline{J}}_z = \begin{pmatrix} 1/2 & 0 \\ 0 & -1/2 \end{pmatrix} \qquad \underline{\underline{J}}^2 = \begin{pmatrix} 3/4 & 0 \\ 0 & 3/4 \end{pmatrix} \tag{15.55}$$

$j = 1$

$$\underline{\underline{J}}_+ = \begin{pmatrix} 0 & \sqrt{2} & 0 \\ 0 & 0 & \sqrt{2} \\ 0 & 0 & 0 \end{pmatrix} \qquad \underline{\underline{J}}_- = \begin{pmatrix} 0 & 0 & 0 \\ \sqrt{2} & 0 & 0 \\ 0 & \sqrt{2} & 0 \end{pmatrix}$$
$$\underline{\underline{J}}_z = \begin{pmatrix} 1 & 0 & 0 \\ 0 & 0 & 0 \\ 0 & 0 & -1 \end{pmatrix} \qquad \underline{\underline{J}}^2 = \begin{pmatrix} 2 & 0 & 0 \\ 0 & 2 & 0 \\ 0 & 0 & 2 \end{pmatrix} \tag{15.56}$$

Since the $|jm\rangle$ kets are eigenkets of $\underline{J}^2$ and $\underline{J}_z$, their matrices are diagonal. The raising and lowering operators, $\underline{J}_+$ and $\underline{J}_-$, have matrix elements one step above and below the principal diagonal, respectively.

D. ORBITAL ANGULAR MOMENTUM AND THE ZEEMAN EFFECT

In dealing with orbital angular momentum, it is conventional to refer to the angular momentum operators as $\underline{L}^2$ and $\underline{L}_z$, rather than $\underline{J}^2$ and $\underline{J}_z$, and the quantum numbers as ℓ and m, rather than j and m. The m quantum number is often written as m_ℓ to indicate that it is associated with orbital angular momentum. In the hydrogen atom problem (Chapter 7), m_ℓ was called the magnetic quantum number. In the absence of a magnetic field, states with different m_ℓ values for a given value of ℓ and principal quantum number, n, have the same energy. When a magnetic field is applied, the m_ℓ states belonging to a particular value of ℓ are no longer degenerate. For a p orbital ($\ell = 1$), the three m_ℓ values, 1, 0, -1, have three different energies. The influence of a magnetic field on the orbital energy is called the Zeeman effect. This effect does not depend on the intrinsic angular momentum of the electron, called electron spin, which gives rise to the magnetic moment of the electron. Electron spin is discussed in Chapter 16.

In addition to the operators for the kinetic energy and the Coulomb interactions in the Hamiltonian for the hydrogen atom (Chapter 7), the helium atom (Chapter 10), and other atoms and molecules, there is an additional term for the interaction of the electron orbital angular momentum with an applied magnetic field. Equation (12.31) gives the Hamiltonian for a charged particle in any combination of electric and magnetic fields in terms of the vector potential. The Zeeman effect involves an external static magnetic field. From Maxwell's equations,

$$\vec{B} = \vec{\nabla} \times \vec{A}. \tag{15.57}$$

For a static magnetic field,

$$\vec{A} = \frac{1}{2}\vec{B} \times \vec{r}. \tag{15.58}$$

This form of the vector potential can be demonstrated to be correct by substituting it into equation (15.57) and evaluating. It is necessary to determine the value of

$$\frac{1}{2}\vec{\nabla} \times (\vec{B} \times \vec{r}). \tag{15.59}$$

The vector identity for the triple cross product

$$\vec{a} \times (\vec{b} \times \vec{c}) = (\vec{a} \cdot \vec{c})\vec{b} - (\vec{a} \cdot \vec{b})\vec{c}, \tag{15.60}$$

can be used. However, if the first vector is a differential operator, it operates on both subsequent vectors, so the identity takes the form

$$\vec{\nabla} \times (\vec{b} \times \vec{c}) = \vec{b}\vec{\nabla} \cdot \vec{c} + \vec{c} \cdot \vec{\nabla}\vec{b} - \vec{c}\vec{\nabla} \cdot \vec{b} - \vec{b} \cdot \vec{\nabla}\vec{c}. \tag{15.61}$$

Then,

$$\vec{\nabla} \times (\vec{B} \times \vec{r}) = \vec{B}\vec{\nabla} \cdot \vec{r} + \vec{r} \cdot \vec{\nabla}\vec{B} - \vec{r}\vec{\nabla} \cdot \vec{B} - \vec{B} \cdot \vec{\nabla}\vec{r}. \tag{15.62}$$

Since $\vec{B}$ is a constant vector, its derivative is 0. Therefore, the second and third terms on the right-hand side of equation (15.62) vanish. The first term on the right-hand side is

$$\begin{aligned} \vec{B}\vec{\nabla} \cdot \vec{r} &= \vec{B}\left(\frac{\partial}{\partial x}x + \frac{\partial}{\partial y}y + \frac{\partial}{\partial z}z\right) \\ &= 3\vec{B}. \end{aligned} \tag{15.63}$$

The fourth term on the right-hand side of equation (15.62), with unit vectors, $\hat{i}$, $\hat{j}$, $\hat{k}$, in the x, y, z directions, respectively, is

$$\begin{aligned} \vec{B} \cdot \vec{\nabla}\vec{r} &= B_x \frac{\partial}{\partial x}(\hat{i}x + \hat{j}y + \hat{k}z) \\ &\quad + B_y \frac{\partial}{\partial y}(\hat{i}x + \hat{j}y + \hat{k}z) \\ &\quad + B_z \frac{\partial}{\partial z}(\hat{i}x + \hat{j}y + \hat{k}z) \\ &= \hat{i}B_x + \hat{j}B_y + \hat{k}B_z \\ &= \vec{B}. \end{aligned} \tag{15.64}$$

Therefore,

$$\frac{1}{2}\vec{\nabla} \times (\vec{B} \times \vec{r}) = \frac{1}{2}(3\vec{B} - \vec{B})$$
$$= \vec{B}, \tag{15.65}$$

which confirms equation (15.58), the form of the vector potential.

The atomic (or molecular) Hamiltonian can be written as

$$\underline{H} = \underline{H}_0 + \underline{H}_M, \tag{15.66}$$

where the first term is

$$\underline{H}_0 = -\frac{\hbar^2}{2m}\nabla^2 + V, \tag{15.67}$$

the sum of the kinetic energy and the Coulomb terms in the potential energy. $\underline{H}_M$ is the additional Zeeman term arising from the magnetic interaction. $\underline{H}_M$ is obtained from equation (12.31), the Hamiltonian for a charged particle in electric and magnetic fields. The first term is the kinetic energy term, which is included as part of $\underline{H}_0$. The second term depends on the square of the vector potential. It is dropped because it is negligible unless the magnetic field is very large or the principal atomic quantum number, n is > 20 (the orbital angular momentum can be very large). For a static magnetic field, the divergence of $\vec{A}$ is zero, so the third term is 0. Since $\vec{B}$ is given in terms of the vector potential, the scalar potential, ϕ, is zero. Therefore,

$$\underline{H}_M = \frac{ie\hbar}{mc}\vec{A} \cdot \vec{\nabla}. \tag{15.68}$$

Substituting equation (15.58) for the vector potential gives

$$\underline{H}_M = \frac{ie\hbar}{2mc}(\vec{B} \times \underline{\vec{r}}) \cdot \vec{\nabla}$$
$$= -\frac{e}{2mc}\vec{B} \cdot \left(\underline{\vec{r}} \times \frac{\hbar}{i}\vec{\nabla}\right)$$
$$= -\frac{e}{2mc}\vec{B} \cdot \left(\underline{\vec{r}} \times \underline{\vec{P}}\right)$$
$$= -\frac{e}{2mc}\vec{B} \cdot \underline{\vec{L}}. \tag{15.69}$$

$\underline{L}$ is the orbital angular momentum operator, and, in the second line, the vector identity, $(\vec{a} \times \vec{b}) \cdot \vec{c} = \vec{a} \cdot (\vec{b} \times \vec{c})$, was used. Since the magnetic field is constant, $\vec{B}$ is not an operator.

The energy eigenvalue equation is

$$\underline{H}|\psi\rangle = \underline{H}_0\,|\psi\rangle - \frac{e}{2mc}\vec{B} \cdot \underline{\vec{L}}\,|\psi\rangle = E\,|\psi\rangle\,. \tag{15.70}$$

Since magnetic field can be chosen to be in any direction, it simplifies the analysis to take it to be along the z axis. The Bohr magneton, μ_0, is the fundamental unit of magnetic moment. It is

$$\mu_0 = \frac{e\hbar}{2mc} = -0.9273 \times 10^{-23} \text{ joules/tesla.} \tag{15.71}$$

With the magnetic field along z and using μ_0, the eigenvalue equation is

$$\underline{H}_0 \left|\psi\right\rangle - \mu_0 B \underline{L}_z \left|\psi\right\rangle = E \left|\psi\right\rangle, \tag{15.72}$$

with the $\hbar$ from $\underline{L}_z$ absorbed into μ_0. B is the magnitude of the field.

The wavefunction solutions to the hydrogen atom eigenvalue problem obtained in Chapter 7 have the form

$$\left|\psi_{n\ell m}(r, \theta, \varphi)\right\rangle = R(r) Y_\ell^m(\theta, \varphi), \tag{15.73}$$

where the $R(r)$ are the solutions to the radial part of the wave equation and the $Y_\ell^m(\theta, \varphi)$ are the spherical harmonics, the solutions to the angular parts of the wave equation. This form applies to any atom or other system with a spherically symmetric potential. As discussed above, in solving the angular parts of the hydrogen atom problem, two quantized parameters emerged; one has values $\ell(\ell+1)$ with $\ell = 0, 1, 2, \ldots$, and the other has values $m = 0, \pm 1, \pm 2, \ldots, \pm \ell$. For integer values of the total angular momentum quantum number, these are identical to the values obtained from the raising and lower operator solution to the angular momentum eigenvalue problem. The spherical harmonics are the Schrödinger representation orbital angular momentum wavefunctions,

$$Y_\ell^m(\theta, \varphi) = \left|\ell m_\ell\right\rangle. \tag{15.74}$$

$\left|\ell m_\ell\right\rangle = \left|jm\right\rangle$ for integer values of j; that is, the $Y_\ell^m(\theta, \varphi)$ are the eigenvectors of $\underline{L}^2$ and $\underline{L}_z$ operators.

$$\underline{L}^2 Y_\ell^m(\theta, \varphi) = \ell(\ell+1) Y_\ell^m(\theta, \varphi) \tag{15.75a}$$

$$\underline{L}_z Y_\ell^m(\theta, \varphi) = m Y_\ell^m(\theta, \varphi) \tag{15.75b}$$

Substituting the wavefunction [equation (15.73)] into the eigenvalue equation (15.72) gives

$$\underline{H} \left|\psi_{n\ell m}\right\rangle = (E_{n\ell} - m\mu_0 B) \left|\psi_{n\ell m}\right\rangle = E \left|\psi_{n\ell m}\right\rangle. \tag{15.76}$$

The wavefunction is an eigenfunction of the atomic Hamiltonian in the absence of a magnetic field. For the hydrogen atom, the energy only depends on the principal quantum number, n; for example, the $2s$ and $2p$ orbitals have the same energy. For many electron atoms, the energy depends on both n and ℓ. When a magnetic field is applied, there is an extra term in the Hamiltonian, $-\mu_0 B \underline{L}_z$. The atomic wavefunctions are also eigenfunctions of the $\underline{L}_z$ operator [equation (15.75b)]. The eigenvalues of $\underline{L}_z$ are m. Therefore, the atomic wavefunctions are also eigenfunctions of the Hamiltonian when a static magnetic field is applied along z. The energies are

$$E_{n\ell m} = E_{n\ell} - m\mu_0 B. \tag{15.77}$$

The $2\ell + 1$ degenerate m states associated with a particular ℓ value are split into $2\ell + 1$ distinct, equally spaced energy levels when a static magnetic field is applied.

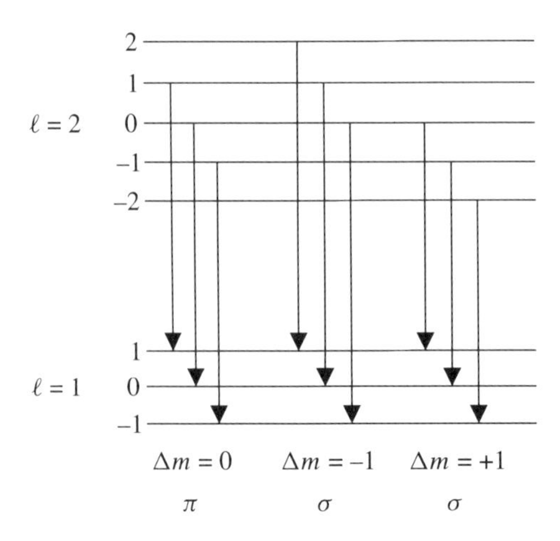

Figure 15.3 Optical emission transitions between a d orbital and a p orbital for a atom in a static magnetic field.

The influence of a magnetic field on the states of atoms can be observed in spectroscopic experiments. The frequency of emission between two states of differing m value is

$$\nu = \nu_0 + \frac{\mu_0 B}{h}\Delta m, \tag{15.78}$$

where $\Delta m = m_I - m_F$ is the difference in the m value between the initial and final states. This type of atomic emission obeys the selection rule

$$\Delta m = \pm 1 \text{ or } 0. \tag{15.79}$$

A d orbital ($\ell = 2$) has five m values, $m = 2, 1, 0, -1, -2$, which are split into five energy levels by a static magnetic field. A p orbital ($\ell = 1$) has three m values, $m = 1, 0, -1$, which are split into three energy levels. Figure 15.3 shows the possible transitions between the d and p multiplets.

The transitions with $\Delta m = 0$ have the same frequency as the transition does in the absence of a magnetic field. The $\Delta m = -1$ transitions are higher in frequency, and the $\Delta m = +1$ transitions are lower in frequency. For a one-tesla field, the shift in frequency is 13.4 GHz. The transitions with $\Delta m = 0$ are said to be π polarized; the light is emitted polarized parallel to the direction of the magnetic field. The transitions with $\Delta m = \pm 1$ are σ polarized; the light is emitted polarized perpendicular to the direction of the magnetic field.

E. ADDITION OF ANGULAR MOMENTUM

In many quantum mechanical problems containing angular momentum, there are two or more distinct angular momenta involved. For example, in NMR, there may be a number of interacting nuclei, such as protons, with $j = 1/2$. Particles with $j = 1/2$ are referred to as spin 1/2 particles. Each particle has $j = 1/2$ with $m = 1/2$ or $-1/2$. In Chapter 16, electron angular momentum (electron spin) is discussed, and the problem of spin–orbit coupling is examined. Spin–orbit coupling describes the interaction between the electron orbital angular momentum and the electron spin. Electron spin resonance spectroscopy frequently involves the interaction of electron spins with nuclear spins—that is, the hyperfine interaction. In such situations, it is often preferable to deal with a total angular momentum vector, a collective vector combining all of the angular momenta, rather than treating the problem in terms of a collection of individual angular momentum vectors.

Non-quantum mechanical vectors, including angular momentum vectors, can be added to obtain a resultant vector. However, quantum angular momentum vectors cannot be added in an arbitrary manner because the resultant vector must still obey the commutator relationships for angular momentum and must still be an eigenket of the both the $\underline{J}^2$ and $\underline{J}_z$ operators. In this section, the general problem of the addition of two angular momentum vectors is considered. The two vectors have total angular momentum quantum numbers, j_1 and j_2. The resulting total angular momentum vectors are useful because they are the eigenkets of operators that couple the individual vectors. This is illustrated in Chapter 16 for spin–orbit coupling. Coupling more than two vectors is an extension of the two vector coupling problem.

First, consider two distinct angular momentum degrees of freedom for which j_1 and j_2 are both equal to 1/2. Then

$$j_1 = \frac{1}{2} \qquad j_2 = \frac{1}{2}$$
$$m_1 = \pm\frac{1}{2} \qquad m_2 = \pm\frac{1}{2}.$$

There are four different product states depending on the values of j_1, m_1 and j_2, m_2. The product states are usually written only in terms of the m_1 and m_2 because the values of j_i are constant. The four product states are

$$\begin{array}{ll} j_1\ m_1\ j_2\ m_2 & m_1\ m_2 \end{array}$$

$$\begin{aligned} \left|\frac{1}{2}\,\frac{1}{2}\right\rangle\left|\frac{1}{2}\,\frac{1}{2}\right\rangle &= \left|\frac{1}{2}\,\frac{1}{2}\right\rangle \\ \left|\frac{1}{2}\,\frac{1}{2}\right\rangle\left|\frac{1}{2}-\frac{1}{2}\right\rangle &= \left|\frac{1}{2}-\frac{1}{2}\right\rangle \\ \left|\frac{1}{2}-\frac{1}{2}\right\rangle\left|\frac{1}{2}\,\frac{1}{2}\right\rangle &= \left|-\frac{1}{2}\,\frac{1}{2}\right\rangle \\ \left|\frac{1}{2}-\frac{1}{2}\right\rangle\left|\frac{1}{2}-\frac{1}{2}\right\rangle &= \left|-\frac{1}{2}-\frac{1}{2}\right\rangle \end{aligned} \tag{15.80}$$

On the right-hand side, the j_1 and j_2 values have been omitted because they are do not vary. This form of representing two angular momenta is called the m_1m_2 representation of the product states with a ket $|m_1m_2\rangle$ standing for $|j_1j_2m_1m_2\rangle$.

There exists a different representation of the product states, which is obtained by a unitary transformation. These states are labeled

$$|j_1j_2jm\rangle = |jm\rangle\,. \tag{15.81}$$

On the right-hand side, j_1 and j_2 have been suppressed because they do not vary. The kets $|jm\rangle$ describe the combined angular momentum vectors that are represented individually in the m_1m_2 representation. The representation for the combined vectors is called the jm representation. The kets $|jm\rangle$ are eigenkets of the operators $\underline{J}^2$ and $\underline{J}_z$, that is,

$$\underline{J}^2 \,|jm\rangle = j(j+1)\,|jm\rangle \tag{15.82a}$$

$$\underline{J}_z \,|jm\rangle = m\,|jm\rangle\,, \tag{15.82b}$$

where

$$\underline{J} = \underline{J}_1 + \underline{J}_2 \tag{15.83a}$$

$$\underline{J}_z = \underline{J}_{1z} + \underline{J}_{2z}. \tag{15.83b}$$

The jm kets are a superposition of the m_1m_2 kets,

$$|jm\rangle = \sum_{m_1m_2} C_{m_1m_2}\,|m_1m_2\rangle\,, \tag{15.84a}$$

with

$$C_{m_1m_2} = \langle m_1m_2|jm\rangle\,. \tag{15.84b}$$

The coefficients, $C_{m_1m_2}$, are called Clebsch–Gordon coefficients, Wigner coefficients, or vector coupling coefficients. To change from the m_1m_2 representation to the jm representation, it is necessary to determine the coefficients, which together comprise the unitary transformation.

A unitary transformation takes one orthonormal basis set into another orthonormal basis set (see Chapter 13, Section B). If there are N states in the original basis, there will be N states in the transformed basis. Then, starting with N states in the m_1m_2 representation, there will be N states in the jm representation. Since the states in m_1m_2 are orthonormal, the states in jm representation will be orthonormal. The number of states in either representation is $(2j_1+1)(2j_2+1)$.

The operators $\underline{J}^2$ and $\underline{J}_z$ obey the same commutation relations as any other angular momentum operators. This can be seen by making the substitution $\underline{J} = \underline{J}_1 + \underline{J}_2$ into equations (15.15)–(15.18) and working through commutator relations using the fact that $\underline{J}_1$ and $\underline{J}_2$ and their components commute since they operate on different state spaces.

The first step in the development of the Clebsch–Gordon coefficients is to show that, since $\underline{J}_z = \underline{J}_{1z} + \underline{J}_{2z}$,

$$m = m_1 + m_2, \tag{15.85}$$

or the coefficient vanishes. To prove this, consider operating $\underline{J}_z$ on the left-hand side of (15.84a) and operating the equivalent operator, $\underline{J}_{1z} + \underline{J}_{2z}$, on the right-hand side of the equation:

$$\underline{J}_z \,|jm\rangle = m\,|jm\rangle \tag{15.86a}$$

$$= (\underline{J}_{1z} + \underline{J}_{2z}) \sum_{m_1 m_2} C_{m_1 m_2} |m_1 m_2\rangle \tag{15.86b}$$

$$= \sum_{m_1 m_2} (m_1 + m_2) C_{m_1 m_2} |m_1 m_2\rangle . \tag{15.86c}$$

The left-hand side of equation (15.86a) is equal to equation (15.86b). Therefore, the right-hand side of equation (15.86a) is equal to equation (15.86c). Since the sum over the coefficients times the kets in equation (15.86c) is equal to the ket $|jm\rangle$, for the right-hand side of equation (15.86a) to be equal to equation (15.86c), $m = m_1 + m_2$. Terms in the sum in equation (15.86c) that do not have $m = m_1 + m_2$ must be zero. Because the ket is nonzero and $m_1 + m_2$ is nonzero, the coefficient, $C_{m_1 m_2}$ must be zero when $m \neq m_1 + m_2$.

Since $m = m_1 + m_2$, it follows that the largest value of m in the jm representation is

$$m = j_1 + j_2, \tag{15.87}$$

because in the $m_1 m_2$ representation, the largest value of m_1 equals j_1 and the largest value of m_2 equals j_2. $m = j_1 + j_2$ is the only way to obtain the largest value of m. Thus, in the jm representation, the largest value of j is $j_1 + j_2$, and there is only one such state. Associated with this value of j will be $2j + 1$ m states.

The next largest value of m is $m - 1$, which corresponds to $j_1 + j_2 - 1$. There are two ways to obtain this value of m consistent with the requirement $m = m_1 + m_2$:

$$m_1 = j_1 \qquad \text{and} \qquad m_2 = j_2 - 1$$

or

$$m_1 = j_1 - 1 \qquad \text{and} \qquad m_2 = j_2.$$

Two independent linear combinations of these can be formed, one of which belongs to the jm state, $j = j_1 + j_2$, because it will have m values:

$$m = (j_1 + j_2), (j_1 + j_2 - 1), \ \ldots, (-j_1 - j_2)$$

The other combination with $m = j_1 + j_2 - 1$ must belong to a different value of j, that is,

$$j = j_1 + j_2 - 1,$$

and the m value $m = j_1 + j_2 - 1$ is the largest m value for this j state. The state $j = j_1 + j_2 - 1$ has m values

$$m = (j_1 + j_2 - 1), (j_1 + j_2 - 2), \ \ldots, (-j_1 - j_2 + 1).$$

In the same manner, examining smaller and smaller values of m that are consistent with $m = m_1 + m_2$, it is found that the possible j values in the jm representation range from

$$j = j_1 + j_2 \ to \ |j_1 - j_2| \tag{15.88}$$

in integer steps. Each of these j values has associated with it $2j + 1$ m values.

As an example, consider the m_1m_2 state for $j_1 = 1/2$, $j_2 = 1/2$, where the four m_1m_2 kets are given in equation (15.80). The possible j values are determined using equation (15.88). They are

$$\begin{aligned} j &= \frac{1}{2} + \frac{1}{2} = 1 \\ j &= \frac{1}{2} - \frac{1}{2} = 0. \end{aligned} \tag{15.89}$$

There are two different j states in the coupled system. As with any angular momentum, if $j = 1$ then $m = 1, 0, -1$, and if $j = 0$ then $m = 0$. The four m_1m_2 kets [equation (15.80)], when transformed into the jm representation, yield the four jm kets

$$|1\ 1\rangle\,,\ |1\ 0\rangle\,,\ |0\ 0\rangle\,,\ |1\ -1\rangle\,. \tag{15.90}$$

Like the m_1m_2 kets, there are actually four quantum numbers. The kets are $|j_1 j_2 jm\rangle$. Since j_1 and j_2 never change, they are suppressed, and the kets are written as $|jm\rangle$.

Each of the four jm kets given in (15.90) is a superposition of the m_1m_2 kets of equation (15.80). The specific superposition of m_1m_2 kets corresponding to a particular jm ket is found using a procedure involving lowering operators. Finding the particular superposition is finding the Clebsch–Gordon coefficients. Beginning with the ket with the largest j and largest m, $|11\rangle$, applying $\underline{J}_z$ gives

$$\underline{J}_z\,|1\ 1\rangle = 1\,|1\ 1\rangle\,,$$

that is, $m = 1$. Since $m = m_1 + m_2$, the values of m_1 and m_2 are

$$m_1 = \frac{1}{2}, \qquad m_2 = \frac{1}{2};$$

there is no other way to have $m_1 + m_2 = 1$. Therefore,

$$|11\rangle = \left|\frac{1}{2}\ \frac{1}{2}\right\rangle. \tag{15.91}$$

The Clebsch–Gordon coefficient $= 1$.

The jm lowering operator is the sum of the m_1m_2 lowering operators,

$$\underline{J}_- = \underline{J}_{1-} + \underline{J}_{2-}. \tag{15.92}$$

The jm lowering operator is applied to the left-hand side of equation (15.91), and the equivalent m_1m_2 lowering operators are applied to the right-hand side. In applying the lowering operators, the lowering operator relation, equation (15.49) is used. [To obtain the coefficient in the lowering operator formula, it is necessary to substitute in values for j and m. For the jm lowering operator, these values are j and m. For the m_1m_2 lowering operator, $\underline{J}_{1-}$ operates on m_1, and the values j_1 (implicit in the m_1m_2 ket) and m_1 are used in the formula. The $\underline{J}_{2-}$ operates on m_2, and the values j_2 (implicit in the m_1m_2 ket) and m_2 are used in the formula.]

$$\underline{J}_-\,|11\rangle = \sqrt{2}\,|10\rangle \tag{15.93}$$

$$(\underline{J}_{1-} + \underline{J}_{2-}) \left|\frac{1}{2}\,\frac{1}{2}\right\rangle = \underline{J}_{1-} \left|\frac{1}{2}\,\frac{1}{2}\right\rangle + \underline{J}_{2-} \left|\frac{1}{2}\,\frac{1}{2}\right\rangle \tag{15.94a}$$

$$= 1\left|-\frac{1}{2}\,\frac{1}{2}\right\rangle + 1\left|\frac{1}{2} - \frac{1}{2}\right\rangle \tag{15.94b}$$

The left-hand side of equation (15.93) is equal to the left-hand side of equation (15.94a); therefore, the right-hand side of equation (15.93) is equal to the right-hand side of equation (15.94b). Equating these terms and dividing by $\sqrt{2}$ gives

$$|10\rangle = \frac{1}{\sqrt{2}}\left|\frac{1}{2} - \frac{1}{2}\right\rangle + \frac{1}{\sqrt{2}}\left|-\frac{1}{2}\,\frac{1}{2}\right\rangle. \tag{15.95}$$

Equation (15.95) expresses the jm ket $|10\rangle$ as a superposition of the two m_1m_2 kets. The $1/\sqrt{2}$ that precedes the two m_1m_2 kets are the Clebsch–Gordon coefficients.

Applying lowering operators to both sides of equation (15.95) yields

$$\underline{J}_- |1\ 0\rangle = \sqrt{2}\,|1 - 1\rangle \tag{15.96a}$$

$$= (\underline{J}_{1-} + \underline{J}_{2-})\frac{1}{\sqrt{2}}\left(\overset{m_1\ m_2}{\left|\frac{1}{2} - \frac{1}{2}\right\rangle} + \overset{m_1 m_2}{\left|-\frac{1}{2}\,\frac{1}{2}\right\rangle}\right) \tag{15.96b}$$

$$= \left[\frac{1}{\sqrt{2}}\left|-\frac{1}{2} - \frac{1}{2}\right\rangle + 0 + 0 + \frac{1}{\sqrt{2}}\left|-\frac{1}{2} - \frac{1}{2}\right\rangle\right]. \tag{15.96c}$$

In equation (15.96b), m_1m_2 labels were put above the kets to make it clear which ket is operated on by each lowering operator. The two lowering operators operate on the two kets, so there are four terms. The four terms that result are shown in equation (15.96c). The two operators were applied to the first ket and then applied to the second ket. Two of the terms are zero, shown explicitly, because $m = -1/2$ is the smallest m value for $j = 1/2$. Applying a lowering operator to $m = -1/2$ gives 0. Equating the right-hand sides of equations (15.96a) and (15.96c) gives

$$|1 - 1\rangle = \left|-\frac{1}{2} - \frac{1}{2}\right\rangle; \tag{15.97}$$

the Clebsch–Gordon coefficient is 1.

Equations (15.91), (15.95), and (15.97) give the three the three jm kets with $j = 1$ as superpositions of the m_1m_2 kets. It remains to find the jm ket $|0\ 0\rangle$ in terms of the m_1m_2 kets. Since $m = 0 = m_1 + m_2$, there are two kets that meet this condition:

$$\left|\frac{1}{2} - \frac{1}{2}\right\rangle, \left|-\frac{1}{2}\,\frac{1}{2}\right\rangle.$$

The ket $|0\ 0\rangle$ is a superposition of these that must be orthogonal to the other kets and normalized. The ket $|1\ 0\rangle$ [equation (15.95)] is already a combination of these two kets. Therefore,

$$|0\ 0\rangle = \frac{1}{\sqrt{2}}\left|\frac{1}{2} - \frac{1}{2}\right\rangle - \frac{1}{\sqrt{2}}\left|-\frac{1}{2}\,\frac{1}{2}\right\rangle \tag{15.98}$$

since this is the only combination that is orthogonal to the right-hand side of equation (15.95) and normalized.

Equations (15.19), (15.95), (15.97), and (15.98) give the four jm kets in terms of the four $m_1 m_2$ kets for the two angular momenta with $j_1 = 1/2$ and $j_2 = 1/2$. The superpositions are actually only determined within an arbitrary phase factor. By convention, the Clebsch–Gordon coefficients are taken to be real. Even with this condition, each superposition giving the jm kets could be multiplied by -1 [a phase factor $\exp(i\pi)$]. Again, by convention the superpositions are taken so that those with a single sign are positive. As long as the set of superpositions is internally consistent in the choice of signs, calculations using the set will be independent of the choice.

The Clebsch–Gordon coefficients can be collected into a table. For $j_1 = 1/2$ and $j_2 = 1/2$ the table has the following form:

$j_1 = 1/2$, $j_2 = 1/2$		1	1	0	1	j
		1	0	0	-1	m
$\frac{1}{2}$	$\frac{1}{2}$	1				
$\frac{1}{2}$	$-\frac{1}{2}$		$\frac{1}{\sqrt{2}}$	$\frac{1}{\sqrt{2}}$		
$-\frac{1}{2}$	$\frac{1}{2}$		$\frac{1}{\sqrt{2}}$	$\frac{-1}{\sqrt{2}}$		
$-\frac{1}{2}$	$-\frac{1}{2}$				1	
m_1	m_2					

There is also a convention for tables of Clebsch–Gordon coefficients. The j and m values are written across the top of the table starting with the largest j value on the left. The largest m value is also written on the top left. The j and m values are written sequentially moving to the right, with preference given to having the m values descend in order but with the largest possible j value written first. Thus the order is 11, 10, but then 00 rather than $1\ -1$. The $m_1 m_2$ values are written down the left side of the table, with the largest m_1 and largest m_2 value at the top and the smallest m_1 and the smallest m_2 value on the bottom. The $m_1 m_2$ values are written in descending order with preference given a maximum value of $m_1 + m_2$ and to a positive m_1. The jm ket with the largest j and largest m always corresponds to the single $m_1 m_2$ with the largest m_1 and the largest m_2. Thus, the upper left-hand corner position of the table will always be 1. The jm ket with the smallest j and smallest m always corresponds to the single $m_1 m_2$ with the smallest m_1 and the smallest m_2, so the lower right-hand corner position of the table will be 1. To obtain a particular jm ket in terms of a superposition of $m_1 m_2$ kets, the table is read down from the jm ket, and the coefficient in the table multiplies the $m_1 m_2$ ket to its left. So the jm ket $|0\,0\rangle$ (written as 0 0 on the top of the table) is $1/\sqrt{2}\,|1/2\ {-1/2}\rangle - 1/\sqrt{2}\,|{-1/2}\ 1/2\rangle$ (with the $m_1 m_2$ kets written as $1/2\ {-1/2}$ and ${-1/2}\ 1/2$). This table of Clebsch–Gordon coefficients is a matrix that is the unitary transformation that takes the $m_1 m_2$ kets into the jm kets, and vice versa.

The next largest possible combination of j_1 and j_2 values is

$$
\begin{array}{ll}
j_1 = 1 & j_2 = \frac{1}{2} \\
m_1 = 1, 0, -1 & m_2 = \frac{1}{2}, -\frac{1}{2}.
\end{array}
$$

The six $m_1 m_2$ kets are

$$
\left|1 \frac{1}{2}\right\rangle \left|1 -\frac{1}{2}\right\rangle \left|0 \frac{1}{2}\right\rangle \left|0 -\frac{1}{2}\right\rangle \left|-1 \frac{1}{2}\right\rangle \left|-1 -\frac{1}{2}\right\rangle.
$$

Implicit in each of these kets are the j_1 and j_2 values. Transforming into the jm representation will yield six states with j and m values:

$$
\begin{array}{ll}
j = j_1 + j_2 = \frac{3}{2} & m = \frac{3}{2}, \frac{1}{2}, -\frac{1}{2}, -\frac{3}{2} \\
j = j_1 - j_2 = \frac{1}{2} & m = \frac{1}{2}, -\frac{1}{2}.
\end{array}
$$

The jm kets are

$$
\left|\frac{3}{2}\ \frac{3}{2}\right\rangle \left|\frac{3}{2}\ \frac{1}{2}\right\rangle \left|\frac{3}{2}\ -\frac{1}{2}\right\rangle \left|\frac{3}{2}\ -\frac{3}{2}\right\rangle \left|\frac{1}{2}\ \frac{1}{2}\right\rangle \left|\frac{1}{2}\ -\frac{1}{2}\right\rangle.
$$

The table of Clebsch–Gordon coefficients is

$j_1 = 1$, $j_2 = ½$	J	$\frac{3}{2}$	$\frac{3}{2}$	$\frac{1}{2}$	$\frac{3}{2}$	$\frac{1}{2}$	$\frac{3}{2}$
m_1	m_2 \\ m	$\frac{3}{2}$	$\frac{1}{2}$	$\frac{1}{2}$	$-\frac{1}{2}$	$-\frac{1}{2}$	$-\frac{3}{2}$
1	$\frac{1}{2}$	1					
1	$-\frac{1}{2}$		$\sqrt{\frac{1}{3}}$	$\sqrt{\frac{2}{3}}$			
0	$\frac{1}{2}$		$\sqrt{\frac{2}{3}}$	$\sqrt{\frac{1}{3}}$			
0	$-\frac{1}{2}$				$\sqrt{\frac{2}{3}}$	$\sqrt{\frac{1}{3}}$	
-1	$\frac{1}{2}$				$\sqrt{\frac{1}{3}}$	$\sqrt{\frac{2}{3}}$	
-1	$-\frac{1}{2}$						1

For example,

$$
\overset{j\,m}{\left|\tfrac{1}{2}\ \tfrac{1}{2}\right\rangle} = \sqrt{\tfrac{2}{3}}\,\overset{m_1\ \ m_2}{\left|1 - \tfrac{1}{2}\right\rangle} - \sqrt{\tfrac{1}{3}}\,\overset{m_1\ \ m_2}{\left|0\ \ \tfrac{1}{2}\right\rangle}.
$$

There is no general formula for the Clebsch–Gordon coefficients. The Clebsch–Gordon coefficients can be obtained using the lower operator procedure illustrated for the system $j_1 = 1/2$, $j_2 = 1/2$. The general procedure is as follows. The jm ket with the largest j and largest m is always equal to the $m_1 m_2$ ket with the largest m_1 and the largest m_2. To obtain the jm ket with one unit smaller m value as a superposition of $m_1 m_2$ kets, the jm lowering operator is applied to the jm ket and the $m_1 m_2$ lowering operators are applied to the $m_1 m_2$

to ket. This procedure is repeated until all of the $2j+1$ m states associated with the largest value of j are obtained. The next series is started with j one unit less than the maximum and the corresponding m value using $m = m_1 + m_2$ and orthonormality. Lowering operators are then used to obtain the rest of this series. The series for the next lower value of j is again started using $m = m_1 + m_2$ and orthonormality, and the series is completed using the lowering operator procedure. The process continues until all jm kets have been obtained. In the system $j_1 = 1$, $j_2 = {}^1\!/_2$, the procedure begins by recognizing that the jm ket $\left|{}^3\!/_2\, {}^3\!/_2\right\rangle$ equals the $m_1 m_2$ ket $\left|1\, {}^1\!/_2\right\rangle$. Lowering operators are used to find all of the $j = {}^3\!/_2$ kets. The next largest values of j and m are $j = {}^1\!/_2$ and $m = {}^1\!/_2$. It is necessary to find the ket $\left|{}^1\!/_2\, {}^1\!/_2\right\rangle$. Since $m = m_1 + m_2$, there are only two $m_1 m_2$ kets for which $m_1 + m_2 = 1/2$, $\left|1\, -{}^1\!/_2\right\rangle$ and $\left|0\, {}^1\!/_2\right\rangle$. Looking at the Clebsch–Gordon coefficient table above, it is seen that a superposition of these two kets was used to express the jm ket $\left|{}^3\!/_2\, {}^1\!/_2\right\rangle$. Therefore, the jm ket $\left|{}^1\!/_2\, {}^1\!/_2\right\rangle$ is the superposition of these two $m_1 m_2$ kets that is normalized and orthogonal to the $\left|{}^3\!/_2\, {}^1\!/_2\right\rangle$ superposition. Lowering operators are then used to complete the series by finding the $\left|{}^1\!/_2\, -{}^1\!/_2\right\rangle$ as a superposition of $m_1 m_2$ kets.

Although there is no formula for the Clebsch–Gordon coefficients, there are books that contain tables of the coefficients for a given j_1, j_2. In addition, some commercial math application computer programs will calculate Clebsch–Gordon coefficients. Clebsch–Gordon coefficients can also be used for adding more than two angular momenta. Two of the angular momentum vectors are added. The resulting jm vectors are then added to a third vector. This procedure can be continued, adding any number of angular momentum vectors to produce a set of jm vectors representing the combined system. The utility of the jm representation is illustrated in Chapter 16.

CHAPTER 16

ELECTRON SPIN

In the treatments of the hydrogen atom and the helium atom (Chapters 7 and 10) the intrinsic angular momentum of the electron was not included in the descriptions of the electronic states. Orbital angular momentum, which gives rise to the s, p, d, and f hydrogen atom orbitals, arose naturally out of the Schrödinger treatment of the hydrogen atom. Three quantum numbers were obtained, n, ℓ, and m_ℓ (the subscript ℓ on the magnetic quantum number m_ℓ shows that this m value is associated with orbital angular momentum). Dirac solved the hydrogen atom problem with an approach that was consistent with both quantum mechanics and the theory of relativity. The results produced an additional quantum number, m_s, which is associated with the intrinsic angular momentum of the electron. The m_s are the projections of the intrinsic electron angular momentum, called s, on the z axis. Because of its intrinsic angular momentum, the electron has a permanent magnetic dipole. The angular momentum of the electron is referred to as electron spin because a classical charge distribution with angular momentum (a rotating or spinning charge distribution) gives rise to a magnetic moment.

In addition to Dirac's relativistic treatment of the hydrogen atom, many experiments can only be explained if the electron has intrinsic angular momentum. For example, alkali metals show splittings of spectral lines, even in the absence of an external magnetic field. The sodium D line, which is the intense orange emission from excited Na at 589 nm, is a doublet with a splitting of 17 cm^{-1}. The D-line fluorescence is an electronic transition from the Na excited $3p$ state to the $3s$ ground state. The multiplicity of the p state is $2\ell + 1 = 3$. For a single electron promoted to the $3p$ state in Na, the three p orbitals are degenerate. Even in the presence of an external magnetic field, three transitions should be observed, not a doublet (see Chapter 15, Section D). As discussed below, the doublet splitting is a consequence of the coupling between the intrinsic electron spin angular momentum and the $3p$ state's orbital angular momentum.

A dramatic demonstration of the existence of electron spin was provided by the Stern–Gerlach experiment (Figure 16.1). In the Stern–Gerlach experiment, an oven containing metallic silver is heated until silver vapor is produced. Some of the vapor escapes through a small hole in the oven. Baffles skim off silver atoms that are not moving in the proper direction. The silver atoms that pass through the baffles form an atomic beam. The silver atom beam passes through an inhomogeneous magnetic field, and subsequently strikes a glass plate. The silver atoms stick to the glass plate, producing a silver deposit. Observing the location of the silver deposit permits the trajectory of the silver atoms to be determined. It was found that the glass plate contained two well-separated lines of silver.

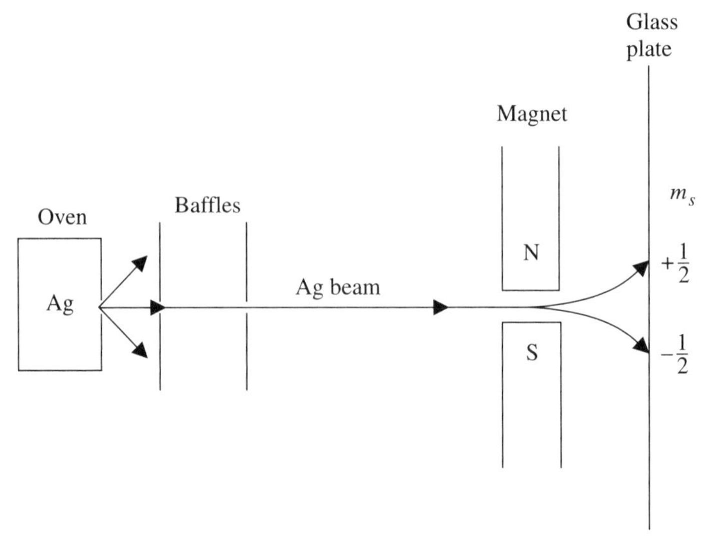

Figure 16.1 A schematic of the Stern–Gerlach experiment. Silver atoms pass out of a heated oven and are formed into an atomic beam by baffles. The beam passes through an inhomogeneous magnetic field, exerting a force on atoms that depends on the electrons m_s quantum number.

A magnetic dipole in an inhomogeneous magnetic field experiences a force that deflects the dipole in a direction that depends on the orientation of the dipole relative to the magnetic field. Silver atoms have one $5s$ electron, which is unpaired (see Section C below for a discussion of the Pauli Principle and electron spin pairing). The direction of the magnetic field defines the z axis. If the electron has a total angular momentum $s = {}^1/_2$, then it can have two possible projections on the z axis determined by the values of $m_s = \pm{}^1/_2$. Thus, the two silver lines on the glass plate arise from two m_s values of the electron. From the separation of the lines on the glass plate, the magnetic moment of the electron was determined. Since the s electron has no orbital angular momentum, the deflection cannot arise from the a magnetic field produced by orbital angular momentum. The size of the deflection is too large to arise from the much smaller proton magnetic moment. The Stern–Gerlach experiment is one of many experiments that demonstrates the existence of intrinsic electron angular momentum.

A. THE ELECTRON SPIN HYPOTHESIS

Although Dirac's quantum mechanical and relativistic treatment of the hydrogen atom gave a theoretical treatment of electron spin, the mathematics of the treatment are so complex that they do not provide a useful approach for the quantum mechanical analysis of atomic and molecular problems. Rather, nonrelativistic quantum theory is used, and electron spin is added as a hypothesis.

It is assumed that the electron has an intrinsic angular momentum (spin), with the total angular momentum

$$s = \frac{1}{2}\hbar, \tag{16.1}$$

that is,

$$\underline{S}^2 |sm_s\rangle = \frac{1}{2}\left(\frac{1}{2}+1\right)\hbar^2 |sm_s\rangle , \tag{16.2}$$

where $\underline{S}^2$ is the square of the total angular momentum operator (called $\underline{J}^2$ in Chapter 15). The kets $|sm_s\rangle$ are also eigenkets of $\underline{S}_z$,

$$\underline{S}_z |sm_s\rangle = \pm\frac{1}{2}\hbar |sm_s\rangle . \tag{16.3}$$

As in Chapter 15, angular momentum is generally expressed in units of $\hbar$, and $\hbar$ is not usually written explicitly. The magnitude of the magnetic moment of the electron is

$$\mu_s = \frac{e\hbar}{2mc}, \tag{16.4}$$

which is one Bohr magneton. The ratio

$$\frac{\mu_s}{s} = \frac{e}{mc}$$

is twice the ratio for orbital angular momentum,

$$\frac{\mu_\ell}{\ell} = \frac{e}{2mc}.$$

1. Electronic Wavefunctions with Spin

In the treatment of the hydrogen atom (Chapter 7), the wavefunctions describing the electronic states are functions of three spatial coordinates only.

$$\psi = \psi(x, y, z).$$

The inclusion of electron spin adds another degree of freedom or "coordinate," the m_s value:

$$\psi = \psi(x, y, z, m_s), \tag{16.5a}$$

with

$$m_s = \pm\frac{1}{2}. \tag{16.5b}$$

There is actually a fifth coordinate s, the total angular momentum of the electron. Since $s = 1/2$ for any electron and the value does not change, it does not have to be specified explicitly. In a many-electron wavefunction, each electron has four coordinates, three spatial coordinates, and a value of m_s.

The angular momentum eigenvectors $|sm_s\rangle$ of the operators $\underline{S}^2$ and $\underline{S}_z$ representing the square of the total angular momentum and the projection of the angular momentum on the z axis are

$$|\chi(m_s)\rangle = \left|\frac{1}{2}\,\frac{1}{2}\right\rangle, \left|\frac{1}{2}\,-\frac{1}{2}\right\rangle. \tag{16.6}$$

These are frequently called α and β spin states with

$$\alpha = \left| \frac{1}{2}\ \frac{1}{2} \right\rangle \tag{16.7a}$$

$$\beta = \left| \frac{1}{2}\ -\frac{1}{2} \right\rangle . \tag{16.7b}$$

In the state α, the spin is parallel to the z axis, and in the state β the spin is antiparallel to the z axis.

2. Electronic States in a Central Field

Consider an electron in a state of orbital angular momentum ℓ. In a centrally symmetric system such as an atom, the three spatial coordinates are written in spherical polar coordinates, r, θ, and φ. The θ and φ angular dependence can be expressed in terms of spherical harmonics, $Y_\ell^{m_\ell}$ (see Chapter 7), where m_ℓ is used to identify the magnetic quantum number with orbital angular momentum ℓ. The spin wavefunctions are $\chi(^1/_2)$ and $\chi(-^1/_2)$, where only the m_s value in equation (16.6) has been written. The part of the wavefunction that describes the angular momentum state of the electron is

$$|\psi(\ell m_\ell m_s)\rangle = \left| Y_\ell^{m_\ell} \right\rangle |\chi(m_s)\rangle . \tag{16.8}$$

The ket $|\psi(\ell m_\ell m_s)\rangle$ defines a product space for the eigenvectors of the electron. This ket is in the $m_1 m_2$ representation of the two types of angular momenta, orbital, and spin (see Chapter 15, Section E).

The kets $|\psi(\ell m_\ell m_s)\rangle$ are simultaneous eigenfunctions of orbital angular momentum and spin angular momentum operators $\underline{L}^2$, $\underline{L}_z$, $\underline{S}^2$, and $\underline{S}_z$, that is,

$$\underline{L}^2 |\psi\rangle = \ell(\ell+1) |\psi\rangle$$

$$\underline{L}_z |\psi\rangle = m_\ell |\psi\rangle$$

$$\underline{S}^2 |\psi\rangle = \frac{3}{4} |\psi\rangle$$

$$\underline{S}_z |\psi\rangle = m_s |\psi\rangle .$$

The third equation arises because the eigenvalues of $\underline{S}^2$ are $s(s+1)$ with s always equal to $^1/_2$. There are $2(2\ell+1)$ linearly independent $|\psi(\ell m_\ell m_s)\rangle$ functions.

B. SPIN–ORBIT COUPLING

In the treatment of the hydrogen and helium atoms (Chapters 7 and 10), only Coulomb interactions were considered in the potential energy term of the Hamiltonian. The Coulomb interactions determine the energies of the atomic states of atoms and molecules on a gross scale. However, as mentioned at the beginning of this chapter, there are many phenomena that can be explained only if electron spin is included. Interactions that depend on the intrinsic angular momentum and magnetic dipole moment of the electron are responsible for the fine structure of the electronic states of atoms and molecules. One such interaction, which plays an important role in many aspects of atomic and molecular properties and processes, is spin–orbit coupling, the coupling between the orbital angular momentum of an electron and its spin angular momentum. In this section, spin–orbit coupling will be treated for the specific case of an electron in a p orbital—for example, the $3p$ excited state of the Na atom. An electron undergoing a radiative transition from the Na $3p$ state to the $3s$ state emits orange light (589 nm). Observation of fluorescence from an ensemble of excited Na atoms reveals two closely spaced frequencies. The splitting in the Na $3p$ to $3s$ emission can be explained by adding a spin–orbit coupling term to the Hamiltonian.

1. p States of an Electron

The angular functions for the $\ell = 1$ states of a single electron in an atom are the three spherical harmonics

$$\begin{aligned} Y_1^1 &= |11\rangle \\ Y_1^0 &= |10\rangle \\ Y_1^{-1} &= |1\;-1\rangle\,. \end{aligned} \tag{16.9}$$

The electron spin states are

$$\begin{aligned} \alpha &= \left|\frac{1}{2}\;\frac{1}{2}\right\rangle \\ \beta &= \left|\frac{1}{2}\;-\frac{1}{2}\right\rangle. \end{aligned} \tag{16.10}$$

The combined angular momenta states of the system can be written in the $m_1 m_2$ representation (see Chapter 15, Section E):

$$j_1 m_1 = \ell m_\ell$$

and

$$j_2 m_2 = s m_s\,.$$

The combined states of the system in the $m_1 m_2$ representation are

$$\begin{aligned} &\quad\;\; \ell\, s m_\ell m_s \quad m_\ell\; m_s \\ Y_1^1\alpha &= \left|1\frac{1}{2}\;1\;\frac{1}{2}\right\rangle = \left|1\;\frac{1}{2}\right\rangle \\ Y_1^1\beta &= \left|1\frac{1}{2}\;1\;-\frac{1}{2}\right\rangle = \left|1\;-\frac{1}{2}\right\rangle \\ Y_1^0\alpha &= \left|1\frac{1}{2}\;0\;\frac{1}{2}\right\rangle = \left|0\;\frac{1}{2}\right\rangle \end{aligned}$$

$$Y_1^0\beta = \left|1\frac{1}{2}\,0 - \frac{1}{2}\right\rangle = \left|0 - \frac{1}{2}\right\rangle$$

$$Y_1^{-1}\alpha = \left|1\frac{1}{2} - 1\,\frac{1}{2}\right\rangle = \left|-1\frac{1}{2}\right\rangle$$

$$Y_1^{-1}\beta = \left|1\frac{1}{2} - 1 - \frac{1}{2}\right\rangle = \left|-1 - \frac{1}{2}\right\rangle. \tag{16.11}$$

In the right-hand set of kets, the $j_1 j_2 = \ell s$ values are not shown explicitly because they are constants. The six $m_1 m_2$ kets in equation (16.11) describe the orbital and spin angular momentum components of the states. Each of these is multiplied by the same radial function, $R(r)$. Since the radial part of the wavefunction is not operated on by angular momentum operators, it will be left implicit until it is needed.

The six states of the p electron can also be written in terms of the total angular momentum vectors in the jm representation (Chapter 15, Section E). The total angular momentum, j, has two possible values

$$j = j_1 + j_2 = \ell + s = \frac{3}{2} \tag{16.12a}$$

$$j = j_1 + j_2 - 1 = |j_1 - j_2| = \ell - s = \frac{1}{2}. \tag{16.12b}$$

The six jm kets are

$$\left|\frac{3}{2}\,\frac{3}{2}\right\rangle \left|\frac{3}{2}\,\frac{1}{2}\right\rangle \left|\frac{3}{2} - \frac{1}{2}\right\rangle \left|\frac{3}{2} - \frac{3}{2}\right\rangle \left|\frac{1}{2}\,\frac{1}{2}\right\rangle \left|\frac{1}{2} - \frac{1}{2}\right\rangle. \tag{16.13}$$

The jm representation kets can be obtained from the $m_1 m_2$ kets using the table of Clebsch–Gordon coefficients for $j_1 = 1$, $j_2 = 1/2$ given in Chapter 15. For example, the jm ket $\left|\frac{1}{2}\,\frac{1}{2}\right\rangle$ is the superposition of $m_1 m_2$ kets

$$\left|\frac{1}{2}\,\frac{1}{2}\right\rangle = \sqrt{\frac{2}{3}}\left|1 - \frac{1}{2}\right\rangle - \sqrt{\frac{1}{3}}\left|0\,\frac{1}{2}\right\rangle. \tag{16.14}$$

2. Spin–Orbit Coupling Term in the Hamiltonian

In Hamiltonian for hydrogen like atoms (or many electron atoms) terms in Hamiltonian arising from electron spin have not been included. Thus the six states of a single p electron in sodium, for example, would be degenerate. However, the intrinsic angular momentum of the electron gives rise to a magnetic dipole that is moving in the electric field of the atom (or molecule). The interaction of a moving dipole with an electric field produces changes in the electron energy.

Classically the energy, W, of a magnetic dipole, $\vec{\mu}$, moving in an electric field, $\vec{E}$, is

$$W = -(\vec{E} \times \vec{V}) \cdot \vec{\mu}, \tag{16.15}$$

where $\vec{V}$ is the velocity. In units of $\hbar$, the magnetic moment of the electron is

$$\vec{\mu} = -\frac{|e|\,\vec{\underline{S}}}{2mc} \tag{16.16}$$

where $\vec{\underline{S}}$ is the electron spin angular momentum operator. Then

$$W = \frac{|e|}{2mc}(\vec{E} \times \vec{V}) \cdot \vec{\underline{S}}. \tag{16.17}$$

In terms of the potential, ϕ,

$$|e|\,\vec{E} = \overrightarrow{\text{grad}\phi},$$

where grad is the gradient, and

$$\vec{p} = m\vec{V}.$$

Substituting, along with replacing the classical momentum with the quantum mechanical operator, gives

$$\underline{W} = \frac{1}{2m^2c}(\overrightarrow{\text{grad}\phi} \times \vec{\underline{p}}) \cdot \vec{\underline{S}}. \tag{16.18}$$

($\underline{W}$ is now a quantum mechanical operator rather than the classical energy.)

For hydrogen-like atoms (one electron atoms)

$$\phi = -\frac{ze^2}{4\pi\varepsilon_o r} \tag{16.19}$$

where z is the nuclear charge and r is the distance. Then

$$\overrightarrow{\text{grad}\phi} = \frac{ze^2}{4\pi\varepsilon_o r^2}\frac{\vec{r}}{r} = \frac{ze^2}{4\pi\varepsilon_o r^3}\vec{r}, \tag{16.20}$$

where $\vec{r}/r$ is a unit vector. Substituting equation (16.20) into equation (16.18) gives $\underline{H}_{so}$, the spin–orbit coupling piece of the Hamiltonian for a hydrogen-like atom,

$$\underline{H}_{so} = \frac{ze^2}{8\pi\varepsilon_o m^2 c}\frac{1}{\underline{r}^3}(\vec{\underline{r}} \times \vec{\underline{p}}) \cdot \vec{\underline{S}}, \tag{16.21}$$

and since $(\vec{\underline{r}} \times \vec{\underline{p}})$ is the orbital angular momentum, $\vec{\underline{L}}$,

$$\underline{H}_{so} = \frac{ze^2}{8\pi\varepsilon_o m^2 c}\frac{1}{\underline{r}^3}\vec{\underline{L}} \cdot \vec{\underline{S}}. \tag{16.22}$$

$\vec{\underline{S}}$ operates on the electron spin wavefunctions, $\vec{\underline{L}}$ operates on the orbital angular momentum wavefunctions (the spherical harmonics), and $1/\underline{r}^3$ operates on the radial part of the wavefunction.

For one electron in a central field ($3p$ electron of Na in the field of nucleus and core electrons),

$$\phi = V(r) \tag{16.23}$$

and

$$\overrightarrow{\text{grad}V}(r) = \frac{\partial V(r)}{\partial r}\frac{1}{r}\vec{r}. \tag{16.24}$$

Then,

$$\underline{H}_{so} = \frac{1}{2m^2c}\frac{1}{r}\frac{\partial V(r)}{\partial r}\vec{\underline{L}} \cdot \vec{\underline{S}} = a(r)\vec{\underline{L}} \cdot \vec{\underline{S}}, \tag{16.25}$$

where $a(r)$ is a number that results from operating the term dependent on r on the radial component of the wavefunction. For a many-electron system,

$$\underline{H}_{so} = \frac{1}{2m^2c}\sum_i \left[\frac{1}{r_i}\frac{\partial V(r_i)}{\partial r_i}\right]\vec{\underline{L}}_i \cdot \vec{\underline{S}}_i = \sum_i a_i(r)\vec{\underline{L}}_i \cdot \vec{\underline{S}}_i, \tag{16.26}$$

where i labels the individual electrons, and terms involving $\vec{\underline{S}}_i$ with $\vec{\underline{L}}_j$, which are small, have been neglected.

For hydrogen-like atoms, the value of the coefficient of the $\vec{\underline{L}} \cdot \vec{\underline{S}}$,

$$a(r) \propto z^4, \tag{16.27}$$

and for other types of atoms, the $a(r)$ scales approximately as z^4. The z^4 dependence gives rise to what is called the "heavy atom effect." Spin–orbit coupling increases rapidly with the nuclear charge of the atom. This is also true for molecules. Bromobenzene will have greatly enhanced spin–orbit coupling compared to benzene. Molecules in a solvent containing heavy atoms can have increased spin–orbit coupling because of the "external heavy atom effect." Since atomic and molecular wavefunctions are spatially extended, there is some probability of finding a molecule's electrons on the heavy atoms of the solvent. Electron delocalization is responsible for the external heavy atom effect.

In $\underline{H}_{so}$, $a(r)$ will be treated as a parameter because it is independent of the $\vec{\underline{L}}$ and $\vec{\underline{S}}$ operators. $\vec{\underline{L}} \cdot \vec{\underline{S}}$ can be written in terms of its components,

$$\vec{\underline{L}} \cdot \vec{\underline{S}} = \underline{L}_x\underline{S}_x + \underline{L}_y\underline{S}_y + \underline{L}_z\underline{S}_z. \tag{16.28}$$

The x and y components can be expressed in terms of angular momentum raising and lowering operators (Chapter 15, Section B):

$$\underline{S}_x = \frac{1}{2}(\underline{S}_+ + \underline{S}_-) \tag{16.29a}$$

$$\underline{S}_y = \frac{1}{2i}(\underline{S}_+ - \underline{S}_-) \tag{16.29b}$$

$$\underline{L}_x = \frac{1}{2}(\underline{L}_+ + \underline{L}_-) \tag{16.29c}$$

$$\underline{L}_y = \frac{1}{2i}(\underline{L}_+ - \underline{L}_-). \tag{16.29d}$$

Substituting the raising and lowering operator expressions into equation (16.28) gives

$$\underline{\vec{L}} \cdot \underline{\vec{S}} = \underline{L}_z \underline{S}_z + \frac{1}{2}(\underline{L}_+ \underline{S}_- + \underline{L}_- \underline{S}_+). \tag{16.30}$$

3. Spin–Orbit Coupling Energies of the Six p States of Na

a. The $m_1 m_2$ Representation

As written, $\underline{\vec{L}} \cdot \underline{\vec{S}}$ is an operator in the $m_1 m_2$ representation. The L operators operate on the orbital angular momentum part of the kets in equations (16.11), and the S operators operate on the spin angular momentum part of the kets. The two types of angular momenta are treated separately—that is, in the $m_1 m_2$ representation.

The $\underline{\vec{L}} \cdot \underline{\vec{S}}$ operator can be represented as a 6×6 matrix with the $m_1 m_2$ kets [the right-hand side of equation (16.11)] as the basis set. The $m_1 m_2$ kets are eigenkets of the $\underline{L}_z \underline{S}_z$ operator. Operating on a ket, $\underline{S}_z$ brings out the eigenvalue m_s and $\underline{L}_z$ brings out the eigenvalue m_ℓ. For example,

$$\underline{L}_z \underline{S}_z \left| 1 \frac{1}{2} \right\rangle = \frac{1}{2} \left| 1 \frac{1}{2} \right\rangle. \tag{16.31}$$

The $m_1 m_2$ kets are not eigenkets of the $\frac{1}{2}(\underline{L}_+ \underline{S}_- + \underline{L}_- \underline{S}_+)$ part of the $\underline{\vec{L}} \cdot \underline{\vec{S}}$ operator. Therefore, the matrix will have off-diagonal matrix elements, and the $m_1 m_2$ kets are not eigenkets of $\underline{H}_{so}$. $\frac{1}{2}(\underline{L}_+ \underline{S}_- + \underline{L}_- \underline{S}_+)$ must be applied to each of the six $m_1 m_2$ kets to determine the off-diagonal matrix elements. Consider its application to the first two kets of equation (16.11):

$$\frac{1}{2}(\underline{L}_+ \underline{S}_- + \underline{L}_- \underline{S}_+) \left| 1 \frac{1}{2} \right\rangle = 0. \tag{16.32}$$

Applying the first term of the operator gives 0 because the $\underline{L}_+$ can't raise 1 above 1, and applying the second term gives 0 because $\underline{S}_+$ can't raise $1/2$ above $1/2$. Therefore, the ket $\left|1 \frac{1}{2}\right\rangle$ is not coupled to other kets by off-diagonal matrix elements. Applying the operator to the second ket gives

$$\frac{1}{2}(\underline{L}_+ \underline{S}_- + \underline{L}_- \underline{S}_+) \left| 1 -\frac{1}{2} \right\rangle = \frac{\sqrt{2}}{2} \left| 0 \frac{1}{2} \right\rangle. \tag{16.33}$$

Applying the first term of the operator gives 0 because the $\underline{L}_+$ can't raise 1 above 1. However, applying the second term to the ket, $\underline{L}_-$ lowers 1 to 0, and $\underline{S}_+$ raises -1/2 to 1/2. The result is the ket $\left|0\, 1/2\right\rangle$ multiplied by a constant. The constant is obtained from the raising and lowering operator formulas given in Chapter 15, Section B. Left-multiplying equation (16.33) by the bra $\left\langle 0\, 1/2\right|$ gives the matrix element

$$\left\langle 0 \frac{1}{2} \right| \frac{1}{2}(\underline{L}_+ \underline{S}_- + \underline{L}_- \underline{S}_+) \left| 1 - \frac{1}{2} \right\rangle = \frac{\sqrt{2}}{2}. \tag{16.34}$$

In a similar manner, the other matrix elements are determined.

Then the total 6×6 matrix has diagonal elements obtained from the operator $\underline{L}_z \underline{S}_z$ and off-diagonal elements obtained from the operator $\frac{1}{2}(\underline{L}_+\underline{S}_- + \underline{L}_-\underline{S}_+)$. The matrix has the form

$$\underline{\underline{H}}_{so} = a(r) \begin{array}{c|cccccc} & \left|1\tfrac{1}{2}\right\rangle & \left|1-\tfrac{1}{2}\right\rangle & \left|0\tfrac{1}{2}\right\rangle & \left|0-\tfrac{1}{2}\right\rangle & \left|-1\tfrac{1}{2}\right\rangle & \left|-1-\tfrac{1}{2}\right\rangle \\ \hline \left\langle 1\tfrac{1}{2}\right| & \frac{1}{2} & & & & & \\ \left\langle 1-\tfrac{1}{2}\right| & & -\frac{1}{2} & \sqrt{2}/2 & & & \\ \left\langle 0\tfrac{1}{2}\right| & & \sqrt{2}/2 & 0 & & & \\ \left\langle 0-\tfrac{1}{2}\right| & & & & 0 & \sqrt{2}/2 & \\ \left\langle 1\tfrac{1}{2}\right| & & & & \sqrt{2}/2 & -\frac{1}{2} & \\ \left\langle 1-\tfrac{1}{2}\right| & & & & & & \frac{1}{2} \end{array} . \tag{16.35}$$

The matrix is multiplied by $a(r)$, which is determined by the radial part of the wavefunction. Since the matrix is not diagonal, the $m_1 m_2$ representation kets are not eigenkets of the $\vec{\underline{L}} \cdot \vec{\underline{S}}$ operator. The matrix is block diagonal. There are four blocks, namely, two 1×1 blocks and two 2×2 blocks. The 1×1 blocks are eigenvalues of $\underline{H}_{so}$. Therefore, two of the eigenvalues are $^1\!/_2 a(r)$.

The two 2×2 blocks are equal. The eigenvalues can be found by matrix diagonalization (Chapter 13, Section E). The secular determinant for the first 2×2 block is

$$\begin{vmatrix} -\frac{1}{2} - \lambda & \sqrt{2}/2 \\ \sqrt{2}/2 & -\lambda \end{vmatrix} = 0, \tag{16.36}$$

where the values of λ are the desired eigenvalues. Expanding the determinant and solving for the λs, which are multiplied by $a(r)$, give

$$\lambda = \frac{1}{2} a(r)$$
$$\lambda = -1a(r).$$

The other 2×2 block gives the same eigenvalues.

The six eigenvalues of $\underline{H}_{so}$ are four values of $0.5a(r)$ and two values of $-1a(r)$. Figure 16.2 shows an energy level diagram. The zero of energy is the energy of the p electron in the absence of spin–orbit coupling. The total splitting between the two sets of energy levels is $1.5a(r)$. As mention above, the Na $3p$ to $3s$ fluorescence (Na D line) is composed of two spectroscopic lines split by 17 cm^{-1}. The splitting of the Na D line is caused by the spin–orbit interaction. For Na, $1.5a(r) = 17$ cm^{-1}, and $a(r) = 11.3$ cm^{-1}. The ratio of the splitting of the line, 17 cm^{-1}, to the energy of the transition, $E = 16{,}980$ cm^{-1}, is

$$\frac{a(r)}{E} = 7 \times 10^{-4}.$$

While this is a very small correction to the energy, without consideration of the spin–orbit interaction, it is not possible to explain the observed fluorescence from Na. Spin–orbit coupling is also responsible for many other phenomena.

The spin–orbit interaction splits the six states of a p electron into a set of four degenerate states and a set of two degenerate states. The reason for the groupings of energy levels can be understood by determining the eigenvectors. The eigenvectors can be obtained using the procedure given in Chapter 13, Section E. Instead, the problem will be considered again in the jm representation of the angular momentum states rather than in the m_1m_2.

b. The jm Representation

The spin–orbit coupling problem can be treated in the jm representation (Chapter 15, Section E) with the orbital and spin angular momentum vectors combined into a single total vector. The jm kets for a single p electron, $\ell = 1, s = {}^1\!/_2$, are given in (16.13). These can be written as superpositions of the m_1m_2 kets using the table of Clebsch–Gordon coefficients given in Chapter 15.

The $\underline{H}_{so}$ matrix elements of the jm kets can be obtained by operating $a(r)\underline{\vec{L}} \cdot \underline{\vec{S}}$ on each of them. Consider the first ket in (16.13), $\left|{}^3\!/_2{}^3\!/_2\right\rangle$. $\underline{\vec{L}} \cdot \underline{\vec{S}}$, is an m_1m_2 operator. Therefore, it is necessary to convert the jm ket into the m_1m_2 representation:

$$a(r)\underline{\vec{L}} \cdot \underline{\vec{S}} \left| \frac{3}{2}\,\frac{3}{2} \right\rangle = a(r)\underline{\vec{L}} \cdot \underline{\vec{S}} \left| 1\,\frac{1}{2} \right\rangle = \frac{1}{2}a(r) \left| 1\,\frac{1}{2} \right\rangle. \qquad (16.37)$$

Therefore,

$$a(r)\underline{\vec{L}} \cdot \underline{\vec{S}} \left| \frac{3}{2}\,\frac{3}{2} \right\rangle = \frac{1}{2}a(r) \left| \frac{3}{2}\,\frac{3}{2} \right\rangle. \qquad (16.38)$$

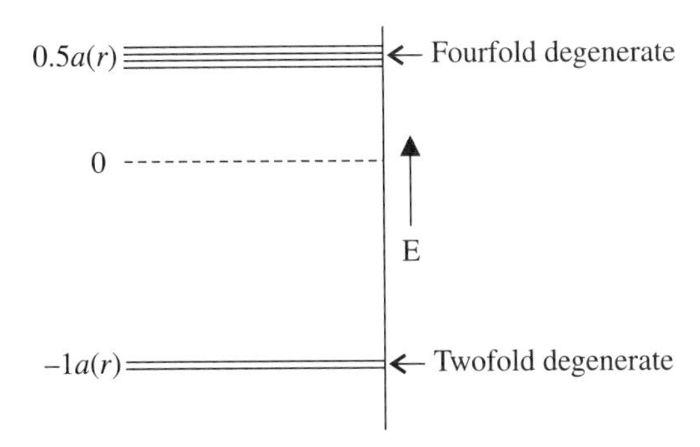

Figure 16.2 Energy level diagram showing the six energy levels of a p electron when spin–orbit coupling is included in the Hamiltonian. The zero of energy is the p electron energy in the absence of the spin–orbit interaction. $a(r)$ is determined by the radial part of the wavefunction.

$|{}^3\!/_2\,{}^3\!/_2\rangle$ is an eigenket of $a(r)\vec{\underline{L}} \cdot \vec{\underline{S}}$ with eigenvalue 1/2 $a(r)$. From the table of Clebsch–Gordon coefficients in Chapter 15, it can be seen that

$$\left|\frac{3}{2}\,\frac{3}{2}\right\rangle = \left|1\,\frac{1}{2}\right\rangle.$$

In the matrix in equation (16.35), $|1\,\frac{1}{2}\rangle$ is the basis for a 1×1 block; that is, it is an eigenvector; therefore, $|{}^3\!/_2\,{}^3\!/_2\rangle$ is also an eigenvector. Application of $a(r)\vec{\underline{L}} \cdot \vec{\underline{S}}$ to each of the jm kets shows that they are the six eigenvectors of $a(r)\vec{\underline{L}} \cdot \vec{\underline{S}}$. For example, the jm ket $|{}^1\!/_2\,{}^1\!/_2\rangle$ can be written in the $m_1 m_2$ representation as

$$\left|\frac{1}{2}\,\frac{1}{2}\right\rangle = \left[\sqrt{\frac{2}{3}}\left|1\,-\frac{1}{2}\right\rangle - \sqrt{\frac{1}{3}}\left|0\,\frac{1}{2}\right\rangle\right].$$

Operating with $a(r)\vec{\underline{L}} \cdot \vec{\underline{S}}$ gives

$$a(r)\vec{\underline{L}} \cdot \vec{\underline{S}}\left|\frac{1}{2}\,\frac{1}{2}\right\rangle = a(r)\left(\underline{L}_z\underline{S}_z + \frac{1}{2}\left(\underline{L}_+\underline{S}_- + \underline{L}_-\underline{S}_+\right)\right)\left[\sqrt{\frac{2}{3}}\left|1\,-\frac{1}{2}\right\rangle - \sqrt{\frac{1}{3}}\left|0\,\frac{1}{2}\right\rangle\right] \tag{16.39}$$

$$= -\frac{1}{2}1a(r)\sqrt{\frac{2}{3}}\left|1\,-\frac{1}{2}\right\rangle - 0 + \frac{1}{2}a(r)\left[0 - \sqrt{2}\sqrt{\frac{1}{3}}\left|1\,-\frac{1}{2}\right\rangle + \sqrt{2}\sqrt{\frac{2}{3}}\left|0 + \frac{1}{2}\right\rangle + 0\right]. \tag{16.40}$$

The first term in equation (16.40) comes from operating $\underline{L}_z\underline{S}_z$ on the first ket. $-{}^1\!/_2$ and 1 are the m_s and m_l values, respectively, brought out by the operation. The -0 comes from operating $\underline{L}_z\underline{S}_z$ on the second ket; $\underline{L}_z$ brings out 0. The first term in the square brackets is 0, which comes from the application $\underline{L}_+\underline{S}_-$ on $|1-{}^1\!/_2\rangle$ since $m_\ell = 1$ can't be raised and $m_s = -{}^1\!/_2$ can't be lowered. The next term in equation (16.40) comes from applying $\underline{L}_+\underline{S}_-$ to $|0\,{}^1\!/_2\rangle$. $\underline{L}_+$ raises 0 to 1, and $\underline{S}_-$ lowers ${}^1\!/_2$ to $-{}^1\!/_2$. The $\sqrt{2}$ coefficient is found using the lowering and raising operator formulas in Chapter 15, Section B. The other two terms are found in analogous manner. Collecting terms in equation (16.40) gives

$$= -1a(r)\left[\sqrt{\frac{2}{3}}\left|1\,-\frac{1}{2}\right\rangle - \sqrt{\frac{1}{3}}\left|0\frac{1}{2}\right\rangle\right] = -1a(r)\left[\sqrt{\frac{2}{3}}\left|1\,-\frac{1}{2}\right\rangle - \sqrt{\frac{1}{3}}\left|0\frac{1}{2}\right\rangle\right]. \tag{16.41}$$

Therefore,

$$a(r)\vec{\underline{L}} \cdot \vec{\underline{S}}\left|\frac{1}{2}\,\frac{1}{2}\right\rangle = -1a(r)\left|\frac{1}{2}\,\frac{1}{2}\right\rangle. \tag{16.42}$$

$|{}^1\!/_2\,{}^1\!/_2\rangle$ is an eigenket of $a(r)\vec{\underline{L}} \cdot \vec{\underline{S}}$ with eigenvalue $-1a(r)$.

The jm kets are the eigenkets of the $a(r)\vec{\underline{L}} \cdot \vec{\underline{S}}$ operator. The interaction between the orbital angular momentum and the spin angular momentum couples the two angular momentum vectors. The jm representation describes coupled angular momentum vectors. Generally, the jm representation kets are the eigenstates for a system involving two or more coupled angular momenta. The four jm kets with $j = {}^3\!/_2$ have eigenvalues ${}^1\!/_2 a(r)$. The two

jm kets with $j = {}^1\!/_2$ have eigenvalues $-1a(r)$. In Figure 16.2, the eigenkets corresponding to the four states with energy ${}^1\!/_2\, a(r)$ are

$$\left|\frac{3}{2}\ \frac{3}{2}\right\rangle \left|\frac{3}{2}\ \frac{1}{2}\right\rangle \left|\frac{3}{2}\ -\frac{1}{2}\right\rangle \left|\frac{3}{2}\ -\frac{3}{2}\right\rangle.$$

The two states with energy $-1a(r)$ are

$$\left|\frac{1}{2}\ \frac{1}{2}\right\rangle \left|\frac{1}{2}\ -\frac{1}{2}\right\rangle.$$

The system divides into sets of energy levels based on the magnitude of the total angular momentum.

Although the spin–orbit induced splitting of the Na D line is small, it is readily observable. With very high resolution spectroscopy, it is found that the Na D line is split into even more very finely spaced energy levels. The additional splittings are caused by the hyperfine interaction, which arises because of the interaction of the magnetic field of the electron with the magnetic field of the nucleus. For Na, the nuclear spin state is $I = {}^3\!/_2$. The hyperfine term in the Hamiltonian can have the form $\gamma \underline{\vec{I}} \cdot \underline{\vec{S}}$, where γ is independent of the angular momentum, $\underline{\vec{I}}$ operates on the nuclear spin, and $\underline{\vec{S}}$ operates on the electron spin. In the Na ground state, the electron is in a $3s$ orbital. Thus there is no orbital angular momentum. The eigenkets of $\gamma \underline{\vec{I}} \cdot \underline{\vec{S}}$ are the jm kets with the two j values 2 and 1 (${}^3\!/_2 + {}^1\!/_2$ and ${}^3\!/_2 - {}^1\!/_2$). The eight energy levels divide into two groups corresponding to five levels with $j = 2$ and three levels with $j = 1$. The splitting is 1.7 GHz. In the $3p$ excited state there are three types of angular momentum: orbital spin, electron spin, and nuclear spin. The nuclear spin states couple to the jm eigenkets of $\underline{H}_{so}$. $j = {}^3\!/_2$ combines with $I = {}^3\!/_2$ to give states with total angular moment values, $F = 3, 2, 1$, and 0. Each of these has corresponding m_F values. There are a total of 16 states. $j = {}^1\!/_2$ combines with $I = {}^3\!/_2$ to give states with total angular moment values, $F = 2$ and 1. Each of these has corresponding m_F values. There are a total of 8 states. The $3p$ state of Na is split into a total of 6 sets of states. The observed spectroscopic splittings in the $3p$ to $3s$ transition are in the range of 100 MHz.

The matrix in equation (16.35) is not diagonal because the m_1m_2 representation kets are not eigenkets of $\underline{H}_{so}$. The eigenvalues were found by solving the secular determinants. The matrix can also be diagonalized with a similarity transformation (Chapter 13, Section E). The matrix used in a similarity transformation is the matrix that takes the initial basis set into the eigenvector basis. In the spin–orbit problem, the initial basis set contained the m_1m_2 kets. The eigenkets are the jm kets. The matrix of Clebsch–Gordon coefficients takes the m_1m_2 kets into the jm kets. Therefore, the similarity transformation that takes the nondiagonal matrix in equation (16.35) into the diagonal matrix with the eigenvalues on the diagonal is

$$\underline{\underline{H}}_{so}^{jm} = \underline{\underline{U}}\, \underline{\underline{H}}_{so}^{m_1m_2}\, \underline{\underline{U}}^{-1}, \tag{16.43}$$

where $\underline{\underline{H}}_{so}^{m_1m_2}$ is the initial nondiagonal matrix in the m_1m_2 representation, $\underline{\underline{H}}_{so}^{jm}$ is the diagonal matrix of eigenvalues in the jm representation, and $\underline{\underline{U}}$ is the matrix of Clebsch–Gordon coefficients.

C. ANTISYMMETRIZATION AND THE PAULI PRINCIPLE

Equation (16.10) gives the kets that represent the intrinsic angular momentum states of the electron. In treating the spin–orbit coupling of a p electron, only a single electron was dealt with. However, in most atomic and molecular problems, more than one electron is involved. In this section, the nature of many electron wavefunctions that have both orbital and spin components will be developed. It will be shown that antisymmetrization of many electron total wavefunctions (orbital and spin) builds the Pauli Principle into the quantum mechanics mathematical formalism. The Pauli Principle states that at most two electrons can occupy an atomic or molecular orbital, and the two electrons must have opposite spin. Alternatively, in a system, no two electrons can have all four quantum numbers, n, ℓ, m_ℓ, and m_s, identical. Since $m_s = \pm 1/2$, it is possible for only two electrons to have the same n, ℓ, m_ℓ. To illustrate the role played by symmetric and antisymmetric superpositions of states and their relationship to the Pauli Principle, the excited states of helium, first without spin and then with spin, are discussed.

1. Excited States of Helium Neglecting Spin

In Chapter 10, Section A, nondegenerate first-order perturbation theory was used to calculate the ground-state ($1s$) energy of the helium atom. The electron–electron repulsive interaction, $e^2/4\pi\varepsilon_0 r_{12}$, was taken to be the perturbation. In performing the calculation, the product of two $1s$ hydrogen-like wavefunctions was used as the zeroth wavefunction. The zeroth-order wavefunction for any state of He is the product function

$$\left|\psi_{n_1\ell_1m_1}(1)\right\rangle\left|\psi_{n_2\ell_2m_2}(2)\right\rangle = \left|\psi_{n_1\ell_1m_1}(1)\psi_{n_2\ell_2m_2}(2)\right\rangle. \tag{16.44}$$

The first electron is labeled 1, and the second electron is labeled 2. (1) represents coordinates of first electron, $(r_1, \theta_1, \varphi_1)$, and (2) represents the coordinates of the second electron, $(r_2, \theta_2, \varphi_2)$. The zeroth-order energy is

$$E^0_{n_1n_2} = -4Rhc\left(\frac{1}{n_1^2} + \frac{1}{n_2^2}\right), \tag{16.45}$$

where

$$R = \frac{\mu e^4}{8\varepsilon_o^2 h^3 c}$$

is the Rydberg constant, and μ is the reduced mass of the hydrogen atom.

The zero-order energy of the first excited state of He is

$$E^0 = -5Rhc. \tag{16.46}$$

There are two combinations of the quantum numbers that give this energy, $n_1 = 1$ and $n_2 = 2$, or $n_2 = 1$ and $n_1 = 2$. These two possibilities must have the same energy since the electrons are indistinguishable. It does not matter if electron 1 is in the lowest energy level and electron 2 is excited or if electron 2 is in the lowest energy level and electron 1 is excited.

Consider the first excited configuration in detail. The first excited zeroth-order energy level [equation (16.45)] is eightfold degenerate. The eight states are

$$\begin{array}{ll} |1s(1)\rangle\,|2s(2)\rangle & |1s(1)\rangle\,|2p_y(2)\rangle \\ |2s(1)\rangle\,|1s(2)\rangle & |2p_y(1)\rangle\,|1s(2)\rangle \\ |1s(1)\rangle\,|2p_x(2)\rangle & |1s(1)\rangle\,|2p_z(2)\rangle \\ |2p_x(1)\rangle\,|1s(2)\rangle & |2p_z(1)\rangle\,|1s(2)\rangle\,. \end{array} \tag{16.47}$$

Each of these has either $n_1 = 1$ and $n_2 = 2$, or $n_2 = 1$ and $n_1 = 2$. Therefore, treating the first excited state of He is a degenerate perturbation theory problem (Chapter 9, Section C). The degenerate perturbation theory problem is solved by finding the matrix elements of the perturbation operator for the eight states and diagonalizing. The secular determinant obtained from the matrix is given in equation (16.48).

$$\begin{array}{lll} & & \begin{array}{cccccccc} \quad 1 \quad & \quad 2 \quad & \quad 3 \quad & \quad 4 \quad & \quad 5 \quad & \quad 6 \quad & \quad 7 \quad & \quad 8 \quad \end{array} \\ \begin{array}{l} 1 \\ 2 \\ 3 \\ 4 \\ 5 \\ 6 \\ 7 \\ 8 \end{array} & \begin{array}{l} 1s(1)2s(2) \\ 2s(1)1s(2) \\ 1s(1)2p_x(2) \\ 2p_x(1)1s(2) \\ 1s(1)2p_y(2) \\ 2p_y(1)1s(2) \\ 1s(1)2p_z(2) \\ 2p_z(1)1s(2) \end{array} & \begin{bmatrix} J_s - E' & K_s & & & & & & \\ K_s & j_s - E' & & & & & & \\ & & J_{p_x} - E' & K_{p_x} & & & & \\ & & K_{p_x} & J_{p_x} - E' & & & & \\ & & & & J_{p_y} - E' & K_{p_y} & & \\ & & & & K_{p_y} & J_{p_y} - E' & & \\ & & & & & & J_{p_z} - E' & K_{p_z} \\ & & & & & & K_{p_z} & J_{p_z} - E' \end{bmatrix} = 0. \end{array} \tag{16.48}$$

The left-hand side of the determinant has been labeled with the bras. The bras have been numbered 1 to 8. Along the top of the determinant are the numbers 1 to 8, which stand for the corresponding kets. The E's are the eigenvalues of the matrix. The E's are the first-order corrections to the zeroth-order energy, equation (16.46). The J and K matrix elements are

$$\begin{aligned} J_s &= \int\!\!\int 1s(1)2s(2)\frac{e^2}{4\pi\varepsilon_o r_{12}}1s(1)2s(2)d\tau_1 d\tau_2 \\ K_s &= \int\!\!\int 1s(1)2s(2)\frac{e^2}{4\pi\varepsilon_o r_{12}}2s(1)1s(2)d\tau_1 d\tau_2 \\ J_{p_x} &= \int\!\!\int 1s(1)2p_x(2)\frac{e^2}{4\pi\varepsilon_o r_{12}}1s(1)2p_x(2)d\tau_1 d\tau_2 \\ K_{p_x} &= \int\!\!\int 1s(1)2p_x(2)\frac{e^2}{4\pi\varepsilon_o r_{12}}2p_x(1)1s(2)d\tau_1 d\tau_2. \end{aligned} \tag{16.49}$$

The other matrix elements, J_{p_y}, K_{p_y}, J_{p_z}, K_{p_z}, are obtained by replacing x with y and z in equation (16.49).

The J integrals are called Coulomb integrals. They can be viewed as representing the average Coulomb interaction energy of two electrons with charge distribution of each electron given by $|\psi|^2$. For example, if electron 1 is in the $1s$ orbital and electron 2 is in the $2s$ orbital, J_s is the Coulomb interaction between the two charge distributions described by $|1s(1)|^2$ and $|2s(2)|^2$.

The K integrals are called exchange integrals and do not have a classical counterpart or interpretation. The K integrals get their names because the two wavefunctions involved differ from each other by the interchange of electrons. In the K_s integral, the function $1s(1)2s(2)$ is on the left-hand side of the operator, and $1s(2)2s(1)$ is on the right-hand side. These only differ by the interchange of which electron occupies which state. The K integrals, which are the off-diagonal matrix elements, arise because the product basis set used to construct the matrix is not the correct zeroth-order set of functions.

The matrix elements not shown in equation (16.48) are zero. For example, an integral such as

$$\int\int 1s(1)2s(2)\frac{e^2}{4\pi\varepsilon_o r_{12}}1s(1)2p_z(2)\,d\tau_1 d\tau_2 = 0$$

because the function $2p_z(2)$ changes sign upon inversion through the origin while the other functions remain unchanged. This is the equivalent of performing an integral of an even function times an odd function over all space, which yields zero.

The values E' are found by setting the determinant of the matrix equal to zero [equation (16.48)], expanding the determinant, and solving the resulting equations. The 8×8 determinant is composed of four 2×2 blocks. The three blocks involving the p orbitals are identical. Therefore, the first order corrections to the energy are

$$\begin{aligned}
&J_s + K_s \\
&J_s - K_s \\
&J_{p_i} + K_{p_i} \qquad \text{(triple root)} \\
&J_{p_i} - K_{p_i} \qquad \text{(triple root)},
\end{aligned} \tag{16.50}$$

where the subscript i is x, y, or z. Figure 16.3 is an energy level diagram. In addition to the first excited states, the ground-state energy level is shown. The left-hand side of the diagram shows the zeroth-order energy levels. The middle of the figure shows the results of the calculation if only the Coulomb integrals are included, but the off-diagonal exchange integrals are not taken into account. The right-hand side of the diagram shows the energy levels calculated to first order. The states involving electrons in the $1s$ and $2s$ orbitals are lower in energy because the $2s$ orbital puts more electron density close to the nucleus than $2p$ orbitals, which increases the Coulombic attraction and lowers the energy. The symbols at the right of the diagram (e.g., ^{3}S) are called term symbols. They will be discussed below after electron spin is included.

The eigenfunctions associated with the diagonalized matrix are:

For $E' = J_s + K_s$

$$\frac{1}{\sqrt{2}}\{1s(1)2s(2) + 2s(1)1s(2)\} \tag{16.51a}$$

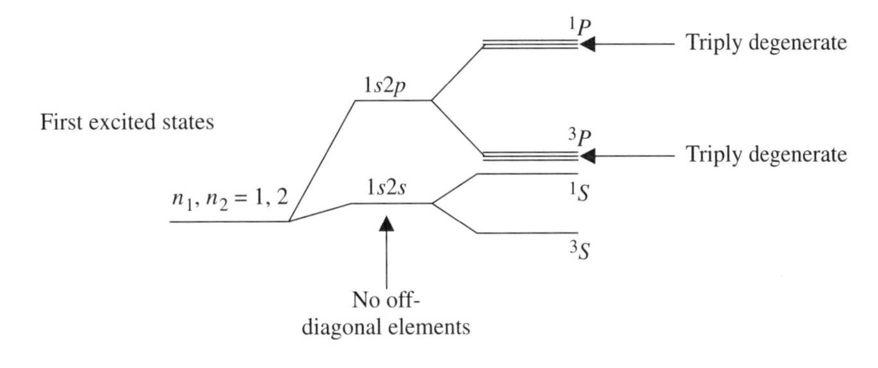

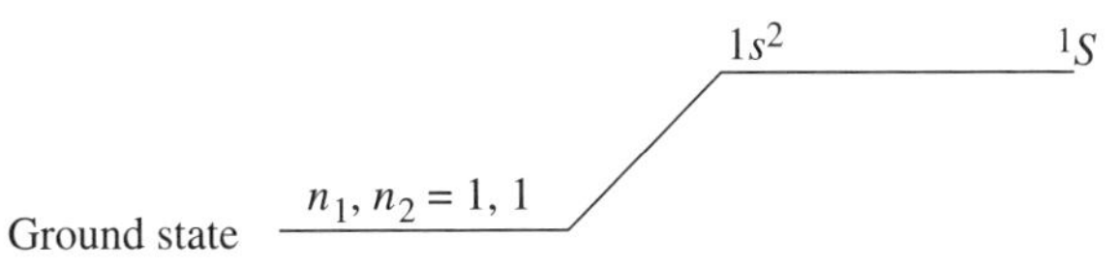

Figure 16.3 Energy level diagram for the ground state and first excited states of helium. The left-hand side of the figure shows the zeroth-order energy levels. The middle of the diagram shows the energy levels including only the Coulomb integrals, J (no off-diagonal matrix elements). The right-hand side of the figure shows energy levels correct to first order.

For $E' = J_s - K_s$

$$\frac{1}{\sqrt{2}}\{1s(1)2s(2) - 2s(1)1s(2)\} \tag{16.51b}$$

For $E' = J_{p_i} + K_{p_i}$

$$\frac{1}{\sqrt{2}}\{1s(1)2p_i(2) + 2p_i(1)1s(2)\} \tag{16.51c}$$

For $E' = J_{p_i} - K_{p_i}$

$$\frac{1}{\sqrt{2}}\{1s(1)2p_i(2) - 2p_i(1)1s(2)\}, \tag{16.51d}$$

where in the last two functions, $i = x$, then y, then z.

2. Symmetric and Antisymmetric Combinations and the Permutation Operator

The functions in equation (16.51) with the + sign are said to be symmetric in the position coordinates of the electrons because interchange of the electron labels yields the identical wavefunction. Since the electron labels represent the coordinates associated with the electron—that is, $(1) \Rightarrow r_1, \theta_1, \phi_1$—interchanging labels is the same as interchanging the positions of two electrons.

The act of interchanging the positions of two electrons is done formally by the application of the permutation operator, $\underline{P}$, to the wavefunction. If a ket $|x\rangle$ is symmetric, then

$$\underline{P}\,|x\rangle = 1\,|x\rangle\,. \tag{16.52}$$

$|x\rangle$ is an eigenket of the permutation operator with eigenvalue 1.

The functions in equation (16.51) with the $-$ sign are antisymmetric because interchange of the electron labels gives the function multiplied by -1. If the ket $|x\rangle$ is the function (16.51b), then application of the permutation operator gives

$$\begin{aligned}\underline{P}\,|x\rangle &= P\frac{1}{\sqrt{2}}\{1s(1)2s(2) \;-\; 2s(1)1s(2)\}\\ &= \frac{1}{\sqrt{2}}\{1s(2)2s(1) \;-\; 2s(2)1s(1)\}\\ &= -1\frac{1}{\sqrt{2}}\{1s(1)2s(2) \;-\; 2s(1)1s(2)\}\\ &= -1\,|x\rangle\,.\end{aligned} \tag{16.53}$$

The function is an eigenfunction of the permutation operator with eigenvalue -1.

Both the $+$ functions and the $-$ functions are eigenfunctions of the permutation operator. Symmetric functions have eigenvalues of 1, and antisymmetric functions have eigenvalues of -1 when operated on by the permutation operator. Since the two electrons are identical particles, interchanging their positions cannot change the state of the system. Therefore, application of the permutation operator must give the same function back. Since a system is defined by the direction of a ket in an abstract vector space, not its length (so long as its length is greater than zero), multiplying the function by -1 does not change the state of the system. All wavefunctions that properly describe a system containing two or more identical particles are either symmetric or antisymmetric with respect to exchange of a pair of particle labels (particle coordinates). Good wavefunctions for systems containing two or more identical particles must be eigenfunctions of the permutations operator.

3. Excited States of Helium Including Spin

Each electron in the He atom is a spin $\frac{1}{2}$ particle. For two spin $\frac{1}{2}$ particles there are four possible states associated with the intrinsic electron angular momentum. In the $m_1 m_2$ representation the four states are

$$\begin{aligned}&\alpha(1)\alpha(2)\\ &\alpha(1)\beta(2)\\ &\beta(1)\alpha(2)\\ &\beta(1)\beta(2).\end{aligned}$$

The two functions $\alpha(1)\beta(2)$ and $\beta(1)\alpha(2)$ are neither symmetric nor antisymmetric. Application of the permutation operator to these two functions does not return the same functions times ± 1. The $m_1 m_2$ representation is not the proper representation for the two electron spins.

The states of the two electron spins can also be written in the jm representation. The necessary table of Clebsch–Gordon coefficients is given in Chapter 15. The four spin states in the jm representation are

$$\begin{gathered}\alpha(1)\alpha(2)\\ \frac{1}{\sqrt{2}}\{\alpha(1)\beta(2)+\beta(1)\alpha(2)\}\\ \beta(1)\beta(2)\\ \\ \frac{1}{\sqrt{2}}\{\alpha(1)\beta(2)-\beta(1)\alpha(2)\}.\end{gathered} \tag{16.54}$$

The first three spin functions correspond to the states with a total angular momentum, $s = 1$, and $m_s = 1, 0, -1$ (from top to bottom). The fourth spin function has $s = 0$, and $m_s = 0$. The first three functions are symmetric with respect to interchange of electron labels, and the fourth is antisymmetric. These functions are the appropriate spin functions because they are eigenfunctions of the permutation operator.

The total wavefunction for an electron is composed of a spatial function times a spin function. There are eight spatial basis functions given in equation (16.47) and four spin functions given in equation (16.54). Forming the product functions, there are 32 independent states associated with the first excited configuration of the He atom. The 32 total basis functions (spatial function times spin function) have the form

$$\begin{aligned}&1s(1)2s(2)\alpha(1)\alpha(2)\\ &2s(1)1s(2)\alpha(1)\alpha(2)\\ &1s(1)2p_x(2)\alpha(1)\alpha(2)\\ &\quad\vdots\\ &1s(1)2s(2)\cdot\frac{1}{\sqrt{2}}(\alpha(1)\beta(2)+\beta(1)\alpha(2))\\ &\quad\vdots\end{aligned} \tag{16.55}$$

Since the Hamiltonian is independent of spin (spin–orbit coupling and other spin-dependent terms have not been included), multiplying a spatial function by each of the four spin functions does not change the energy. The first excited state is 32-fold degenerate. The secular determinant has the form shown in equation (16.56). There are four 8×8 blocks. Each is identical to the 8×8 determinant in equation (16.48) except for multiplication of the spatial basis functions by the spin functions. The matrix elements outside of the blocks vanish because of the orthogonality of the spin functions, which is not removed by the operator $1/r_{12}$ because it is spin-independent.

The eigenfunctions of the matrix, which are the correct zeroth-order functions, are

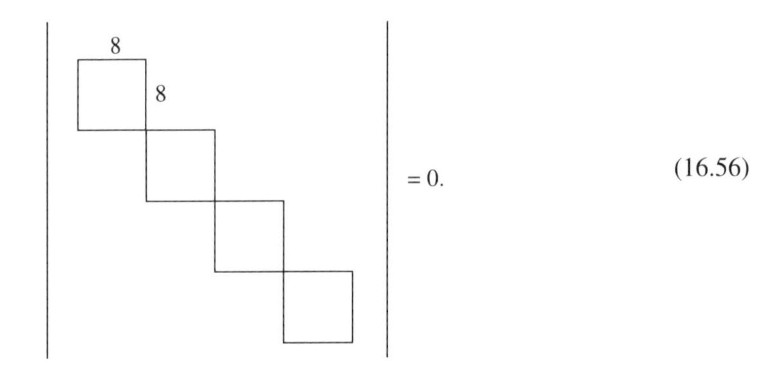

(16.56)

obtained by multiplying each of the spatial functions given in equation (16.51) by the four spin functions given in equation (16.54). Consider one of the 8×8 blocks, the one involving the $1s$ and $2s$ orbitals. The 8 total wavefunctions, spatial $\times$ spin, are

$$\begin{aligned}
&\frac{1}{\sqrt{2}}[1s(1)2s(2) + 2s(1)1s(2)] \cdot \alpha(1)\alpha(2)\\
&\frac{1}{\sqrt{2}}[1s(1)2s(2) + 2s(1)1s(2)] \cdot \frac{1}{\sqrt{2}}[\alpha(1)\beta(2) + \beta(1)\alpha(2)]\\
&\frac{1}{\sqrt{2}}[1s(1)2s(2) + 2s(1)1s(2)] \cdot \beta(1)\beta(2)\\
&\frac{1}{\sqrt{2}}[1s(1)2s(2) - 2s(1)1s(2)] \cdot \frac{1}{\sqrt{2}}[\alpha(1)\beta(2) - \beta(1)\alpha(2)]
\end{aligned} \qquad (16.57a)$$

$$\begin{aligned}
&\frac{1}{\sqrt{2}}[1s(1)2s(2) - 2s(1)1s(2)] \cdot \alpha(1)\alpha(2)\\
&\frac{1}{\sqrt{2}}[1s(1)2s(2) - 2s(1)1s(2)] \cdot \frac{1}{\sqrt{2}}[\alpha(1)\beta(2) + \beta(1)\alpha(2)]\\
&\frac{1}{\sqrt{2}}[1s(1)2s(2) - 2s(1)1s(2)] \cdot \beta(1)\beta(2)\\
&\frac{1}{\sqrt{2}}[1s(1)2s(2) + 2s(1)1s(2)] \cdot \frac{1}{\sqrt{2}}[\alpha(1)\beta(2) - \beta(1)\alpha(2)].
\end{aligned} \qquad (16.57b)$$

The functions have been grouped so that the first four (16.57a) are totally symmetric with respect to interchange of electron labels. Functions 1–3 have spatial and spin parts that are symmetric. Applying the permutation operator, $\underline{P}$, to the spatial part brings out a 1, and applying $\underline{P}$ to the spin part brings out a 1. Then the total function is an eigenfunction of the permutation operator with eigenvalue 1. Function 4 has antisymmetric spatial and spin parts. Applying $\underline{P}$ to each part will bring out a -1; so applying $\underline{P}$ to the total function will yield an eigenvalue of 1. Therefore, function 4 is also totally symmetric; that is, the total function, spatial $\times$ spin, is an eigenfunction of $\underline{P}$ with eigenvalue 1.

Functions 5–8 (16.57b) are totally antisymmetric. Functions 5–7 have antisymmetric spatial functions times symmetric spin functions. Function 8 has a symmetric spatial function and an antisymmetric spin function. Each total function is an eigenfunction of $\underline{P}$ with eigenvalue -1. The functions that give rise to the other 8×8 blocks in equation (16.56), which involve p_x, p_y, and p_z orbitals, can be grouped in the same manner. For each block, four of the total functions (spatial times spin) are totally symmetric and four of the total functions are totally antisymmetric.

Figure 16.4 is a diagram showing qualitatively the relative energies of the ground and first excited states of He. The diagram has been divided so that the states are segregated by the permutation symmetry of the total wavefunction. The states are symmetric if the spatial times spin function is symmetric. The states are antisymmetric if the spatial times spin function is antisymmetric. A filled circle indicates a symmetric spin function, and an open circle indicates an antisymmetric spin function. Totally antisymmetric functions with filled circles have spatial functions that are antisymmetric since the spin parts are symmetric. Totally antisymmetric functions with open circles have spatial functions that are symmetric since the spin parts are antisymmetric. For the totally symmetric functions, filled circles mean that the spatial functions are symmetric, and open circles mean that the spatial functions are antisymmetric. Three circles in a line (e.g., the $1s2s$ totally antisymmetric state) indicate that the energy level is triply degenerate. This can occur because there are three spin states associated with one spatial configuration or because there is a single spin state but the level involves the three different p orbitals. The groups of nine circles represent a level that is ninefold degenerate because there are three spin states associated with each of the three different p orbitals. The orbitals that give rise to each level are listed in the diagram; and on the right-hand side of the diagram are term symbols, which are discussed below.

No additional term in the Hamiltonian can mix the symmetric and antisymmetric states. For the states to mix, a nonzero matrix element must exist. The Hamiltonian, including any additional terms in the Hamiltonian, is symmetric. The product of a symmetric function and an antisymmetric function is antisymmetric. To calculate the value of a matrix element, an integral is performed over all space—that is, a symmetrical region. If one wavefunction is symmetric and one is antisymmetric, the product is antisymmetric, and the integral will vanish. Therefore, the totally symmetric functions and the totally antisymmetric functions are two completely independent sets of functions which never mix; that is, they can have no interactions with each other.

Since totally antisymmetric states can have no interactions with totally symmetric states, only one of these two types of states can exist in nature as we know it. The question is: Which types of functions actually occur in nature? This question has been answered by experiments. All experimental observations demonstrate that the states occurring in nature are totally antisymmetric. For example, in the diagram, the ground state of He is the triply degenerate $s = 1$ spin state (electron spins are unpaired) if the wavefunction is totally symmetric while the ground state is the nondegenerate $s = 0$ spin state (electron spins paired) if the wavefunction is totally antisymmetric. The $s = 1$ state would be paramagnetic because of the combined magnetic dipole moments of the two electrons. In contrast, the $s = 0$ state is diamagnetic. Experiments show that the ground state of

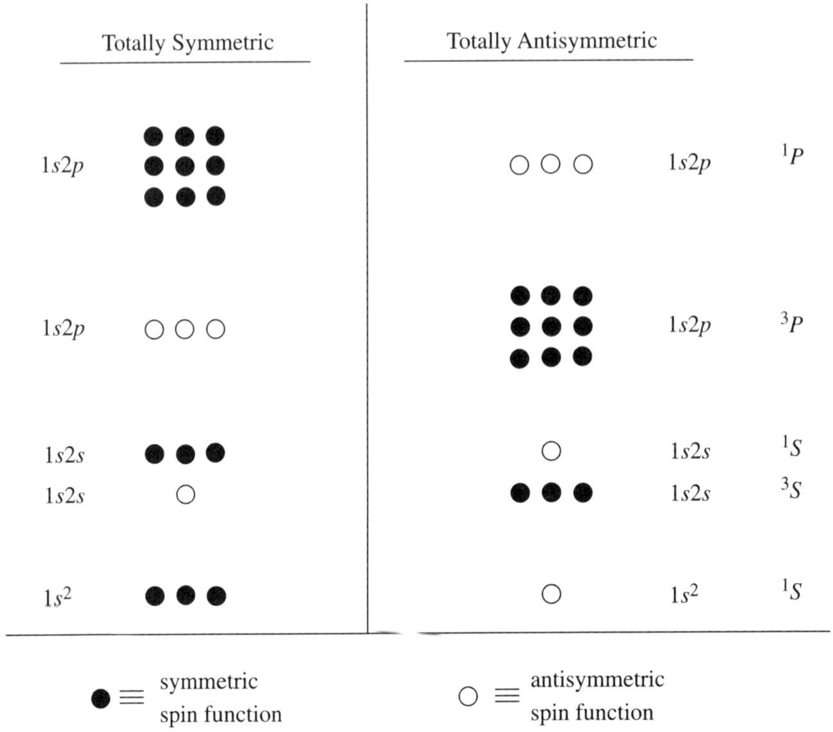

Totally Symmetric		Totally Antisymmetric		
$1s2p$	●●● ●●● ●●●	○○○	$1s2p$	^1P
$1s2p$	○○○	●●● ●●● ●●●	$1s2p$	^3P
$1s2s$	●●●	○	$1s2s$	^1S
$1s2s$	○	●●●	$1s2s$	^3S
$1s^2$	●●●	○	$1s^2$	^1S

● ≡ symmetric spin function ○ ≡ antisymmetric spin function

Figure 16.4 An energy level diagram for the ground state and first excited states of He. A filled circle indicates a symmetric spin function, and an open circle indicates an antisymmetric spin function. The left-hand side of the diagram shows the states that are totally symmetric, and the right-hand side shows the states that are totally antisymmetric. Totally symmetric states do not occur in nature.

He is not paramagnetic. A glass container of liquid helium cannot be picked up with a magnet. Figure 16.4 also shows that for the totally symmetric states the $s = 0$ (singlet) states are lower in energy than the $s = 1$ (triplet) states for states involving the same orbitals. The opposite is true for the totally antisymmetric states. Spectroscopic experiments show that, given the same orbitals, triplets states are always lower in energy than singlet states. The reason for this is discussed below. As will also be discussed below, the Pauli Exclusion Principle is a qualitative statement that all total many-electron wavefunctions must be antisymmetric. The spatial part can be symmetric or antisymmetric. The spin part can be symmetric or antisymmetric, but the total function must be antisymmetric. The fact that the wavefunctions must be antisymmetric (Pauli Exclusion Principle) is responsible for the periodic table of the elements. Returning to Figure 16.4, only the totally antisymmetric states exist. The totally symmetric functions are not valid wavefunctions.

Since the permutation symmetry of many electron wavefunctions can only be determined experimentally, it is necessary to bring the antisymmetrization property into quantum theory as an assumption.

❖ **Assume:** The wavefunction representing an actual state of a system containing two or more electrons must be completely antisymmetric in the coordinates of the electrons; that is, on interchanging the coordinates of any two electrons, the sign of the wavefunction must change.

This assumption is the quantum mechanical statement of the Pauli Exclusion Principle. To see this, consider the fact that antisymmetric wavefunctions can be written as determinants. If $A(1)$ represents an orbit $\times$ spin function for one electron such as $1s\alpha$, and $B, C, \ldots, N$ are others, then the wavefunction

$$\psi = \begin{vmatrix} A(1)B(1) & \cdots & N(1) \\ A(2)B(2) & \cdots & N(2) \\ A(N)B(N) & \cdots & N(N) \end{vmatrix} \tag{16.58}$$

is a completely antisymmetric in the N electrons because the interchange of any two rows changes the sign of the determinant. The number in each row is the electron label. Each electron is in every orbital. For example, the ground-state wavefunction of He can be written as a determinant. The ground state will involve the $1s$ orbital and α and β spin states. If electron 1 is in the $1s$ state with β spin, the function is written as $1s(1)\ \beta(1) = \overline{1s(1)}$, where the bar over the function indicates that the spin state is β. If electron 1 is in the $1s$ state with α spin, the function is written as $1s(1)\ \alpha(1) = 1s(1)$, where the absence of a bar over the function indicates α spin. The 2×2 determinant is formed and expanded to obtain the totally antisymmetric wavefunction:

$$\begin{aligned} \begin{vmatrix} 1s(1)\overline{1s(1)} \\ 1s(2)\overline{1s(2)} \end{vmatrix} &= 1s(1)\overline{1s(2)} - 1s(2)\overline{1s(1)} \\ &= 1s(1)\alpha(1)1s(2)\beta(2) - 1s(2)\alpha(2)1s(1)\beta(1) \\ &= 1s(1)1s(2)[\alpha(1)\beta(2) - \beta(1)\alpha(2)]. \end{aligned} \tag{16.59}$$

The last line is the correct antisymmetric ground state function. The spatial part is symmetric and the spin part is antisymmetric. The spin part corresponds to $s = 0, m_s = 0$; that is, the electron spins are paired.

Determinants have another important property. If two columns of a determinant are equal, it vanishes. This property of determinants builds the Pauli Exclusion Principle into the mathematical formalism. For a given one-electron orbital, there are only two possible spatial $\times$ spin functions—that is, those obtained by multiplying the spatial function by either of the two spin functions, α and β. Thus, no more than two electrons can occupy the same orbital in an atom or molecule, and these two must have their spins opposed; no two electrons can have the same values of all four quantum numbers n, ℓ, m_ℓ, and m_s. If all of the quantum numbers are the same, two columns in the determinant form of the wavefunction will be equal, and the wavefunction (determinant) vanishes. Therefore, the requirement that all many electron total wavefunctions (spatial $\times$ spin) must be antisymmetric with respect to the interchange of all pairs of electrons forces the wavefunctions to satisfy the Pauli Exclusion Principle.

For example, a wavefunction for the ground state of the He atom in which both electrons are in the $1s$ orbital and both have α spins is written as

$$\psi = \begin{vmatrix} 1s(1) & 1s(1) \\ 1s(2) & 1s(2) \end{vmatrix}. \tag{16.60}$$

Expanding the determinant

$$\begin{vmatrix} 1s(1) & 1s(1) \\ 1s(2) & 1s(2) \end{vmatrix} = 1s(1)\alpha(1)1s(2)\alpha(2) - 1s(2)\alpha(2)1s(1)\alpha(1) = 0. \tag{16.61}$$

The attempt to write a wavefunction that violates the Pauli Principle (a wavefunction that is not totally antisymmetric) gives 0.

D. SINGLET AND TRIPLET STATES

The right-hand side of Figure 16.4 shows the energy level diagram for the totally antisymmetric states that actually occur as states of the He atom. There are two types of states, those with open circles and those with filled circles. The states with open circles are obtained by multiplying a symmetric spatial function by the single antisymmetric spin function. For each spatial configuration, there is only one such state with $s = 0$ and $m_s = 0$. These states are called singlet states.

The states with filled circles are obtained by multiplying an antisymmetric spatial function by a symmetric spin function. Each antisymmetric spatial function can be multiplied by three symmetric spin functions with $s = 1$ and $m_s = 1, 0,$ or -1. Because each antisymmetric spatial function gives rise to three total spatial $\times$ spin functions, these states are called triplet states. The three triplet states differ by the value of m_s associated with each one. The three triplet states are not necessarily degenerate; for example, spin–orbit coupling can split the degeneracy of the triplet states, giving rise to multiplets in spectra.

The symbols at the far right in Figure 16.4—$^1S, ^3S, ^1P$, and ^3P, are called term symbols. ^3S means a triplet state derived from s states, while ^1P means a singlet state derived from s and p states.

Triplet states have inherently lower energies than the singlet states that arise from the same orbital configurations. For example, the He excited triplet states involving $1s2s$ orbitals are lower in energy than the $1s2s$ singlet state (see Figure 16.4) in spite of the fact that the spatial wavefunctions contain the same orbitals. The difference in energy arises from the permutation symmetry of the spatial wavefunctions. The triplet states have antisymmetric spatial functions, while the singlet states have symmetric spatial functions. The antisymmetric spatial function has a node (vanishes) if the two electrons are at the same spatial location, while the symmetric spatial function does not vanish. Consider the antisymmetric spatial function,

$$\frac{1}{\sqrt{2}}[1s(1)2s(2) - 2s(1)1s(2)]. \tag{16.62}$$

Since the electron labels, 1 and 2, represent the coordinates of the two electrons, if the two electrons are at the same location, q, the function is

$$\frac{1}{\sqrt{2}}[1s(q)2s(q) - 2s(q)1s(q)] = 0.$$

The triplet electrons are anti-correlated. A plot of a function like (16.62) around the point q shows that the probability of finding the two electrons near each other is small and vanishes at q for all q. In contrast, if the electrons are located at q, the symmetric spatial function is

$$\frac{1}{\sqrt{2}}[1s(q)2s(q) + 2s(q)1s(q)] \neq 0.$$

The symmetric spatial function permits the two electrons to be at the same point in space, and the probability is not necessarily small for finding the two electrons at the same location or near each other.

The anticorrelation of the triplet electrons, which causes them to be further apart than the singlet electrons on average, reduces the triplet-state energy by reducing the magnitude of electron–electron repulsion. Electron–electron repulsion is an interaction that increases the energy of a state. Therefore, the energy of the state is not solely determined by the orbitals that make up the spatial wavefunction. The permutation symmetry of the spatial wavefunction can play a significant role in determining the energy of a state.

CHAPTER

17 THE COVALENT BOND

In this chapter the nature of the covalent bond will be addressed. Chemical bonds range from purely ionic, for example, the bonding in a sodium chloride crystal, to purely covalent, for example, the bond that is responsible for the hydrogen molecule. Ranging from the very smallest molecule, H_2, to very large molecules, such as DNA, covalent bonds are primarily responsible for the formation of molecules. Ionic bonds arise from electrostatic interactions, which can be explained with classical mechanics. However, the covalent bond is a purely quantum mechanical phenomenon. The explanation of the covalent bond is a major accomplishment of quantum theory.

In this chapter, bonding in the two simplest molecules, H_2^+ and H_2, will be examined. A very simple treatment will be used that permits the calculations to be done analytically. The treatments of the two molecules will introduce the ideas of molecular structure calculations. However, the aim is to explicate the basis of the covalent bond rather than to present a description of the many powerful methods that have been developed to perform quantum mechanical calculations of molecular structure.

A. SEPARATION OF ELECTRONIC AND NUCLEAR MOTION: THE BORN–OPPENHEIMER APPROXIMATION

The complete Schrödinger equation for a molecule with r nuclei and s electrons is

$$\sum_{j=1}^{r} \frac{1}{M_j} \nabla_j^2 \psi + \frac{1}{m_0} \sum_{i=1}^{s} \nabla_i^2 \psi + \frac{2}{\hbar^2}(W - V)\psi = 0, \tag{17.1}$$

where M_j is the mass of the jth nucleus, m_0 is the mass of the electron, ∇_j^2 is the Laplace operator in terms of coordinates of the jth nucleus, and ∇_i^2 is the Laplace operator in terms of the coordinates of the ith electron. V is the potential,

$$V = \sum_{i,i'} \frac{e^2}{4\pi\varepsilon_0 r_{ii'}} + \sum_{j,j'} \frac{Z_j Z_{j'} e^2}{4\pi\varepsilon_0 r_{jj'}} - \sum_{i,j} \frac{Z_j e^2}{4\pi\varepsilon_0 r_{ij}}. \tag{17.2}$$

The first term is the electron–electron Coulomb repulsion between electrons i and i'. The second term is the nucleus–nucleus Coulomb repulsion between nuclei j and j' with charges Z_j and $Z_{j'}$. The third term is the Coulomb attraction between electron i and nucleus j. In equation (17.1), the first term is the kinetic energy operator for the nuclei; the second term is the kinetic energy operator for the electrons; and W is the energy.

Even in the simplest cases, equation (17.1) cannot be solved exactly. The problem of calculating the energy of a molecule and determining its molecular wavefunctions is made more tractable (but still not exactly solvable) by invoking the Born–Oppenheimer approximation. The Born–Oppenheimer approximation is based on the difference in mass between electrons and atomic nuclei. Electrons are very light relative to even the smallest nucleus, the proton. Because electrons are very light, they move very fast compared to nuclear motion. In the time it takes the nuclei to change position a significant amount (vibrational motions), electrons will have sampled their full configuration space. Because of this disparity between electrons and nuclei in their rates of change in position, under the Born–Oppenheimer approximation, the nuclei are placed in fixed positions, and the electronic energies and wavefunctions are calculated for the fixed nuclear positions. The nuclei are then moved, and the electronic energies and wavefunctions are calculated again. This procedure is continued until the electronic energy is mapped out as a function of nuclear position. A minimum in this energy surface corresponds to a possible stable molecular structure. The depth of the energy minimum is a measure of the stability of the structure.

Mathematically, the Born–Oppenheimer approximation permits a separation of the full wave equation into two separate problems, one for the electronic energy of the molecule as a function of the nuclear coordinates and the other for the vibrational states of the molecule in terms of the electronic energy surface, which provides the potential energy for the nuclear motion.

Let γ represent the $3r$ coordinates of the nuclei relative to an axis system fixed in space, and let x represent the $3s$ coordinates of the electrons relative to axes determined by the nuclei. v are quantum numbers associated with the motions of the nuclei. n are quantum numbers associated with the motions of the electrons. The total approximate wavefunction has the form

$$\psi_{nv}(x, \gamma) = \psi_n(x, \gamma)\psi_{nv}(\gamma). \tag{17.3}$$

$\psi_n(x, \gamma)$ is the electronic wavefunction. $\psi_{nv}(\gamma)$ is the nuclear wavefunction—that is, the vibrational–rotational wavefunction. $\psi_n(x, \gamma)$ depends on the fixed nuclear coordinates, γ. Unlike the exact separation of the hydrogen atom wave equation into three equations, which is accomplished by rigorously separating the wavefunction into the product of three functions (see Chapter 7, Section A), the product function in equation (17.3) is approximate. The product function permits an approximate separation of the wave equation into two equations.

$\psi_n(x, \gamma)$ is obtained by solving the approximate electronic wave equation for fixed nuclear positions, γ:

$$\sum_{ii}^{S} \nabla_i^2 \psi_n(x, \gamma) + \frac{2M_0}{\hbar^2}[U_n(\gamma) - V(x, \gamma)]\psi_n(x, \gamma) = 0. \tag{17.4}$$

$U_n(\gamma)$ is the energy. It depends on the nuclear coordinates and the electronic quantum number. The potential function, $V(x, \gamma)$, is the complete potential function [equation (17.2)], but with fixed nuclear positions. The important simplification contained in equation

(17.4) is the absence of the nuclear kinetic energy term. Thus, under the Born–Oppenheimer approximation, the electron motion is decoupled from the nuclear motion. Solution of equation (17.4) gives the electron state energies $U_n(\gamma)$ as a function of the nuclear coordinates. A minimum in $U_n(\gamma)$ gives the energy and nuclear positions of a potentially stable molecular configuration. For a diatomic system, there will be at most one minimum. For polyatomic molecules, there may be many minima. In some situations, different minima correspond to different stable configurations of the molecule. Alternatively, the molecule may assume the structure associated with the lowest energy minimum.

The nuclear wave equation is

$$\sum_{j=1}^{r} \frac{1}{M_j} \nabla_j^2 \psi_{nv}(\gamma) + \frac{2}{\hbar^2}[E_{n,v} - U_n(\gamma)]\psi_{nv}(\gamma) = 0. \tag{17.5}$$

$U_n(\gamma)$, the electronic energy as a function of nuclear coordinates, γ, acts as the potential function for the nuclear wave equation. Once the electronic wave equation is solved, $U_n(\gamma)$ is known. Equation (17.5) describes the motion of the nuclei with a potential energy that is determined by the electronic energy for fixed nuclear positions. The wave equation does not contain a term for the kinetic energy of the electrons. In Chapter 6, the harmonic oscillator problem was solved. The parabolic potential is the simplest model for molecular vibrations. In Chapter 9, Section B, an anharmonic term in the potential was shown to change the harmonic oscillator energy levels. The potential, $U_n(\gamma)$, is anharmonic. It provides an accurate potential function for the determination of the vibrational states of a molecule.

The Born–Oppenheimer approximation, which permits the approximate separation of the full wave equation into an electronic equation and a nuclear equation, is obtained by expanding the complete wave equation in powers of $(m_0/M)^{1/4}$, where M is the average nuclear mass. The Born–Oppenheimer approximation provides a very useful method for a wide variety of problems. However, many important effects can be traced to the breakdown of the Born–Oppenheimer approximation. There are a variety of methods that extend the Born–Oppenheimer approximation and provide improved results. However, at a fundamental level, electrons are, in fact, coupled to the motion of the nuclei; and in some situations, it is necessary to consider this coupling explicitly.

B. THE HYDROGEN MOLECULE ION

The simplest molecule is the hydrogen molecular ion, which consists of two protons and one electron. After making the Born–Oppenheimer approximation, it is possible to solve the H_2^+ electronic energy problem with fixed nuclear coordinates exactly by separation of the wave equation in confocal elliptical coordinates. Rather, an approximate procedure is employed that is more general and will be used below to treat the hydrogen molecule.

Figure 17.1 is a diagram showing the labeling of the distance parameters. Hydrogen nuclei (protons) A and B are separated by the distance, r_{AB}. The distance from nucleus A to the electron is r_A, and the distance from nucleus B to the electron is r_B.

Within the context of the Born–Oppenheimer approximation, the H_2^+ electronic wave equation is

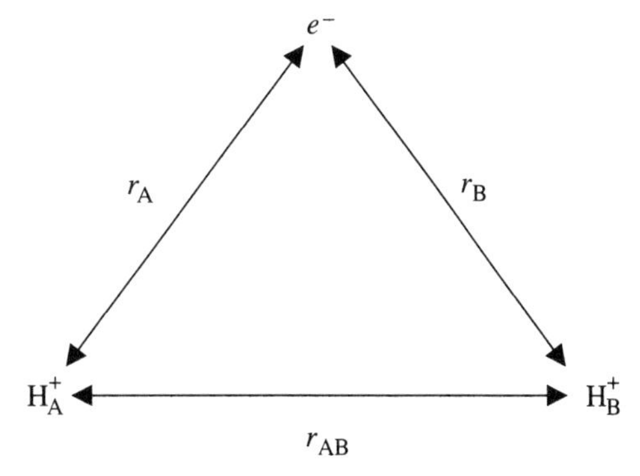

Figure 17.1 Diagram of the distance parameters used in treating the hydrogen molecule ion problem.

$$\nabla^2\psi + \frac{2m_0}{\hbar^2}\left(E + \frac{e^2}{4\pi\varepsilon_0 r_A} + \frac{e^2}{4\pi\varepsilon_0 r_B} - \frac{e^2}{4\pi\varepsilon_0 r_{AB}}\right)\psi = 0. \tag{17.6}$$

The first term is the electron kinetic energy (multiplied by $-2m_0/\hbar^2$). E is the electronic energy, which will be a function of the internuclear separation, r_{AB}. The second term in brackets is the attraction of the electron for nucleus A; the third term is the attraction of the electron for nucleus B, and the last term is the repulsion between nuclei A and B.

For very large internuclear separations r_{AB}, the system will be a hydrogen atom and a hydrogen ion. The hydrogen atom may contain nucleus A with nucleus B the ion, or the hydrogen atom may contain nucleus B with nucleus A the ion. For sufficiently large separation (no interaction), the electronic energy of these two states is

$$E = E_H, \tag{17.7}$$

the energy of the ground state ($1s$ state) of a hydrogen atom. $E_H = -13.6$ eV. The wavefunctions for two states are U_{1s_A} or U_{1s_B}, a 1s hydrogen wavefunction with nucleus A at its center or a $1s$ hydrogen wavefunction with nucleus B at its center. These two states are degenerate.

The two states of the system when the nuclei are well-separated suggests a simple approximate treatment involving U_{1s_A} and U_{1s_B} as basis functions. As a hydrogen atom is brought close to a proton, interactions may result in the formation of a chemical bond. The wavefunctions, U_{1s_A} and U_{1s_B}, are two possible initial states that are equivalent. Therefore, both need to be considered as a minimum basis set for the problem. In terms of these basis states, the Hamiltonian matrix has the form

$$\underline{\underline{H}} = \begin{pmatrix} H_{AA} & H_{AB} \\ H_{BA} & H_{BB} \end{pmatrix}, \tag{17.8}$$

where the matrix elements are given by

$$H_{AA} = \int U_{1s_A}^* \underline{H} U_{1s_A} d\tau \tag{17.9a}$$

$$H_{\mathrm{BA}} = \int U^*_{1s_{\mathrm{B}}} \underline{H} U_{1s_{\mathrm{A}}} d\tau \tag{17.9b}$$

$$H_{\mathrm{AA}} = H_{\mathrm{BB}} \tag{17.9c}$$

$$H_{\mathrm{AB}} = H_{\mathrm{BA}} \tag{17.9d}$$

The general problem of matrix diagonalization to find eigenvalues and eigenvectors, and in particular 2×2 matrix diagonalization, was treated in Chapter 13, Section E. In Chapter 13, the basis vectors were taken to be normalized and orthogonal. $U_{1s_{\mathrm{A}}}$ and $U_{1s_{\mathrm{B}}}$ are normalized, but they are not orthogonal. Equation (17.9e)

$$\Delta = \int U^*_{1s_{\mathrm{A}}} U_{1s_{\mathrm{B}}} d\tau, \tag{17.9e}$$

is the "Overlap integral." It is the scalar product of the two vector functions. If the functions were orthogonal, then $\Delta = 0$, which is not the case here. For a normalized but nonorthogonal basis set, the system of equations [equation (13.80)] becomes

$$\sum_{j=1}^{N} (a_{ij} - \alpha \Delta_{ij}) u_j = 0 \quad (i = 1, 2, \; \ldots, N), \tag{17.10}$$

where Δ_{ij} is the overlap between the ith and jth basis functions rather than the Kronecker delta, δ_{ij}. Then the determinant of the coefficients of the vector representatives u_j [equation (13.82)] becomes

$$\begin{vmatrix} (a_{11} - \alpha) & (a_{12} - \Delta_{12}\alpha) & (a_{13} - \Delta_{13}\alpha) & \cdots \\ (a_{21} - \Delta_{21}\alpha) & (a_{22} - \alpha) & (a_{23} - \Delta_{23}\alpha) & \cdots \\ (a_{31} - \Delta_{31}\alpha) & (a_{32} - \Delta_{32}\alpha) & (a_{33} - \alpha) & \\ \vdots & & & \ddots \end{vmatrix} = 0. \tag{17.11}$$

Δ_{ii} is omitted from the diagonal terms because the basis functions are normalized; therefore, Δ_{ii} is 1.

For the 2×2 problem under consideration here, the secular determinant corresponding to the Hamiltonian matrix is

$$\begin{vmatrix} H_{\mathrm{AA}} - E & H_{\mathrm{AB}} - \Delta E \\ H_{\mathrm{BA}} - \Delta E & H_{\mathrm{BB}} - E \end{vmatrix} = 0, \tag{17.12}$$

where the roots E are the desired energy eigenvalues. The eigenvalues and eigenfunctions are

$$E_{\mathrm{S}} = \frac{H_{\mathrm{AA}} + H_{\mathrm{AB}}}{1 + \Delta}$$

$$E_{\mathrm{A}} = \frac{H_{\mathrm{AA}} - H_{\mathrm{AB}}}{1 - \Delta}$$

$$\psi_{\mathrm{S}} = \frac{1}{\sqrt{2 + 2\Delta}} (U_{1s_{\mathrm{A}}} + U_{1s_{\mathrm{B}}})$$

$$\psi_{\rm A} = \frac{1}{\sqrt{2-2\Delta}}(U_{1s_{\rm A}} - U_{1s_{\rm B}}). \tag{17.13}$$

Except for the appearance of Δ in the denominators, the results are the standard solutions for a 2×2 matrix. The wavefunction composed of the positive combination of the basis functions and the associated energy have been labeled with the subscript S, and the negative combination and the associated energy have been labeled with the subscript A. In the hydrogen molecule treated below, similar functions occur. In the hydrogen molecule, the S and A reflect the permutation symmetry of the two electrons (see Chapter 16, Section C) and stand for symmetric and antisymmetric. In the H_2^+ molecule, there is only one electron, but the designations S and A have been used in analogy to the functions in H_2 and other many-electron molecules.

To find the energies $E_{\rm S}$ and $E_{\rm A}$, the matrix elements, equations (17.9a) and (17.9b), and the overlap integral, equation (17.9e), must be evaluated. Calculation of the matrix elements can be simplified. From equation (17.6), the Hamiltonian is

$$\underline{H} = -\frac{\hbar^2}{2m_0}\nabla^2 - \frac{e^2}{4\pi\varepsilon_0 r_{\rm A}} - \frac{e^2}{4\pi\varepsilon_0 r_{\rm B}} + \frac{e^2}{4\pi\varepsilon_0 r_{\rm AB}}. \tag{17.14}$$

The first two terms of $\underline{H}$ are the Hamiltonian for a hydrogen atom about proton A; that is, they are the electron kinetic energy operator and the attraction of the electron to the proton at A. $U_{1s_{\rm A}}$, which is a $1s$ orbital for hydrogen atom A, is an eigenfunction of the first two terms of $\underline{H}$

$$\left(-\frac{\hbar^2}{2m_0}\nabla^2 - \frac{e^2}{4\pi\varepsilon_0 r_{\rm A}}\right)U_{1s_{\rm A}} = E_{\rm H}U_{1s_{\rm A}}, \tag{17.15}$$

where $E_{\rm H}$ is the energy of the hydrogen $1s$ state. Since the matrix elements involve $\underline{H}$ operating on $U_{1s_{\rm A}}$, the first two terms in the Hamiltonian can be replaced with $E_{\rm H}$.

The diagonal matrix elements are

$$H_{\rm AA} = \int U_{1s_{\rm A}}^* \left(E_{\rm H} - \frac{e^2}{4\pi\varepsilon_0 r_{\rm B}} + \frac{e^2}{4\pi\varepsilon_0 r_{\rm AB}}\right) U_{1s_{\rm A}}\, d\tau, \tag{17.16}$$

$$H_{\rm AA} = E_{\rm H} + J + \frac{e^2}{4\pi\varepsilon_0 a_0 D}, \tag{17.17}$$

with

$$J = \int U_{1s_{\rm A}}^* \left(-\frac{e^2}{4\pi\varepsilon_0 r_{\rm B}}\right) U_{1s_{\rm A}}\, d\tau, \tag{17.18a}$$

$$J = \frac{e^2}{4\pi\varepsilon_0 a_0}\left[-\frac{1}{D} + e^{-2D}\left(1 + \frac{1}{D}\right)\right], \tag{17.18b}$$

and

$$D = \frac{r_{\rm AB}}{a_0}. \tag{17.19}$$

D is the internuclear separation in units of a_0, the Bohr radius. In evaluating the matrix element, equation (17.16), the first and last terms in brackets are independent of the coordinate of the electron, so they factor out of the integral. For these terms, the integral reduces to the normalization integral. Since U_{1s_A} is normalized, the integrals yield 1.

The off-diagonal matrix elements, as well as the overlap integral Δ, are evaluated in a similar manner:

$$\begin{aligned} H_{\mathrm{BA}} &= \int U^*_{1s_{\mathrm{B}}} \left(E_{\mathrm{H}} - \frac{e^2}{4\pi\varepsilon_0 r_{\mathrm{B}}} + \frac{e^2}{4\pi\varepsilon_0 r_{\mathrm{AB}}} \right) U_{1s_{\mathrm{A}}} d\tau \\ &= \Delta E_{\mathrm{H}} + K + \frac{\Delta e^2}{4\pi\varepsilon_0 a_0 D} \end{aligned} \tag{17.20}$$

$$K = \int U^*_{1s_{\mathrm{B}}} \left(-\frac{e^2}{4\pi\varepsilon_0 r_{\mathrm{B}}} \right) U_{1s_{\mathrm{A}}} d\tau \tag{17.21a}$$

$$K = -\frac{e^2}{4\pi\varepsilon_0 a_0} e^{-D}(1 + D) \tag{17.21b}$$

and

$$\Delta = e^{-D}\left(1 + D + {}^1\!/_3 D^2\right). \tag{17.22}$$

J is a Coulomb integral. It represents the Coulomb interaction of an electron in a $1s$ orbital around nucleus A with the other nucleus at B. The J integral has a classical interpretation. It is the interaction of a negatively charged, spherically symmetric, charge distribution centered at A with a positive, point charge at B.

K is an exchange integral. It does not have a classical counterpart. It arises because the basis functions, $U_{1s_{\mathrm{A}}}$ and $U_{1s_{\mathrm{B}}}$, are not eigenfunctions of the Hamiltonian. Examining equation (17.21a), it can be seen that the interaction couples the state of the system that has the electron around nucleus A to the state of the system in which the electron is around nucleus B. The K integral represents an exchange or resonance energy of the electron between the two nuclei. Note that K is intrinsically negative.

In terms of the above,

$$\begin{aligned} E_{\mathrm{S}} &= E_{\mathrm{H}} + \frac{e^2}{4\pi\varepsilon_0 a_0 D} + \frac{J + K}{1 + \Delta} \\ E_{\mathrm{A}} &= E_{\mathrm{H}} + \frac{e^2}{4\pi\varepsilon_0 a_0 D} + \frac{J - K}{1 - \Delta}. \end{aligned} \tag{17.23}$$

In addition to these energy eigenvalues, it is informative to also calculate what is, in essence, the classical mechanics description of the hydrogen molecule ion. Classically, the molecule would be bound by electrostatic interactions—that is, the interaction of a hydrogen atom with a hydrogen ion separated at distance r_{AB}. This interaction energy is the diagonal matrix element, H_{AA}, which is the Coulomb interaction with no quantum mechanical exchange interaction (i.e., the electron remains attached to nucleus A). The classical energy, E_{N}, is

$$E_{\mathrm{N}} = H_{\mathrm{AA}} = E_{\mathrm{H}} + \frac{e^2}{4\pi\varepsilon_0 a_0} e^{-2D}\left(1 + \frac{1}{D}\right). \tag{17.24}$$

Figure 17.2 is a plot of the three functions E_{S}, E_{A}, and E_{N} versus D, the internuclear separation in units of a_0, the Bohr radius, where $D = r_{\mathrm{AB}}/a_0$. At very large separation (no interactions), all three curves have the same value, the energy of a hydrogen atom

$$E_H = -1.0\frac{e^2}{8\pi\varepsilon_0 a_0}.$$

As the separation is reduced, E_{A} rapidly increases in energy. E_{A} is the energy of the antibonding molecular orbital ψ_{A} [equation (17.13)]. The antisymmetric combination of the basis $1s$ functions is a state in which the particles are repulsive even at large distances. The classical mechanics energy, E_{N}, also increases rapidly as the internuclear separation is reduced, but the particles can be brought closer together before the increase begins. The classical mechanics description does not lead to the formation of a stable bond. In contrast to E_{A} and E_{N}, E_{S} decreases as the particles are brought closer together. The energy goes through a deep minimum, and then begins to increase again as the nuclei are brought very close together. The appearance of the minimum means that a stable covalent bond is formed. The separation, r_{AB}, at which the minimum occurs is the equilibrium bond length, and the depth of the minimum is the bond strength, the amount of energy required to separate the particles (break the bond). E_{S} is the energy of the bonding molecular orbital, ψ_{S} [equation

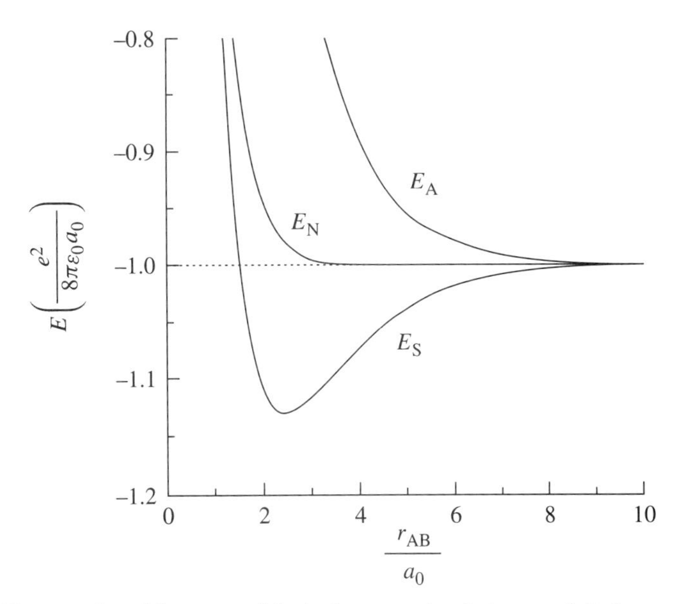

Figure 17.2 The energies of the states of the hydrogen molecule ion are plotted versus the internuclear separation, $D = r_{\mathrm{AB}}/a_0$. E_{S} is the energy associated with the symmetric combination of atomic $1s$ orbitals, and E_{A} is the energy of the antisymmetric combination. E_{N} is the classical mechanics energy. The minimum in the E_{S} curve shows that a stable bond is formed, and the location of the minimum gives the bond length.

(17.13)].

The source of the bonding interaction can be determined by examining equations (17.23) and equation (17.24). In Figure 17.2, the difference between the curve $E_{\rm N}$ and the curves $E_{\rm A}$ and $E_{\rm S}$ is the neglect of the exchange or resonance of the electron between the two nuclei in the classical treatment. The energy of the bond is mainly determined by the resonance energy (exchange interaction). (The are other small contributions not considered in this treatment such as the polarization of the atom in the field of the ion.) The curves in Figure 17.2 show that the resonance interaction becomes significant at much greater distances than the Coulomb interaction of the atom and the ion (curve $E_{\rm N}$). The reason for this is that $H_{\rm AA}$ contains the factor e^{-2D} while the resonance integral K contains the factor e^{-D}. For distances larger than $\sim$2 Å, E_S and E_A are quite accurately given by

$$\begin{aligned} E_{\rm S} &\cong E_{\rm H} + K \\ E_{\rm A} &\cong E_{\rm H} - K \end{aligned} \tag{17.25}$$

with K intrinsically negative. The bonding molecular orbital decreases in energy, and the antibonding molecular orbital increases in energy because of the exchange interaction. The first two terms in equation (17.23) for $E_{\rm S}$ and $E_{\rm A}$ are identical. The J in the third term is the same for $E_{\rm S}$ and $E_{\rm A}$. The differences in the expressions are in the signs of K and Δ. Both of these arise because the basis functions, in which the electron is either localized around A or localized around B, are not the correct description of the system. The eigenkets [equations (17.13)] are superpositions in which the electron is shared or is exchanged between the two atomic centers. Therefore, the quantum mechanical exchange interaction, not the classical Coulomb interaction, is responsible for the covalent bond.

The analytical treatment of H_2^+ explicates the nature of the covalent bond and also provides values for the bond dissociation energy and the equilibrium bond length. The bond dissociation energy, D_e, is the difference in energy between the bottom of the well (actually the vibrational zero point energy) and the energy of the isolated hydrogen atom. From the calculation, $D_e = 1.77$ eV. The calculated equilibrium bond length is $\overline{r_{\rm AB}} = 1.32$ Å. These calculated values can be compared to the experimental values, $D_e = 2.78$ eV and $\overline{r_{\rm AB}} = 1.06$ Å. In spite of its simplicity, the treatment provides reasonable values for H_2^+ bond length and bond strength. If the approach is augmented by the simplest variational calculation (see Chapter 10, Section B), using the nuclear charge Z as the minimization parameter, the calculated values become $D_e = 2.25$ eV and $\overline{r_{\rm AB}} = 1.06$ Å. More sophisticated methods can produce the experimentally measured parameters. In all cases, the main source of bonding comes from the exchange interaction.

C. THE HYDROGEN MOLECULE

The hydrogen molecule ion discussed in Section B is atypical because it has only one electron. In this section, the hydrogen molecule will be treated using the same approach that was applied to H_2^+.

Figure 17.3 is a diagram showing the labeling of the distance parameters. Hydrogen nuclei (protons) A and B are separated by the distance, $r_{\rm AB}$. The electrons are labeled 1 and

2. The distance from electron 1 to nucleus A is r_{A1}, and so on. The distance between the two electrons is r_{12}.

In the Born–Oppenheimer approximation, the electronic wave equation is

$$\nabla_1^2\psi + \nabla_2^2\psi + \frac{2m_0}{\hbar^2}\left[E + \frac{e^2}{4\pi\varepsilon_0 r_{A1}} + \frac{e^2}{4\pi\varepsilon_0 r_{B1}} + \frac{e^2}{4\pi\varepsilon_0 r_{A2}} + \frac{e^2}{4\pi\varepsilon_0 r_{B2}} - \frac{e^2}{4\pi\varepsilon_0 r_{12}} - \frac{e^2}{4\pi\varepsilon_0 r_{AB}}\right]\psi = 0. \quad (17.26)$$

The first two terms are the kinetic energy operators (multiplied by $-2m_0/\hbar^2$) for electron 1 and 2. E is the energy eigenvalue. Terms 2–5 inside the square brackets are the Coulombic attractions of electrons, 1 and 2, to the nuclei, A and B. The sixth term is the electron–electron repulsion, and the last term is the nucleus–nucleus repulsion.

For very large internuclear separation, r_{AB}, the system will consist of two noninteracting hydrogen atoms. The system wavefunctions for very large separation are the products of two $1s$ hydrogen wavefunctions, U_{1s}. The system wavefunctions at large separation suggests the following normalized basis functions for the approximate solution to the H_2 electronic energy problem.

$$\begin{aligned}\psi_I &= U_{1s_A}(1)U_{1s_B}(2)\\ \psi_{II} &= U_{1s_A}(2)U_{1s_B}(1)\end{aligned} \quad (17.27)$$

ψ_I has electron 1 in a $1s$ orbital about nucleus A and electron 2 in a $1s$ orbital about nucleus B. ψ_{II} has electron 2 in a $1s$ orbital about nucleus A and electron 1 in a $1s$ orbital about nucleus B. These two states are degenerate. When the nuclei are widely separated, the energy of the system is $2E_H$, the ground-state energy of two hydrogen atoms.

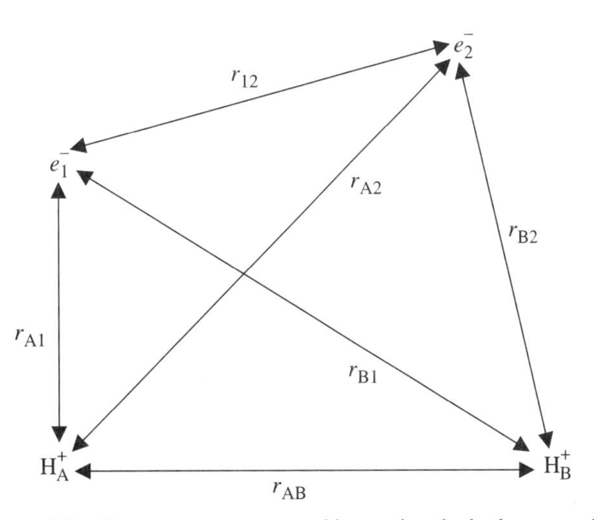

Figure 17.3 Diagram of the distance parameters used in treating the hydrogen molecule problem.

In terms of these basis functions, the Hamiltonian matrix is

$$\underline{\underline{H}} = \begin{pmatrix} H_{\mathrm{I\,I}} & H_{\mathrm{I\,II}} \\ H_{\mathrm{II\,I}} & H_{\mathrm{II\,II}} \end{pmatrix}. \tag{17.28}$$

The terms in the matrix are

$$\begin{aligned} H_{\mathrm{I\,I}} &= \iint \psi_{\mathrm{I}} \underline{H} \psi_{\mathrm{I}} d\tau_1 d\tau_2 \\ H_{\mathrm{I\,II}} &= \iint \psi_{\mathrm{I}} \underline{H} \psi_{\mathrm{II}} d\tau_1 d\tau_2 \\ \Delta^2 &= \iint \psi_{\mathrm{I}} \psi_{\mathrm{II}} d\tau_1 d\tau_2, \end{aligned} \tag{17.29}$$

with $H_{\mathrm{I\,I}} = H_{\mathrm{II\,II}}$ and $H_{\mathrm{I\,II}} = H_{\mathrm{II\,I}}$. As discussed in Section B, the basis functions are normalized but not orthogonal. Equation (17.30) is the "overlap integral."

$$\Delta^2 = \iint \psi_{\mathrm{I}} \psi_{\mathrm{II}} d\tau_1 d\tau_2. \tag{17.30}$$

It is the scalar product of the two vector functions. Since the functions are not orthogonal, $\Delta^2 \neq 0$. Δ^2 is the square of the overlap integral, Δ, that occurred in the H_2^+ problem in Section B because there are now two electrons rather than one. As discussed in Section B, the nonorthogonality of the basis functions leads to additional factors of $\Delta^2 E$ in the secular determinant.

The secular determinant is

$$\begin{vmatrix} H_{\mathrm{I\,I}} - E & H_{\mathrm{I\,II}} - \Delta^2 E \\ H_{\mathrm{II\,I}} - \Delta^2 E & H_{\mathrm{II\,II}} - E \end{vmatrix} = 0, \tag{17.31}$$

where the E are the energy eigenvalues. Expanding the determinant and solving for the two energy eigenvalues and eigenfunctions gives

$$\begin{aligned} E_{\mathrm{S}} &= \frac{H_{\mathrm{I\,I}} + H_{\mathrm{I\,II}}}{1 + \Delta^2} \\ E_{\mathrm{A}} &= \frac{H_{\mathrm{I\,I}} - H_{\mathrm{I\,II}}}{1 - \Delta^2} \\ \psi_{\mathrm{S}} &= \frac{1}{\sqrt{2 + 2\Delta^2}} \left[U_{1s_{\mathrm{A}}}(1)\, U_{1s_{\mathrm{B}}}(2) + U_{1s_{\mathrm{B}}}(1)\, U_{1s_{\mathrm{A}}}(2) \right] \\ \psi_{\mathrm{A}} &= \frac{1}{\sqrt{2 - 2\Delta^2}} \left[U_{1s_{\mathrm{A}}}(1)\, U_{1s_{\mathrm{B}}}(2) - U_{1s_{\mathrm{B}}}(1)\, U_{1s_{\mathrm{A}}}(2) \right]. \end{aligned} \tag{17.32}$$

The S stands for the symmetric combination of the basis functions, and the A stands for the antisymmetric combination. As discussed in Chapter 16, Section C, the wavefunctions are eigenfunctions of the permutation operator, a necessary requirement for all many electron wavefunctions.

The energies, E_{S} and E_{A}, can be evaluated in a manner analogous to that used for the hydrogen molecule ion. First, terms in the Hamiltonian are found that correspond to two hydrogen atom Schrödinger equations, one for a hydrogen atom centered about nucleus A and one for a hydrogen atom centered about nucleus B. The basis functions are eigenfunctions of these terms and give the energies of two hydrogen atoms. Therefore, these terms in $\underline{H}$ can be replaced by $2E_{\mathrm{H}}$. The remaining terms can be evaluated analytically.

$$H_{\mathrm{I\,I}} = 2E_H + 2J + J' + \frac{e^2}{4\pi\varepsilon_0 \mathrm{a}_0 \mathrm{D}}, \tag{17.33}$$

where J is the same as in the H_2^+ problem [equation (17.18b)], and

$$J' = \frac{e^2}{4\pi\varepsilon_0 a_0}\left[\frac{1}{D} - e^{-2D}\left(\frac{1}{D} + \frac{11}{8} + \frac{3}{4}D + \frac{1}{6}D^2\right)\right]. \tag{17.34}$$

D is the same as before [equation (17.19)].

$$H_{\mathrm{I\,II}} = 2\Delta^2 E_{\mathrm{H}} + 2\Delta K + K' + \Delta^2 \frac{e^2}{4\pi\varepsilon_0 \mathrm{a}_0 \mathrm{D}}, \tag{17.35}$$

where K and Δ are the same as in the H_2^+ problem, equations (17.21) and (17.22) respectively, and

$$\begin{aligned} K' = \frac{e^2}{20\pi\varepsilon_0 a_0}&\left[-e^{-2D}\left(-\frac{25}{8} + \frac{23}{4}D + 3D^2 + \frac{1}{3}D^3\right)\right.\\ &\left.+\frac{6}{D}\left[\Delta^2(\gamma + \ln D) + \Delta'^2 Ei(-4D) - 2\Delta\Delta' Ei(-2D)\right]\right]. \end{aligned} \tag{17.36}$$

γ is Euler's constant (0.5772 . . .), and

$$\Delta' = e^D\left(1 - D + \frac{1}{3}D^2\right). \tag{17.37}$$

Ei is the function, the integral logarithm. Ei is related to the exponential integral, E_1 by

$$E_1(x) = -\mathrm{Ei}(-x).$$

Values of $E_1(x)$ are tabulated. For the range of the variables of interest in equation (17.36), a very good approximation is

$$E_1(x) = \left(\frac{x^2 + a_1 x + a_2}{x^2 + b_1 x + b_2}\right)\frac{e^{-x}}{x}, \tag{17.38}$$

$1 \le x < \infty$, with

$$a_1 = 2.334733, \quad a_2 = 0.2500621, \quad b_1 = 3.330657, \quad b_2 = 1.681534.$$

J and K have same meanings as in the H_2^+ problem. J' is a Coulomb integral describing the interaction of an electron in a 1s orbital on $\mathrm{H_A}$ with an electron in a 1s orbital on $\mathrm{H_B}$. J' has a classical interpretation. It is the Coulomb interaction of a negatively charged, spherically symmetric, charge distribution centered at A with an identical negatively charged, spherically symmetric, charge distribution centered at B. K' is the corresponding exchange or resonance integral.

The final expressions for the energies of the H_2 molecular orbitals are

$$E_S = 2E_H + \frac{e^2}{4\pi\varepsilon_0 a_0 D} + \frac{2J + J' + 2\Delta K + K'}{1 + \Delta^2} \tag{17.39a}$$

$$E_A = 2E_H + \frac{e^2}{4\pi\varepsilon_0 a_0 D} + \frac{2J + J' - 2\Delta K - K'}{1 - \Delta^2}. \tag{17.39b}$$

As in the H_2^+ problem, it is informative also to calculate the classical mechanics energy of two interacting hydrogen atoms. The classical energy involves the Coulomb interactions of a hydrogen atom at A with a hydrogen atom at B, but without the exchange interactions. The classical energy, E_N, is

$$E_N = H_{I\,I}, \tag{17.40}$$

that is, the energy of the basis function, $\psi_I = U_{1s_A}(1)\, U_{1s_B}(2)$. Figure 17.4 is a plot of the three functions, E_S, E_A, and E_N versus D, the internuclear separation in units of a_0, the Bohr radius, $D = r_{AB}/a_0$.

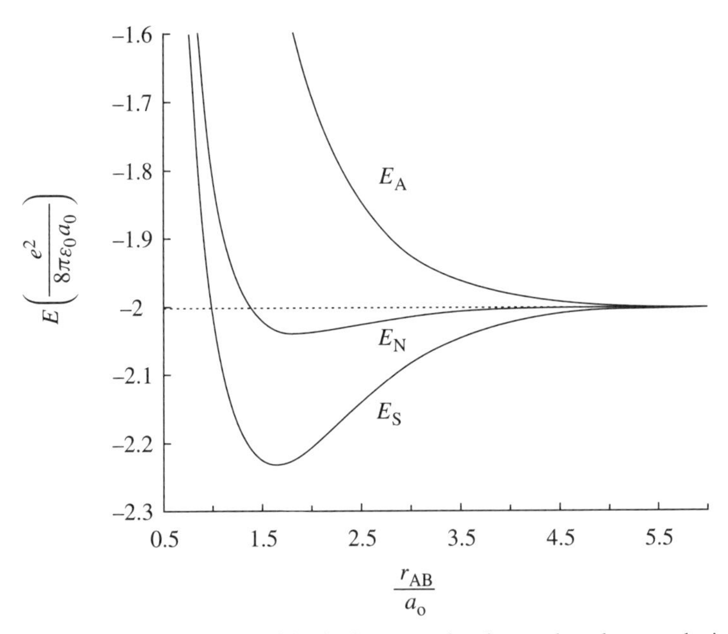

Figure 17.4 The energies of the states of the hydrogen molecule are plotted versus the internuclear separation, $D = r_{AB}/a_0$. E_S is the energy associated with the symmetric combination of atomic 1s orbitals, E_A is the energy of the antisymmetric combination, and E_N is the classical mechanics energy. The minimum in the E_S curve shows that a stable bond is formed, and the location of the minimum gives the bond length.

At very large separation (no interactions), all three curves have the same value, the energy of two hydrogen atoms, $2E_{\mathrm{H}}$. As the separation is reduced, E_{A} rapidly increases in energy. E_{A} is the energy of the antibonding molecular orbital ψ_{A} [equation (17.32)]. The antisymmetric combination of the basis $1s$ functions is a state in which the particles are repulsive even at large distances. In contrast to H_2^+ (Figure 17.2), the classical mechanics energy, E_{N}, shows a small broad minimum. This minimum is not nearly deep enough to account for the experimental strength of the H_2 bond. It is the weak attraction that arises between nonbonded atoms or molecules from van der Waals interactions. E_{S} decreases as the particles are brought closer together. The energy goes through a deep minimum, and then it begins to increase again as the nuclei are brought very close together. The appearance of the minimum means that a stable covalent bond is formed. The separation, r_{AB}, at which the minimum occurs is the equilibrium bond length, and the depth of the minimum is the bond strength. E_{S} is the energy of the bonding molecular orbital, ψ_{S} [equation (17.32)].

The nature of the curves in Figure 17.4 shows that the covalent bond is a consequence of the nonclassical exchange interactions that reflect the sharing of the electron between the two atomic centers. The curve E_{N} accounts for the Coulomb interactions. Coulomb interactions alone do not produce a minimum in the energy curve that can account for the strength of the H_2 bond. The exchange interactions, reflected in the off-diagonal matrix elements, are a strictly quantum mechanical effect. The basis functions, in which each electron remains localized about its own atomic center, are not the eigenfunctions. The molecular orbitals, which are the symmetric and antisymmetric combinations of the basis functions, are the eigenfunctions. Thus, the covalent bond is a quantum mechanical phenomenon.

The calculated results provide values for the bond dissociation energy and the equilibrium bond length. The calculated bond dissociation energy is $D_e = 3.15$ eV, and the calculated equilibrium bond length is $\overline{r_{\mathrm{AB}}} = 0.80$ Å. These calculated values can be compared to the experimental values, $D_e = 4.72$ eV and $\overline{r_{\mathrm{AB}}} = 0.74$ Å. If a variational treatment using an effective nuclear charge, $Z'e$, in the wavefunction Ψ_s [equation (17.32)] is employed, values of $D_e = 3.76$ eV and $\overline{r_{\mathrm{AB}}} = 0.76$ Å are obtained.

Another property that can be calculated is the H_2 vibrational frequency. One approach is to substitute the potential function, equation (17.39a), into the nuclear wave equation, equation (17.5), and solve for the vibrational eigenvalues. A simple approximate approach is to fit a parabola to the potential function (E_s in Figure 17.4) near its minimum and determine the vibrational energies for a harmonic oscillator from the parabolic potential obtained from the fit. Fitting the potential from $D = 1.3$ to 2.2 by varying the location of the minimum, varying the harmonic oscillator force constant (see Chapter 6), but fixing the depth of the minimum, yields a vibrational frequency of $\nu = \sim 3400\ \mathrm{cm}^{-1}$. The measured value is $\nu = 4318\ \mathrm{cm}^{-1}$. The calculated value is ~80% of the measure value.

The calculated values for the various properties of the H_2 molecule have nonnegligible errors; nonetheless, the simple, analytical treatment gives reasonable values. The calculated values can be improved by extensive molecular structure calculations. For the H_2 molecule, essentially exact results can be obtained using numerical methods. However, the calculations as presented above demonstrate the important point. Covalent molecular bonding arises predominantly from nonclassical exchange interactions.

The eigenfunction of the Hamiltonian operator in equation (17.32) are also eigenfunctions of the permutation operator. As discussed in Chapter 16, Section C, all total wavefunctions, spatial $\times$ spin, must be totally antisymmetric with respect to interchange of electron labels (coordinates). The bonding molecular orbital, ψ_S, is symmetric. Therefore, its spin wavefunction must be the antisymmetric combination of the two $s = 1/2$ electron spin wavefunctions. This corresponds to a singlet state; the electron spins are paired. The antibonding molecular orbital, ψ_A, is antisymmetric. Therefore, the spin function must be symmetric. The symmetric combination has $s = 1$ and is a triplet state. Therefore, there are three antibonding molecular orbitals that differ by the value of m_s, which can equal 1, 0, or -1.

Problems

CHAPTER 2

1. A Hermitian operator can be defined by the following relationship:

$$\left\langle a \left| \underline{\gamma} \right| b \right\rangle = \left\langle a \left| \overline{\underline{\gamma}} \right| b \right\rangle \equiv \overline{\left\langle b \left| \underline{\gamma} \right| a \right\rangle}.$$

This, in turn, gives:

$$\langle a \left| \gamma \right| a \rangle = \left\langle a \left| \overline{\underline{\gamma}} \right| a \right\rangle \equiv \overline{\left\langle a \left| \underline{\gamma} \right| a \right\rangle}.$$

The first equation is more general than necessary.

Given: $\langle S \left| \underline{F} \right| S \rangle = \langle S \left| \overline{\underline{F}} \right| S \rangle \equiv \overline{\langle S \left| \underline{F} \right| S \rangle}$

Prove: $\langle T \left| \underline{F} \right| S \rangle = \langle T \left| \overline{\underline{F}} \right| S \rangle$

It is useful to define and use the following kets:

$$|\phi\rangle = |T + S\rangle = |T\rangle + |S\rangle$$
$$|\theta\rangle = |T + iS\rangle = |T\rangle + i\,|S\rangle$$

where $i = \sqrt{-1}$.

2. Take $\psi(x)$ to be a well-behaved function of x. Assume it vanishes at $\pm\infty$. In the Schrödinger representation of quantum mechanics, such vector functions are used as the ket vectors. (The Schrödinger representation is introduced in Chapters 3 and 4 and discussed in detail in Chapter 5.) In linear algebra, the scalar product for vector functions is defined as

$$\langle \psi(x) |\, \underline{A}(x)\, | \psi(x) \rangle = \int_{-\infty}^{\infty} \psi^*(x) A(x) \psi(x)\, dx,$$

where $\underline{A}(x)$ is an operator and $\psi^*(x)$ is the complex conjugate of $\psi(x)$. $\psi(x)$ is a ket vector and $\psi^*(x)$ is the corresponding bra.

Given the above, determine whether or not the following operators are Hermitian:

a. $i\dfrac{d}{dx}$

b. $\dfrac{d^2}{dx^2}$

c. $i\dfrac{d^2}{dx^2}$

3. Given a set of kets $|A_m\rangle$, $m = 1$ to n, which are linearly independent but not orthogonal or normalized, formulate a procedure for obtaining a new set of kets $|B_m\rangle$ which are orthonormal. (This is called the Gram–Schmidt orthonormalization procedure. The Gram–Schmidt procedure can be found in textbooks on linear algebra.)

4. Consider the set of functions:

$$f_0 = 1, \quad f_1 = x, \quad f_2 = x^2, \quad f_3 = x^3,$$

defined on the interval $x = -1$ to $x = +1$. These functions are linearly independent but not orthogonal or normalized. Construct the set of functions g_i from the f_i such that the g_i are orthonormal. Use the definition of the scalar product of two functions $A(x)$ and $B(x)$ defined on the interval γ to δ, that is,

$$\langle A(x)|B(x)\rangle = \int_\gamma^\delta A(x)B(x)\,dx.$$

5. Given a normalized ket, $|\psi\rangle$, that is a linear combination of orthonormal kets, $|A_n\rangle$,

$$|\psi\rangle = \sum_n c_n\,|A_n\rangle\,.$$

a. Show that $\displaystyle\sum_n c_n^* c_n = 1.$

b. Show that $\underline{P}_n = |A_n\rangle\langle A_n|$ is an operator.

c. Show that $\underline{P}_n^2 = \underline{P}_n.$

CHAPTER 3

1. In the optical region of the electromagnetic spectrum, a material has a wavelength-dependent index of refraction, $n(\lambda)$, given by

$$n(\lambda) = ae^{b\lambda},$$

with $a \equiv 1.84$ and $b \equiv -2.56 \times 10^3\ \text{cm}^{-1}$.

a. Find expressions for the group velocity ($V_g(k)$) of a photon (optical wave packet) in the material and for the phase velocity ($V_p(k)$) of the associated plane wave whose wave vector corresponds to the center in k-space of the wave packet.

b. Calculate $V_p(k)$ and $V_g(k)$ for $\lambda = 1.0\ \mu$m, 500 nm, and 200 nm.

2. Use the relation $\Delta x \Delta p = \hbar/2$ for a Gaussian wave packet to determine the spectral width (FWHM, i.e., full width at half-maximum height) of a $\lambda = 1.064\ \mu$m, $\Delta t =$ 100 ps Nd:YAG laser pulse in air (take $n(\lambda)$ to be equal to one at all wavelengths, even though it is actually slightly different from unity). For the purposes of this problem, assume that the value given for Δt is σ, the standard deviation of the duration of the Gaussian wave packet. You may neglect the fact that in real life, measurements are normally made at the intensity (probability) level, rather than at the electric field (probability amplitude) level. This difference would result in an extra factor of $1/\sqrt{2}$ in your answer for the FWHM of the spectrum. It is necessary to convert from σ to FWHM for the final answer.
3. A wave packet with a dispersion that does not depend linearly on k will spread as time goes on. Explain why.
4. For an object of mass m, a particularly simple result can be obtained to describe the spreading of a wave packet for reasonably long times, t, if the packet is a Gaussian, that is,

$$\Delta x = \frac{\hbar t}{2m\Delta x_0},$$

where Δx_0 is the width of the packet at $t = 0$. To see why the spreading of wave packets is not important for macroscopic objects, calculate how long (in years) it will take a 1-g, 1-cm-long object to double in size?
5. Derive the equation in 4. (There are many ways to do this.)
6. Discuss the possibility of compensating for the spreading of an optical wave packet in a dispersive system by having the wave packet travel in one dispersive medium and continue into another.

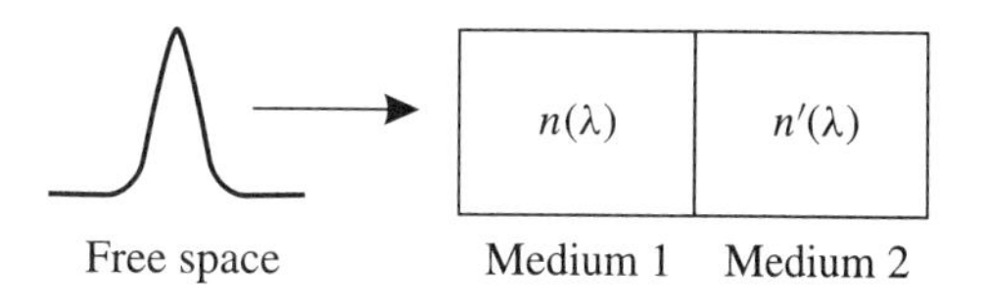

CHAPTER 4

1. Given the definition of the classical Poisson bracket and that the classical Hamiltonian is

$$H = \frac{p^2}{2m} + V(x),$$

verify the classical equations of motion and $\dot{x} = \{x, H\}$ and $\dot{p} = \{p, H\}$, and prove $\dot{f} = \{f, H\}$, where $f = f(x, p)$ is any function of x and p.
2. a. Prove $[\underline{x}^n, \underline{p}] = i\hbar n \underline{x}^{n-1}$, where n is any positive integer.

   b. Prove $[\underline{f}(x), \underline{p}] = i\hbar \frac{\partial f}{\partial x}$, where $f(x)$ is any polynomial in x.

   c. Prove $[\underline{x}, \underline{p}^n] = i\hbar n \underline{p}^{n-1}$.

d. Prove $[\underline{x}, \underline{f}(p)] = i\hbar\frac{\partial f(p)}{\partial p}$, where $\underline{f}(p)$ is any polynomial in p.

3. In the derivation of the uncertainty principle, the following commutator was the starting point,

$$[\underline{A}, \underline{B}] = i\underline{C}$$

with $\underline{A}$ and $\underline{B}$ Hermitian. Taking the operator $\underline{D} = \underline{A} + \alpha\underline{B} + i\beta\underline{B}$ with α and β real numbers, derive the inequality:

$$\langle S|\,\overline{\underline{D}}\,\underline{D}\,|S\rangle = \langle \underline{A}^2\rangle + (\alpha^2 + \beta^2)\langle \underline{B}^2\rangle + \alpha\langle \underline{C}'\rangle - \beta\langle \underline{C}\rangle \geq 0,$$

where $\underline{C}'$ is the anticommutator of $\underline{A}$ and $\underline{B}$.

4. Show that the following set of operators, which define the "momentum representation," obeys Dirac's Quantum Condition.

$$p \to \underline{p} = p$$

$$x \to \underline{x} = +i\hbar\frac{\partial}{\partial p}.$$

5. Discuss the difference between the expectation value of an operator and the eigenvalue of an operator.

CHAPTER 5

1. Consider the eigenvalue problem $\underline{H}\,|U\rangle = E\,|U\rangle$. In addition to satisfying the eigenvalue equation, in the Schrödinger representation a specific function U must be well-behaved in order to be an acceptable eigenfunction. Born defined "well-behaved" as a function that has the following properties:

   1. Single-valued
   2. Finite everywhere
   3. Continuous everywhere
   4. Its first derivative is continuous everywhere.

   Which of the following are well-behaved? For those which are not, indicate why they are not.

   a. $u = \begin{cases} x, & x \geq 0 \\ 0, & x < 0 \end{cases}$

   b. $u = x^2$

   c. $u = e^{|x|}$

   d. $u = e^{-x}$

   e. $u = e^{-x^2}$

   f. $u = \sin\,|x|$

   g. $u = \cos\,x$

2. The operator that represents the time derivative of a quantum mechanical dynamical

variable is $\underline{\dot{A}} = (i/\hbar)[\underline{H}, \underline{A}]$. Using this relationship show for the operators $\underline{p}$ and $\underline{x}$ that

$$\underline{\dot{p}} = -\frac{\partial \underline{V}}{\partial x}$$

and

$$\underline{\dot{x}} = \frac{\underline{p}}{m},$$

where $\underline{H} = \frac{\underline{p}^2}{2m} + \underline{V}(x)$.

3. In addition to the energy eigenvalues, other properties of a particle in a one-dimensional box with infinite walls can be calculated. It is interesting to examine how fast a particle in a box is moving. Calculate the square root of the expectation value of the $\underline{V}^2$ for the lowest state of the particle in a box, where $\underline{V}$ is the velocity operator; that is, $\sqrt{\langle\varphi_1(x)\left|\underline{V}^2\right|\varphi_1(x)\rangle}$. The root-mean-square velocity is calculated because the expectation value of the velocity is zero since the particle will move equally in the positive and negative directions. Equation (5.36) is the Hamiltonian for a free particle. For a free particle, $\underline{H}$ is only the kinetic energy, $mV^2/2$. Therefore, equation (5.36) can be used to obtain the form of the operator $\underline{V}^2$. Using the general result, determine the root-mean-square velocity for an electron in an anthracene size box (6 Å). Note that the velocity is well below velocities for which relativistic considerations are important.

4. Find the energy eigenvalues and the eigenfunctions for a particle in a two-dimensional rectangular box with infinite walls (potential is infinite outside the box). The box is centered at $x = 0$, $y = 0$. Along the x axis, the box extends from $-b$ to b; and along the y axis, the box extends from $-a$ to a. (Note that this two-dimensional problem can be separated into two one-dimensional problems.)

5. Using the eigenfunctions obtained in problem 4, for the lowest energy state of the particle in a two-dimensional box rectangular box, determine the probability of finding the particle in the rectangle with corners (0, 0), (b, 0), (b, a), and (0, a).

CHAPTER 6

1. In the Schrödinger representation, the general form of the harmonic oscillator wave function is

$$\Psi_n(x) = N_n e^{-(\gamma^2/2)} H_n(\gamma),$$

where $\gamma = \alpha^{1/2}x$, $\alpha = 2\pi m\nu_0/\hbar$, and $H_n(\gamma)$ is the nth Hermite polynomial. Using the generating functions $S(\gamma, t)$ and $T(\gamma, t)$ given below [equations (6.38) and (6.41)], show that the set of all $|\Psi_n(x)\rangle$ are orthogonal, and determine the value of the normalization constant, N_n [equation (6.34)].

$$S(\gamma, s) = \sum_n \frac{H_n(\gamma)}{n!} s^n = e^{\gamma^2-(2-\gamma)^2}$$

$$T(\gamma, t) = \sum_n \frac{H_m(\gamma)}{m!} t^m = e^{\gamma^2-(t-\gamma)^2}.$$

Note: The procedure is outlined in Chapter 6.

2. The probability of a spectral transition between two harmonic oscillator states represented by the eigenfunctions $|\Psi_m\rangle$ and $|\Psi_n\rangle$ is, in part, determined by the bracket

$$\left\langle \Psi_n \left| \underline{x} \right| \Psi_m \right\rangle,$$

where the $\underline{x}$ operator represents (in part) the interaction of the harmonic oscillator with x polarized light. (The problem of absorption and emission of light by atoms and molecules is treated in detail in Chapter 12.) Using the procedure analogous to that used in problem 1, evaluate this bracket for all values of m and n. (The same basic problem was worked using raising and lowering operators to obtain equation (6.96). Calculating the bracket with different methods can involve different mathematical operations. However, the method will not change the final result. The transition probability depends on the absolute value squared of the bracket.)

3. The probability of a molecule undergoing an electronic transition depends, in part, on a vibrational overlap factor called the Franck–Condon factor. If the potential energy surfaces of the two electronic states are identical, a transition can only be made from a vibrational state of the ground electronic state to a vibrational state of the excited electronic state with the same vibrational quantum number. In this case, the Franck–Condon factor is unity. However, the potential energy surfaces of the two electronic states are, in general, not the same. Therefore, the vibrational potentials, which are actually anharmonic, are not identical. The result is that the vibrational wavefunctions of the ground electronic state are not orthogonal to the vibrational wavefunctions of the excited electronic state. This makes it possible to have electronic transitions in which the vibrational quantum number changes. The vibrational potential of an excited electronic state can be different from that of the ground electronic state in two ways: (1) The force constant can change, or (2) the equilibrium position can change. Either of these two effects will reduce the transition probability for the transition in which the vibrational quantum number does not change by decreasing the Franck–Condon factor below unity. For a transition from the $n = 0$ vibrational state of one electronic state Ψ_0 to the $n = 0$ vibrational state of another electronic state Ψ_0' (where the subscript denotes the vibrational state):

a. Calculate the vibrational overlap

$$\left\langle \Psi_0' | \Psi_0 \right\rangle,$$

where the vibrational frequency is ν_0 in the state $|\Psi_0\rangle$ and is ν_0' in the state $\left|\Psi_0'\right\rangle$.

b. Calculate the vibrational overlap if the equilibrium position changes from $x = 0$ in the state $|\Psi_0\rangle$ to $x = x_0$ in the state $\left|\Psi_0'\right\rangle$.

c. Demonstrate that the condition

$$\langle \Psi_0' | \Psi_0 \rangle \neq 1$$

is sufficient to ensure that for at least one excited vibrational state of the electronic excited state $|\Psi_i'\rangle$, $i \neq 0$, that

$$\langle \Psi_0 | \Psi_i' \rangle \neq 0;$$

that is, the vibrational states of the primed and unprimed electronic states are not orthogonal, and transitions can occur that do not preserve the original vibrational quantum number. These results are responsible for "vibronic progressions" in the spectroscopy of electronic states. (It is not necessary to do any integrals to solve this problem. Use an expansion in terms of a complete set of kets and assume the coefficients are real.)

4. Obtain the following commutation relations between the harmonic oscillator Hamiltonian in the Dirac representation and the lowering and raising operators, $\underline{a}$ and $\underline{a}^+$,

$$[\underline{a}, \underline{H}] = \underline{a}$$
$$[\underline{a}^+, \underline{H}] = -\underline{a}^+.$$

5. Use the raising and lowering operator formalism to calculate the uncertainty relation, $\Delta x \Delta p$, for all states of the harmonic oscillator. Recall that

$$(\Delta x)^2 = \langle x^2 \rangle - \langle x \rangle^2$$
$$(\Delta p)^2 = \langle p^2 \rangle - \langle p \rangle^2 .$$

What happens to the $\Delta x \Delta p$ product as the quantum number increases?

6. Calculate the probability of finding a harmonic oscillator outside of the classically allowed region (beyond the classical turning points) for the lowest state of the quantum harmonic oscillator. (Note that this probability is not an observable quantity. It cannot be calculated with the raising and lowering operator formalism, but it can be calculated using the Schrödinger representation. While not an observable, such calculations give valuable insights into the differences between classical and quantum mechanics.)

CHAPTER 7

1. What is the average distance of the electron from the nucleus in the hydrogen atom when the atom is in its $1s$ state and in its $2s$ state? To find the average distance, calculate the expectation value of r for each state. [Note the $1s$ and $2s$ wavefunctions are given in equations (7.68) and (7.69), and it is necessary to use the correct differential operator; see equation (10.14a).]
2. What are the probabilities of finding the electron in the region between a_0 and $2a_0$ for the $1s$ and $2s$ states of the hydrogen atom?
3. Use the generating functions to determine the radial part of the hydrogen wavefunctions for the $2p$, $3p$, and $4d$ states.

CHAPTER 8

1. In solving the degenerate time-dependent two-state problem, the energies of the two states, E_0, were set equal to zero [equation (8.7)]. Rework this problem to obtain the results in equations (8.26a) and (8.26b), but do not set $E_0 = 0$; that is, take the energy of the excited states to be E_0.

2. Obtain the solutions for the nondegenerate time-dependent two-state problem (Chapter 8, Section D) given in equations (8.38a) and (8.38b) and equations (8.40a) and (8.40b). Describe the results by considering the cases in which $\gamma/\Delta E > 1$ and $\gamma/\Delta E < 1$, with $\Delta E = |E_A - E_B|$.

3. Determine the energy levels (band structure) for the one-dimensional exciton problem (Chapter 8, Section E) with nearest-neighbor and next-nearest-neighbor interactions. The nearest-neighbor interactions have strength γ [equation (8.52)], the next-nearest-neighbor interactions have strength β. The next-nearest-neighbor interactions can be included by adding a term to the Hamiltonian [equation (8.48)] of the form $\underline{H}_{j,j\pm 2}$, with

$$\underline{H}_{j,j\pm 2}\left|\Phi_j^e\right\rangle = \beta\left|\Phi_{j+2}^e\right\rangle + \beta\left|\Phi_{j-2}^e\right\rangle.$$

CHAPTER 9

1. Consider a particle in a one-dimensional box. The box has width L and extends from 0 to L with its center at $L/2$. While the box has infinite walls, it has a parabolic bottom rather than a flat bottom. The potential inside the box has the form

$$V = \frac{K}{2}x^2.$$

The parabolic bottom can be considered a perturbation of the normal box problem. In the absence of the perturbation, the energy levels and wavefunctions for the particle in the box are

$$E_n^0 = n^2h^2/8mL^2$$
$$\psi_n^0(x) = (2/L)^{1/2}\,\sin(n\pi x/L).$$

a. Use first-order perturbation theory to calculate the correction to the E_n^0 for all n; that is, calculate the E_n' for all n. What is the n dependence for very large n?

b. Take the perturbation potential V at the box wall $x = L$ to be equal to the lowest unperturbed box energy. What value of K yields this result? Give the expression for the E_n' in terms of this K.

c. Find an expression for the ratio of the first-order correction to the zeroth-order energy for all n; that is, find E_n'/E_n^0 for all n. What is the expression for n very large?

d. Evaluate the ratio in c for $n = 1, 2, 4$, and 20 using the results of b. Why does the influence of the perturbation become negligible as n becomes large?

2. Consider a harmonic oscillator perturbed by a potential of the form bx, where b is a constant.

   a. Use perturbation theory to calculate the first- and second-order corrections to the harmonic oscillator energy levels.

   b. This problem can also be solved exactly—that is, without using perturbation theory—by changing the coordinates from x to $x' = x + b/k$. Find the energy levels by employing this transformation, and compare the exact results to those obtained in part a.

3. In Chapter 9, Section B.1, first-order perturbation theory was used to examine the influence of the addition of anharmonic terms to the Hamiltonian for the harmonic oscillator. The perturbations consisted of a cubic term and a quartic term. It was found that, to first order, the cubic term, $c\underline{x}^3$, did not change the energies of the harmonic oscillator states. Using the raising and lowering operator formalism, calculate the second-order correction [equation (9.29)] to the harmonic oscillator energy levels arising from the first-order addition to the harmonic oscillator Hamiltonian, $c\underline{x}^3$.

4. Perturbation theory gives the wavefunction for the nth state to first order as:

$$|\Psi_n\rangle = \left|\Psi_n^0\right\rangle + \lambda \sum_j{}' C_j \left|\Psi_j^0\right\rangle,$$

   where $\left|\Psi_n^0\right\rangle$ is an eigenfunction of the zeroth-order Hamiltonian. The prime in the summation indicates that the nth term is omitted. Thus, there is no additional first-order term,

$$\lambda C_n \left|\Psi_n^0\right\rangle,$$

   added to the zeroth-order wavefunction. Show that for the first-order corrected wavefunction to be normalized, that is,

$$\langle \Psi_n | \Psi_n \rangle = 1$$

   to first order, the coefficient C_n can, indeed, be taken to be zero because the real part of C_n must be zero for the corrected wavefunction to be normalized to first order, and the imaginary part of C_n, is arbitrary, and, therefore, it can be set equal to zero. (*Note:* When working at first-order, it is important to only keep terms that are zeroth or first order in λ. Higher-order terms must be dropped.)

5. In Chapter 9, Section C.3, second-order degenerate perturbation theory was used to investigate the influence of the degeneracy of the plane rotator rotational states on the Stark-effect-induced changes in the energy levels. Using equation (9.84), obtain the final results given in equations (9.87).

CHAPTER 10

1. In this problem, the lowest energy state of the lithium atom will be calculated using a variational calculation and a very simple trial function.

a. Write out the Hamiltonian operator for the lithium atom, that is, a three-electron atom [the equivalent of equation (10.40)].

b. As a trial function, form a product of three $1s$ orbitals (each electron is in a $1s$ orbital) with the nuclear charge, Z', as the variable; that is, write out the equivalent of equation (10.39). (Be sure that the normalization constant is correct.)

c. Rewrite the Hamiltonian in the form containing Z' [the equivalent of equation (10.41)].

d. Calculate the energy and minimize the result with respect to Z'. Give the answer in eV (electron volts). Note that the ground-state energy of the hydrogen atom, $E_{1s}(H)$, is -13.6 eV. (*Hint:* By examining the treatment of the He atom, it is possible to obtain the result without doing any integrals.)

e. The experimentally determined ground-state energy of Li is -203 eV. If you did part d correctly, you will actually have obtained a result lower than this number, a seeming impossibility given the variational theorem. However, the trial function, with all three electrons in $1s$ orbitals, violates the Pauli Exclusion Principle. The Pauli Exclusion Principle is discussed in detail in Chapter 16, Section C. The Pauli Principle states that no more than two electrons can occupy the same orbital in an atom or molecule, and these two must have their electron spins (see Chapter 16) opposed. The trial function treats the electrons as if they are bosons rather than fermions. Only fermions (e.g., electrons) must obey the Pauli Principle. Discuss why the trial function, which violates the Pauli Principle, seemingly violates the variational theorem.

2. To obtain a reasonable value for the ground-state energy of lithium using a variational calculation, it is necessary to use a trial function that does not violate the Pauli Principle. Here, using the results of Problem 1a, and 1c, the ground-state energy of Li is calculated.

a. As a trial function, form a product of two $1s$ orbitals and a $2s$ orbital with the nuclear charge, Z', as the variable; that is, write out the equivalent of equation (10.39). Make sure that the normalization constant (from all three pieces of the wavefunctions) is correct and that you have all of the Z's in the right places [see equation (7.66) and the definition of ρ following equation (7.63)].

b. Use this trial function to calculate the energy and minimize it with respect to Z'. Give the answer in eV. The ground-state energy of the hydrogen atom, $E_{1s}(H)$, is -13.6 eV. The answer you get should be close to the experimental result 203 eV, within a few percent, but greater than the true energy in accord with the variational theorem. [*Hint:* The only real differences between the Li calculation and the He calculation are the terms that involve the $2s$ orbital. Since the electron–electron repulsion terms containing the $2s$ orbital still only involve s orbitals, the form of the $1/R_{ij}$ used for the ground state of He will be the same in all of the electron–electron repulsion terms in the Li calculation. Thus, there will be calculations equivalent to those in equation (10.21) to (10.23) but with $1s$ and $2s$ orbitals.]

Some useful integrals are

$$\int_0^\infty x^n e^{-ax}\,dx = \frac{n!}{a^{n+1}} \qquad n \text{ a positive integer and } a > 0$$

$$\int xe^{ax}\,dx = \frac{e^{ax}}{a^2}(ax - 1)$$

$$\int x^2 e^{ax}\,dx = \frac{x^2 e^{ax}}{a} - \frac{2e^{ax}}{a^3}(ax-1).$$

CHAPTER 11

1. Consider the time-dependent Hamiltonian, $\underline{H} = \underline{H}^0 + \underline{H}'$, where $\underline{H}^0$ is the harmonic oscillator Hamiltonian and $\underline{H}'$ is given by

$$\underline{H}'(t) = \begin{cases} 0, & -\infty < t < 0 \\ Ax^4, & 0 \le t \le t' \\ 0, & t' < t. \end{cases}$$

For $t < 0$ the system is in the lowest harmonic oscillator state; that is, $C_0^* C_0 = 1$. Use time-dependent perturbation theory to calculate the probability of finding the system in all states $|m\rangle$ as a function of time. How does the probability behave for times $0 < t < t'$? How does the probability behave for times $t > t'$? In working this problem you can assume that $C_0^* C_0 \cong 1$ for all time. Under what circumstances is this a poor assumption?

2. This problem will examine the motion of an electron in a system composed of two identical metal atoms joined by a ligand. In addition to the electrons belonging to the two metal atoms, the electrons belonging to the ligand, and the electrons shared by the metal atoms and the ligand in bonding, there is one extra electron. The state of the system in which the extra electron is on the first metal atom is $|m_1\rangle$, on the ligand is $|\ell\rangle$, and on the second metal atom is $|m_2\rangle$. These kets represent (a) the states in the absence of interactions of the extra electron when it is on one atom or (b) the ligand with the states when it is on other atom or ligand; for example, $|m_1\rangle$ does not "feel" $|m_2\rangle$ or $|\ell\rangle$. (Take these kets to be orthogonal and normalized.) These kets are the time-independent, spatial parts of the total ket vector. When the electron is on either metal atom its energy is E_m and when it is on the ligand, its energy is E_ℓ; that is,

$$\underline{H}_0 |m_1\rangle = E_m |m_1\rangle$$
$$\underline{H}_0 |\ell\rangle = E_\ell |\ell\rangle$$
$$\underline{H}_0 |m_2\rangle = E_m |m_2\rangle\,,$$

where $\underline{H}_0$ is the part of the Hamiltonian which operates on the extra electron.

In reality, when the system is in the state $|m_1\rangle$ (the electron is on the first metal atom), there is an interaction with $|\ell\rangle$, and when the system is in the state $|\ell\rangle$, it will interact with $|m_1\rangle$ and $|m_2\rangle$, and so on. (Any direct interaction of $|m_1\rangle$ and $|m_2\rangle$ is neglected.) In addition to $\underline{H}_0$, there will be terms in the Hamiltonian representing these interactions. The result is that the states are coupled, and operating $\underline{H}$ on the kets yields

$$\underline{H}|m_1\rangle = E_m |m_1\rangle + \gamma |\ell\rangle$$
$$\underline{H}|\ell\rangle = E_\ell |\ell\rangle + \gamma |m_1\rangle + \gamma |m_2\rangle$$

$$\underline{H}\,|m_2\rangle = E_m\,|m_2\rangle + \gamma\,|\ell\rangle\,.$$

γ is the strength of the interaction (the off-diagonal coupling matrix element). Generally, $E_\ell > E_m$ and $\gamma < (E_\ell - E_m)$. To simplify the math in this problem, a special case is considered which obeys the above inequalities but uses the specific situation

$$(E_\ell - E_m) = \sqrt{8}\,\gamma.$$

The eigenvalues are

$$E_1 = E_m$$
$$E_2 = E_m + (2 + \sqrt{2})\gamma$$
$$E_3 = E_m - (2 - \sqrt{2})\gamma,$$

and the corresponding eigenvectors are

$$|1\rangle = 1/\sqrt{2}\;(|m_1\rangle - |m_2\rangle)$$
$$|2\rangle = 0.271\,|m_1\rangle + 0.924\,|\ell\rangle + 0.271\,|m_2\rangle$$
$$|3\rangle = 0.653\,|m_1\rangle - 0.383\,|\ell\rangle + 0.653\,|m_2\rangle$$

The decimal fractions are approximations for various ugly factors that come up such as $\sqrt{1/(8+4\sqrt{2})}$. In a situation where orthogonality is involved, the decimal fractions will yield small numbers that are not exactly zero. These should be set to zero when appropriate. Within this proviso, these kets are orthogonal and normalized. They are the spatial parts of the eigenkets. Each has associated with it a time-dependent phase factor (not shown). (The method for obtaining the eigenvalues and eigenvectors for this type of problem is developed in Chapter 13. Obtaining the eigenvalues and eigenvectors for this problem is problem 2 of Chapter 13.

a. If the electron starts at time $t = 0$ on the first metal atom (this is not an eigenstate), then the couplings, γ, will cause the system to evolve in time. The probability of finding the electron on the first metal atom will decrease from 1, and the probabilities of finding the electron on the ligand and on the second metal atom will increase from 0. At $t = 0$ the electron is on the first metal atom; that is, it appears as if the system is in the uncoupled state $|m_1\rangle$. Write the explicit superposition of the eigenkets $|1\rangle$, $|2\rangle$, and $|3\rangle$, which looks like $|m_1\rangle$ at $t = 0$. (*Hint:* Just add and subtract the expressions for the eigenvectors multiplied by possibly necessary constants until only $|m_1\rangle$ is left. Since it is $t = 0$, there will be no time-dependent phase factors. Make sure that the resulting expression is normalized.)

b. Write the general time-dependent state of the system $|t\rangle$ which has the initial condition of part a, that is, include the time-dependent phase factors into the expression that was obtained in part a. (It is not necessary to solve any differential equations for this part.)

c. Using the projection operator for the uncoupled state $|\ell\rangle$, calculate the time-dependent probability of finding the electron on the ligand given that it is in the time-dependent state $|t\rangle$ that was obtained in part b.

d. As an alternative approach to calculating the short time probability of finding the electron on the ligand, given that it starts on the first metal atom, one can use time-dependent perturbation theory. Consider the system to be uncoupled (the Hamiltonian is $\underline{H}_0$) and initially in the state $|m_1\rangle$. At $t = 0$, the coupling γ turns on (the Hamiltonian suddenly becomes $\underline{H}$). Using time-dependent perturbation theory appropriate for short times and small changes in probability, calculate the time-dependent probability of finding the system in $|\ell\rangle$.

e. Show analytically that the functional form of the time-dependent probability of finding the electron on the ligand obtained in part c is identical to the time-dependent perturbation theory solution of part d in the limit of very short times; that is, show that the solution of part c is the same as the solution of part d in the short time limit.

f. At what time and at what value of the probability calculated with perturbation theory does the exact result, obtained in part c, deviate by 5% from the perturbation theory result obtained in part d?

CHAPTER 12

1. In the treatment of the harmonic oscillator (Chapter 6), the selection rule for transitions between harmonic oscillator states was derived [equation (6.96)]. The results showed that dipole allowed transitions could only change the harmonic oscillator quantum number by 1. Real molecular vibrations involve anharmonic oscillators. The following problems will examine how anharmonicity influences vibrational spectra.

a. Obtain the first- and second-order corrections to the energy and the first-order correction to the eigenstates for a perturbed harmonic oscillator where the perturbation term is $\underline{H}' = Ax^4$.

b. Obtain the dipole transition selection rules for the perturbed harmonic oscillator of part a to first order in A; that is, consider the brackets $\langle n' |\underline{x}| m' \rangle$, where $|m'\rangle$ and $|n'\rangle$ are perturbed harmonic oscillator states and determine which brackets are nonzero to first order. (Note that in calculating the brackets, terms that are of order A^2 or higher are dropped because the corrections to the kets were only done to first order. If terms in, for example, A^2 are kept, all terms in A^2 must be calculated. More terms in A^2 would arise from the second-order correction to the kets. These might identically cancel terms that arose out of the first-order treatment, or they might be much larger than terms that arose at first order. Therefore, in calculating brackets that involve perturbation theory, all terms at a given order must be included or none of the terms should be included. However, once a bracket is calculated to a certain order, if the absolute value squared is used to calculate a probability, all of the terms are kept; for example, first-order terms in A will become A^2.)

c. Indicate which transitions will be highly allowed and which transitions will be less allowed.

d. Assuming that the system is initially in the lowest vibrational state ($m' = 0$), for all transitions that are allowed to first order in A—that is, have nonzero brackets $\langle n' |\underline{x}| m' \rangle$ to first order in A—calculate the ratio of each transition probability to that of the transition with the largest probability using equation (12.69). Take the transitions to be x polarized and the light intensity, I_x, to be constant for all frequencies.

2. To obtain a more realistic feel for the difficulties in sorting out weak transitions in a vibrational spectrum, reconsider the absorption spectrum of problem 1 that would result if all of the population does not begin in the lowest level, but rather the system is in thermal equilibrium at 300 K. Take the intensity of the incident light source to be weak enough (and constant with wavelength) that the $T = 300$ K Boltzmann population distribution is not perturbed by the absorption of light. (Time-dependent perturbation theory holds.) Assume that any level with population 10^{-5} times smaller than the ground-state population has zero population. Use the energy expression from problem 1 to first order only.

   a. Use the condition that the quartic term, Ax^4, is 2.0% of the quadratic term, $\frac{1}{2}kx^2$, at the classical turning point of the *harmonic oscillator* ground state (A is negative), as well as that $\bar{\upsilon}_0 = 800\ \text{cm}^{-1}$ (the harmonic oscillator energy in cm^{-1}), to calculate the energies (in cm^{-1}) of the possible transitions. How many lines will be seen in the spectrum? Calculate the relative absorption intensities of the lines relative to the strongest line. [*Note:* In calculating the absorption intensities, ignore the contribution from stimulated emission, which will be small because the population of the higher level in each transition is a small percentage ($\sim$2%) of the lower level. Absorption reduces the intensity of light passing through a sample. Stimulated emission increases the intensity. Therefore, in practice, stimulated emission makes absorption appear weaker than it actually is.]

   b. Discuss the spectrum calculated above. How many lines would appear in the absence of the quartic perturbation—that is, $A \to 0$? What happens if A is significant but $T \to 0$ K?

   c. The spectrum with the first-order quartic perturbation has a distinct pattern. If this pattern is observed for a molecule at 300 K, how can the anharmonicity be extracted from the splittings between the spectral features?

CHAPTER 13

1. Find the eigenvalues and eigenvectors for the two-state problem for the nondegenerate case; that is, in equations (13.84) replace the E_0's with E_1 and E_2. What are the natures of the eigenvalues and eigenvectors when $\gamma \gg E_1 - E_2$, when $\gamma \ll E_1 - E_2$?

2. Consider two identical metal atoms connected by a ligand. In addition to the electrons belonging to the two metal atoms, the electrons belonging to the ligand, and the electrons shared by the metal atoms and the ligand in bonding, there is one extra electron. The state of the system in which the extra electron is on the first metal atom is $|m_1\rangle$, on the ligand is $|\ell\rangle$, and on the second metal atom is $|m_2\rangle$. These kets represent the states of the system in the absence of interactions (couplings) among them; for example, when $|m_1\rangle$ does not "feel" (is not coupled to) $|m_2\rangle$ or $|\ell\rangle$. (Take these kets to be orthogonal and normalized.) These kets are the time-independent, spatial parts of the total ket vector. When the electron is on either metal atom, its energy is E_m, and when it is on the ligand, its energy is E_ℓ; that is,

$$\underline{H}_0 |m_1\rangle = E_m |m_1\rangle$$
$$\underline{H}_0 |\ell\rangle = E_\ell |\ell\rangle$$
$$\underline{H}_0 |m_2\rangle = E_m |m_2\rangle ,$$

where $\underline{H}_0$ is the part of the Hamiltonian which operates on the extra electron.

In reality, when the system is in the state $|m_1\rangle$ (the electron is on the first metal atom), there is an interaction with $|\ell\rangle$, and when the system is in the state $|\ell\rangle$, it will interact with $|m_1\rangle$ and $|m_2\rangle$, and so on. (Any direct interaction of $|m_1\rangle$ and $|m_2\rangle$ and is neglected.) In addition to $\underline{H}_0$, there will be terms in the Hamiltonian representing these interactions. The result is that the states are coupled, and operating $\underline{H}$ on the kets yields

$$\underline{H} |m_1\rangle = E_m |m_1\rangle + \gamma |\ell\rangle$$
$$\underline{H} |\ell\rangle = E_\ell |\ell\rangle + \gamma |m_1\rangle + \gamma |m_2\rangle$$
$$\underline{H} |m_2\rangle = E_m |m_2\rangle + \gamma |\ell\rangle .$$

γ is the strength of the interaction (the off diagonal coupling matrix element). Generally, $E_\ell > E_m$ and $\gamma < (E_\ell - E_m)$. To simplify the math in this problem, a special case is considered which obeys the above inequalities but uses the specific situation

$$(E_\ell - E_m) = \sqrt{8}\,\gamma .$$

Using the above conditions, derive the eigenvalues and eigenvectors by matrix diagonalization (do this analytically, not on a computer). Discuss the probability of finding the electron on the metal atoms and on the ligands for the three eigenstates, and sketch an energy level diagram. [Note that factors such as $\sqrt{1/(8+4\sqrt{2})}$ will appear in solving this problem.]

CHAPTER 14

1. Consider a superposition of harmonic oscillator eigenstates

$$|t\rangle = \sum_{n=0}^{m} c_n |n\rangle .$$

Use the density matrix formalism to calculate the time-dependent expectation value of the position operator $\langle \underline{x}(t) \rangle$ by employing the relation

$$\langle \underline{A} \rangle = \mathrm{Tr}\ \underline{\underline{\rho}}(t) \underline{\underline{A}} ,$$

[equation (14.50)] for $n = 0$ to 1 with $c_0 = 1/\sqrt{2}$ and $c_1 = 1/\sqrt{2}$.

2. Use the density matrix formalism to calculate the time-dependent expectation value of the position operator $\langle \underline{x}(t) \rangle$ in the same manner as in problem 1 but for $n = 0$, 1, and

2 with $c_0 = 1/\sqrt{3}$, $c_1 = 1/\sqrt{3}$, and $c_2 = 1/\sqrt{3}$. Using the result of the calculation, determine the distance traveled in angstroms (the difference between the maximum and minimum of $\langle \underline{x}(t) \rangle$) for a C-H stretch. Take the vibrational frequency to be 3000 cm^{-1} and the mass to be the reduced mass of ^{12}C and ^{1}H.

CHAPTER 15

1. a. Using

$$\underline{J}_x = -i\hbar \left(y\frac{\partial}{\partial z} - z\frac{\partial}{\partial y} \right)$$

$$\underline{J}_y = -i\hbar \left(z\frac{\partial}{\partial x} - x\frac{\partial}{\partial z} \right)$$

$$\underline{J}_z = -i\hbar \left(x\frac{\partial}{\partial y} - y\frac{\partial}{\partial x} \right)$$

and commutators such as $[\underline{x}, \underline{P}_x] = i\hbar$, prove

$$[\underline{J}_y, \underline{J}_z] = i\hbar \underline{J}_x$$

b. Using the commutator in part a and

$$[\underline{J}_x, \underline{J}_y] = i\hbar \underline{J}_z$$

$$[\underline{J}_z, \underline{J}_x] = i\hbar \underline{J}_y,$$

prove the following relationships among the angular momentum operators. Use the definitions of $\underline{J}_+$ and $\underline{J}_-$ in terms of $\underline{J}_x$ and $\underline{J}_y$ (all in units of $\hbar$). (Note that it is not necessary to use the differential operators of part a. Use the relationships among the operators to obtain the results.)

$$[\underline{J}_+, \underline{J}_z] = -\underline{J}_+$$

$$[\underline{J}_+, \underline{J}_-] = 2\underline{J}_z$$

$$\underline{J}_- \underline{J}_+ = \underline{J}^2 - \underline{J}_z^2 - \underline{J}_z.$$

2. For total angular momentum, $\underline{J} = 3/2$, find the explicit matrices for the operators $\underline{J}^2$, $\underline{J}_z$, $\underline{J}_+$, and $\underline{J}_-$.

3. Consider two particles with angular momenta

$$J_1 = 3/2 \quad \text{and} \quad J_2 = 1/2$$

a. What are the product states in the $m_1 m_2$ representation?

b. What are the coupled states in the Jm representation?

c. Using the lowering operator procedure given in Chapter 15, Section E, generate the table of Clebsch–Gordon coefficients that take the $m_1 m_2$ representation into the

JM representation for $J_1 = {}^3\!/_2$ and $J_2 = {}^1\!/_2$. See the tables of Clebsch–Gordon coefficients in Chapter 15 Section E for the ordering of the kets so that the table appears in the proper form.

CHAPTER 16

1. The trial function used in problem 2 of Chapter 10 to calculate the ground-state energy of the lithium atom does not violate the Pauli Principle, but it is not a proper antisymmetric wavefunction. Include spin and expand the determinant form of the wavefunction to obtain a totally antisymmetric function with two electrons in the $1s$ state and one electron in the $2s$ state.
2. Consider a sodium atom in its ground state. The nuclear spin is $I = {}^3\!/_2$, and Na has one unpaired electron with electron spin $s = {}^1\!/_2$. This electron occupies an s orbital, so it has no orbital angular momentum. In addition to the Coulomb interactions of the type treated in solving the hydrogen atom energy eigenvalue problem and the spin–orbit interaction that was calculated for the p state of the Na atom, there is an interaction between the nuclear magnetic moment and the electron magnetic moment. This is called the hyperfine interaction because it results in very small (hyperfine) splittings in the spectral lines of Na and other atoms and molecules. The Hamiltonian for this interaction is

$$\underline{H}_{\mathrm{HF}} = \alpha \vec{\underline{I}} \cdot \vec{\underline{S}},$$

where $\vec{\underline{I}} = \underline{I}_x + \underline{I}_y + \underline{I}_z$ operates on the nuclear spin states, and $\vec{\underline{S}} = \underline{S}_x + \underline{S}_y + \underline{S}_z$ operates on the electron spin states. (These operators are in the $m_1 m_2$ representation because the two spins are considered separately.)

a. In the $m_1 m_2$ representation, what are the angular momentum states of the Na atom where the $m_1 = m_I$ are the nuclear spin states and where the $m_2 = m_S$ are the electron spin states? (There should be eight states.)

b. Derive the expression for the operator $\alpha\,\vec{\underline{I}} \cdot \vec{\underline{S}}$ in terms of $\underline{I}_z$, $\underline{I}_+$, $\underline{I}_-$, $\underline{S}_z$, $\underline{S}_+$, and $\underline{S}_-$. (Remember, these are angular momentum operators. It doesn't matter what they are called, they still obey all of the rules and relationships that other angular momentum operators do.)

c. Using the kets of part a, form the Hamiltonian matrix $\underline{\underline{H}}_{\mathrm{HF}}$. (This matrix is in the $m_1 m_2$ representation because the two spins are considered separately. It should be an 8×8 matrix. Remember, in using the raising and lowering operator formulas, the kets are actually $|j_1 j_2 m_1 m_2\rangle$. The raising and lowering operator formulas are written in terms of a j and an m. So, when operating on one type of spin, j_1 and m_1 go together and j_2 and m_2 go together.)

d. Find the eigenvalues of the hyperfine Hamiltonian. (*Hint:* If you write the states in the correct order, this matrix will be block diagonal. The largest block is a 2×2.) Sketch an energy level diagram.

e. Use the Clebsch–Gordon coefficients for $J_1 = {}^3\!/_2$, $J_2 = {}^1\!/_2$ to form a unitary matrix and perform a similarity transformation on the hyperfine Hamiltonian matrix of part c from the m_1m_2 representation to the JM representation. (The table of Clebsch–Gordon coefficients was found in problem 3 of Chapter 15, or it can be obtained from published tables and computer programs.) Is the transformed hyperfine Hamiltonian matrix diagonal? What does this tell you about the eigenvectors of $\underline{\underline{H}}_{\text{HF}}$?

f. Write out the eigenvectors of $\underline{\underline{H}}_{\text{HF}}$ as explicit superpositions of the $m_I m_S$ kets. Which set of eigenvectors corresponds to each eigenvalue?

3. This problem illustrates the differences between the spatial configurations of electrons in singlet and triplet states formed from the same pair of one-electron wavefunctions. For simplicity the problem will consider a very elementary model of the π-electron system of ethylene. Ethylene has one double bond, which is composed of a σ and a π bond. Consider the lowest excited states of ethylene in terms of a one-dimensional particle in a box. Take the ground state of the π-electron system to be composed of both electrons in the $n = 1$ particle in the box state. The lowest excited states (singlet or triplet) have one electron in the $n = 1$ state and one electron in the $n = 2$ state. Since the total wavefunction (orbital $\times$ spin) for the lowest excited state must be antisymmetric with respect to electron exchange, the orbital part must be the symmetric (singlet) or antisymmetric (triplet) linear combinations of the product of the particle in the box functions for $n = 1$ and $n = 2$.

For this problem, take the center of the box (midpoint between the two carbon atoms) to be at $x = 0$. Take the length of the box to be 1.30 Å, that is, the box goes from $x = -0.65$ Å to $x = +0.65$ Å).

a. What is the spatial part of the first excited singlet-state wavefunction in terms of particle in a box wavefunctions? What is the spin function that goes with the spatial singlet-state function?

b. What is the spatial part of the first excited triplet-state wavefunction? What are the spin functions that go with the spatial triplet-state function?

c. Plot a series of correlation diagrams for the singlet excited state of this system; that is, pick a fixed location, x_1, for electron 1, and plot the spatial distribution function

$$D(x_1, x_2) = [\psi_{\text{singlet}}(x_1, x_2)]^2 dx_1 dx_2$$

as a function of x_2 of the second electron. Then place electron one at a new location and make another plot. The spatial distribution $D(x_1, x_2)$ is the probability of finding the second electron in a thin segment of thickness dx_2, given that the first electron is in thin segment of thickness dx_1 located at x_1. Make six plots to see a trend developing in the relative locations of the electrons. Place electron 1 at points $x_1 = 0.55$ Å, 0.45 Å,0.35 Å,0.25 Å,0.15 Å, and 0.05 Å. Because the results are symmetric about $x_1 = 0$, only positive values are necessary to see the trends. Note that the differentials, dx_1dx_2, are formally necessary because the probability of having an electron in a thin segment of zero thickness is zero. They also make the units come out correctly. However, the choice of their width is arbitrary in the calculation. The heights of the peaks in the plots will depend on the choice. Only the shapes of the plots are of concern here, not their amplitudes. (These calculations can be done with a spreadsheet program, with a programmable calculator, or by hand.)

d. Repeat part c for the triplet excited state.

e. Discuss the difference in electron correlation in the singlet and triplet excited states. Why does the difference in singlet- and triplet-state electron correlation influence the energies of the states?

CHAPTER 17

1. To see the influence of using a normalized but nonorthogonal basis set in a matrix diagonalization solution to an eigenvalue problem, find the eigenvalues and eigenvectors given in equation (17.13) using equations (17.12) and (17.11).
2. In Figure 17.4, the curve E_s is the energy as a function of the internuclear separation for the bound state of the H_2 molecule. The harmonic oscillator, discussed in Chapter 6, is the simplest model of molecular vibrations. To obtain an estimate of the vibrational frequency of the H_2 molecule, the potential function E_s [Equation (17.39a)] can be approximated by a parabola near the curve's minimum. From the parabola, the harmonic oscillator vibrational frequency can be found as an approximation to the H_2 vibrational frequency.

a. Calculate the E_s curve, equation (17.39a), in Figure 17.4 over the range of values $D = \sim 0.75$ to 6.0.

b. Calculate the E_s curve over the range of D restricted to $D = 1.3$ to $D = 2.2$. This is the H_2 potential near the minimum. Fit a parabola to this portion of E_s using the form

$$E = \frac{1}{2}k(x - x_0)^2 + E_0,$$

where k is the harmonic oscillator force constant, x_0 is the displacement of the potential minimum from $x = 0$, and E_0 is the shift of the potential from $E = 0$ at the minimum. Display the parabola obtained from the fitting on a graph along with the E_s curve. Make the plot span the energy range -2.18 to -2.24 and the distance range 1.3 to 2.2. To simplify the fitting, take $E_0 = -2.23205\ (e^2/8\pi\varepsilon_0 a_0)$, where $-(e^2/8\pi\varepsilon_0 a_0)$ is the energy of the hydrogen atom ground state. Therefore, the fit involves two parameters, k and x_0. What parameters are obtained for k and x_0?

c. Convert k into conventional units; and using

$$\omega = \sqrt{\frac{k}{\mu}},$$

where μ is the reduced mass of the H_2 molecule, determine the vibrational frequency in cm^{-1}. (*Note:* ω is the angular frequency. After finding ω, divide by 2π to obtain the frequency, ν, before finding the vibrational energy in cm^{-1}. The result obtained will depend somewhat on the fitting procedure used in part b.)

Physical Constants and Conversion Factors for Energy Units

Physical Constants

Speed of light	c	$2.99792458 \times 10^{8}\ \mathrm{ms^{-1}}$
Electron charge	e	$1.602177 \times 10^{-19}\ \mathrm{C}$
Planck's constant	h	$6.62608 \times 10^{-34}\ \mathrm{Js}$
	$\hbar$	$1.05457 \times 10^{-34}\ \mathrm{Js}$
Mass of electron	m_e	$9.10939 \times 10^{-31}\ \mathrm{kg}$
Mass of proton	m_p	$1.67262 \times 10^{-27}\ \mathrm{kg}$
Mass of neutron	m_n	$1.67493 \times 10^{-27}\ \mathrm{kg}$
Permittivity of vacuum	ε_o	$8.85419 \times 10^{-12}\ \mathrm{J^{-1}C^2m^{-1}}$
	$4\pi\varepsilon_o$	$1.11265 \times 10^{-12}\ \mathrm{J^{-1}C^2m^{-1}}$
Bohr magnetron	$\mu_B = e\hbar/2m_e$	$9.27402 \times 10^{-24}\ \mathrm{JT^{-1}}$
Hydrogen atom		
Bohr radius for H atom (μ = reduced mass of H atom)	$a_0 = \varepsilon_o h^2/\pi\mu e^2$	$5.288891 \times 10^{-11}\ \mathrm{m}$
Bohr radius, infinite mass	$a_0 = \varepsilon_o h^2/\pi m_e e^2$	$5.291771 \times 10^{-11}\ \mathrm{m}$
Rydberg constant	R_H	$1.0967732 \times 10^{7}\ \mathrm{m^{-1}}$
Rydberg constant, infinite mass	R_∞	$1.0973731 \times 10^{7}\ \mathrm{m^{-1}}$
Ground-state energy	E_H	-13.5983 eV
		$-2.17864 \times 10^{-18}\ \mathrm{J}$
Angstrom	Å	$1 \times 10^{-10}\ \mathrm{m}$

Conversion Factors for Energy Units

1 J	$= 6.241506 \times 10^{18}$ eV	$= 5.03411 \times 10^{22}\ \mathrm{cm^{-1}}$
1 eV	$= 1.602177 \times 10^{-19}$ J	$= 8065.54\ \mathrm{cm^{-1}}$
1 $\mathrm{cm^{-1}}$	$= 1.239842 \times 10^{-4}$ eV	$= 1.986447 \times 10^{-23}$ J

INDEX

A
absolute size 1, 2
absorption 31, 56, 85, 172, 181, 182, 186, 221
absorption probability 185
addition of angular momentum 246–54
adjoint 13
angular momentum 232–54
angular momentum matrices 240–42
angular momentum operator 232–36
angular momentum quantum number 240
anharmonic interactions 114
anharmonic oscillator selection rules 307–8
anharmonic oscillators 307
anharmonic potential 140
anharmonicity 140
annihilation operator 84, 191
anti-bonding molecular orbital 287, 293
anticommutator 42
antisymmetric function 311
antisymmetric wavefunction 272
antisymmetrization 268–78
associated Laguerre polynomials 108, 109
associated Legendre functions 101
associated Legendre polynomials 106
associative law 12
B
band structure 302
beat frequency 229
black body radiation density 187
Bloch theorem of solid-state physics 127
Bloch-Siegert frequency shift 223
Bohr magneton 244, 257
Bohr radius 104–5, 285
bond dissociation energy 228, 293
bond length 228, 293
bond strength 287–88, 293
bonding 280–94
bonding molecular orbital 288, 293
Born Conditions 52–53
Born interpretation 26
Born-Oppenheimer approximation 280–82
boundary conditions 53
bra 9
bracket 8–9
Brillouin zone 130
C
causality 2
center of mass 94
change of basis set 199–203
charged particle in E and M fields 173
classical electrodynamics 2
classical equations of motion 297
classical Poisson bracket 297
classical turning point 67, 74–75
classically forbidden region 59, 75
Clebsch-Gordan coefficients 248
Clebsch-Gordon coefficients 252, 310
coherences 224, 227, 231
coherent components 225
coherent coupling 213, 221
coherent emission 228, 231
coherent oscillation 227–28
coherent transfer 125
coherent transport 132
column matrix 198

column vector 198
commutative law 12
commutator 34, 35, 38, 42, 50, 215
commutators 310
commute 11, 34
commuting operators 37
complete set 39, 193
complex conjugate 9, 13
complex conjugate of a matrix 197
conduction band 132
constructive interference 26
conversion factors 314
correlation diagram 312
correspondence principle 32
Coulomb integral 269, 286, 291
coupled pendulums 112
covalent bond 280–94
creation operator 84, 191
cross product 232
crystal 125–26
cyclic boundary condition 126
D
de Broglie wavelength 32
degeneracy 121
degenerate perturbation theory 145–51, 269
delocalized eigenstates 122
delta function 21–22
density matrix 49, 213–14, 221, 223, 310
density matrix equation of motion 215
density operator 213
destructive interference 26
determinant 197
determinant wavefunction 277–78
diagonal matrix 198, 204, 205
dimer splitting 121
Dirac delta function 20, 21, 183
Dirac representation 79
Dirac's quantum condition 34–36
dispersion relation 30
dispersion relations 33
dispersive medium 30
distance of electron from nucleus 301
distributive law 12
dynamical variables 12–13, 204
E
effective field 225
eigenbra 16
eigenfunction 20
eigenket 13
eigenvalue 7, 13, 14, 18, 46–47
eigenvalue equation 13, 14, 19, 204
eigenvalue problem 13
eigenvector 14, 18
eigenvector representative 209
Einstein 2, 172
Einstein A coefficient 187, 192
Einstein B coefficients 186
electromagnetic field 178
electromagnetic radiation 172
electron angular momentum 255
electron correlation 279
electron spin 255–79
electron transfer 114, 305–6
electron-electron repulsion 279
electronic energy surface 281
electronic excitation transfer 114
electronic wavefunction 281
emission 85, 172, 186–88, 221, 246, 255
energy denominator 137
energy gap 125
energy operator 45, 50
equations of motion 48, 118, 223
equilibrium bond length 287
exchange energy 286
exchange integral 270, 286, 292
exchange interactions 293
excited state lifetime 184
exciton band 130
exciton energy 130, 131
exciton group velocity 131–32
exciton wave packet 131
excitons 130
exclusion principle 276
expectation of the dipole operator 227
expectation value 39, 41, 47, 123, 219
external heavy atom effect 262
F
fine structure 259
flip angle 226
fluctuations of the vacuum state 192
fluorescence 187
Fourier transform NMR 229

Frank-Condon factor 300
free induction decay 229–31
free particle 18, 32, 49
free precession 226–28, 230
G
Gaussian wave packet 27, 297
generating function 77, 106, 108
Gram-Schmidt orthonormalization procedure 296
group velocity 29, 32, 296
H
H_2 molecule 313
H_2 vibrational frequency 313
Hamiltonian 37, 45
harmonic oscillator 66, 79, 236
harmonic oscillator Hamilonian matrix 207
harmonic oscillator matrix representation 205–8
harmonic oscillator selection rule 300
harmonic oscillator wave packet 91
heavy atom effect 262
Heisenberg uncertainty principle 43
Heisenberg uncertainty relation 28
helium atom 152, 160, 268
Hermite polynomials 73, 77
Hermite's Equation 70
Hermitian conjugate of a matrix 197–98
Hermitian linear operators 14
Hermitian matrix 198, 204–5
Hermitian operator 16, 18, 204–5, 295
Hilbert space 23
Hooke's Law 66
hydrogen atom 5, 93–111
hydrogen atom energy levels 104–5
hydrogen atom wavefunctions 105–11
hydrogen molecular ion 282–88
hydrogen molecule 280, 288–94
hyperfine interaction 267, 311
I
I_2 molecule 91
identical particles 272
identity operator 35
imaginary matrix 198
impact parameter 166, 169
incoherent transfer 125
incoherent transport 132
indeterminacy 2
index of refraction 296
induced emission 186
inhomogeneous broadening 229
inhomogeneous line 230
intensity 26
intensity of light 184
interference 2, 3–4, 32
interference pattern 4
intermolecular interactions 125
inverse of a matrix 197, 198
ionization 57, 63–65
ionization energies 162
ion-molecule collision 165
J
J^2 operator 235
jm representation 248, 260, 265
JM representation 312
J_z operator 235
K
ket 7–11
ket vectors 7
kinetic energy 37
Kronecker delta 77
L
Lagrangian 175
Laguerre polynomials 108
Laplacian 46
Laplacian operator 95
lattice symmetry 125–26
lattice wave vector 127
Legendre polynomials 106
linear operator 11–13
linear transformation 194, 202
lithium 304
lithium atom 304
lowering operator 82, 84, 232, 236, 239
lowering operator matrix 206–7
M
m_1m_2 representation 247, 259, 311
magnetic dipole 260
magnetic field selection rule 246
magnetic quantum number 98
matrix 197
matrix diagonalization 205, 208
matrix element 194–95, 220

matrix multiplication 196
matrix representation 193–212
Maxwell's equations 173, 178
mixed density matrix 228, 229
molecular orbital 292, 293
molecular wavefunctions 281
molecule 56, 280, 282, 288
momentum 18
momentum eigenfunction 20, 23, 29–30
momentum operator 19, 35, 36, 85
momentum representation 298
N
nearest neighbor interaction 126, 127
non-commutivity 11
non-degenerate 308
non-orthogonal basis set 284, 313
normalization 14, 20
normalization constant 73
normalize 10
nuclear spin 311
nuclear wavefunction 281
number operator 84, 191
O
observable 1, 13, 14
one dimensional lattice 126
on-resonance optical field 224
operator 11, 180
optical Bloch equations 224, 225
orbital angular momentum 93
orbitals 93
orthogonal 10
orthogonality theorem 16
orthonormal basis set 39, 199
overlap integral 284, 290
P
particle in a box 51
Pauli exclusion principle 276, 304
Pauli principle 268
pendulums 112
periodic potential 127
permutation operator 271, 274
permutation symmetry 275
perturbation 139
perturbation theory 133–62, 302
perturbed harmonic oscillator 139–42
phase factor 11, 29
phase velocity 31, 130, 296
phonon 132
phosphorescence 187
photo-electric effect 2
physical constants 314
pi over two pulse 225
pi pulse 225
plane rotor 142
plane wave 20, 24, 30, 176–77, 178
Poisson bracket 34
polaron 132
population inversion 225
position operator 85
potential energy 37, 45, 66
Poynting vector 184
precession 226
principal diagonal 195
probability amplitude 26
projection operator 12, 119–20
pure density matrix 228
Q
quantum condition 13
quantum number 54, 75, 84
quasi-particles 131
R
Rabi frequency 222, 224
radial quantum number 104
radiation density 185
raising operator 83, 84, 232, 236, 239
raising operator matrix 207
real matrix 198
reciprocal lattice vector 130
recursion formula 71, 100
recursion relation 103
reduced mass 92
resonance energy 286
resonance integral 292
rigid plane rotor 142
rotating wave approximation 182, 223
rotation matrix 201
row vector 199
Rydberg constant 105, 268
S
scalar product 9, 10, 21
Schrödinger equation 45–48, 49
Schrödinger representation 45

selection rule 86, 188–89
semi-classical 172
semi-classical treatment 190
similarity transformation 203, 204, 211, 267
simultaneous diagonalization 205
simultaneous eigenfunctions 36
simultaneous eigenvectors 37, 235
singlet 56
singlet excited state 114
singlet state 276, 278, 294, 312
singular matrix 198
sodium atom 311
sodium D line 255, 265
spectral transition 300
spectral width 297
spherical Bessel function 22, 25
spherical harmonics 93, 245, 258, 259
spherical polar coordinates 93
spin 256
spin states 258
spin wavefunctions 258
spin-orbit coupling 258–67, 278
spin-orbit coupling Hamiltonian 261
spontaneous emission 187–88, 192, 230
square well potential 64
standard deviation 27, 43
Stark effect 139, 142–45
state 12
state of a system 5
stationary states 120–23
stimulated emission 181, 186
superconductivity 132
superposition 3, 5, 7, 8, 24
superposition of eigenstates 30
superposition of harmonic oscillator eigenstates 309
superposition principle 1, 2, 3, 5, 7, 24, 29, 39, 193
symmetric matrix 198
symmetric wavefunction 272

T

table of Clebsch-Gordon coefficients 252
term symbols 270, 275, 278
thermal equilibrium 308
thermal fluctuations 123
time-dependent perturbation theory 163–71, 172, 305
time-dependent phase factor 29, 47
time-dependent two-state problem 217–19
time derivative of an operator 48
total quantum number 104
totally antisymmetric states 276
totally antisymmetric wavefunction 275
trace 214, 220
transient nutation 224
transition dipole 181
transition dipole bracket 181, 186
transition dipole interaction 114
transition dipole moment operator 181
transition dipole operator 222
transition probability 181, 183
transpose of a matrix 197
triple cross product 243
triplet state 276, 278, 294, 312
tunneling 57, 58, 63
tunneling parameter 62
tunneling probability 62
turning points 74
two dimensional particle in a box 299
two state problem 112

U

uncertainty 26, 43
uncertainty principle 28, 42–44, 89
uncertainty relation 42, 43, 44, 87
unit matrix 195
unitary matrix 198, 200, 211
unitary transformation 200, 204, 205

V

vacuum state 192
variational method 160
variational theorem 152, 158–60
vector 7, 23
vector coupling coefficients 248
vector functions 21
vector potential 173, 177, 178
vector representative 194, 200
vector space 193, 199
velocity operator 299
vibrational energy transfer 114
vibrational spectra of molecules 87
vibrational states 282

vibrational wave packet 90
vibron 132
W
wave packet 18, 23–29, 31, 61, 90
wave packet spreading 297
wave vector 20
wavefunction 4, 5, 20, 45, 52
wave-particle duality 33
Wigner coefficients 248
Z
Zeeman effect 242
zero matrix 195–96
zero point energy 73, 192